D1700882

# HOW WE GOT HERE

## From Bows and Arrows to the Space Age

C. R. Hallpike

authorHOUSE

*AuthorHouse™ UK Ltd.*
*500 Avebury Boulevard*
*Central Milton Keynes, MK9 2BE*
*www.authorhouse.co.uk*
*Phone: 08001974150*

*© 2008 C. R. Hallpike. All rights reserved.*

*No part of this book may be reproduced, stored in a retrieval system, or transmitted by any means without the written permission of the author.*

*First published by AuthorHouse 10/29/2008*

*ISBN: 978-1-4389-0833-5 (sc)*

*Library of Congress Control Number: 2008909104*

*Printed in the United States of America*
*Bloomington, Indiana*

*This book is printed on acid-free paper.*

# Contents

Preface ..................................................................... vii
Acknowledgements ................................................. xi
Chapter I: How Social Evolution Works ..................... 1
Chapter II: The Simplest Societies ............................ 28
Chapter III: The Agricultural Revolution ................... 52
Chapter IV: The New World of Tribal Society ............ 66
Chapter V: Economics, War, and Politics
    In Tribal Society .................................................. 94
Chapter VI: Primitive Thought ............................... 130
Chapter VII: The State and Civilisation ................... 156
Chapter VIII: Technology and Invention ................. 190
Chapter IX: Writing, Mathematics, and High Culture . 209
Chapter X: Social Crisis and the Need to Think ....... 236
Chapter XI: The New Religions ............................... 266
Chapter XII: Natural Philosophy ............................. 288
Chapter XIII: Ancient Sciences ............................... 320
Chapter XIV: The Uniqueness of Western Society .... 354
Chapter XV: How We Learned to Experiment ......... 396
Chapter XVI: Modern Science and Industrialism .... 429
Conclusions .......................................................... 463
Notes .................................................................... 470
References ............................................................ 565
Index .................................................................... 585

# Preface

In the 1960s and 1970s I was privileged to spend several years living as an anthropologist with tribesmen in the mountains of Ethiopia and Papua New Guinea. Sharing the daily lives of people who are so different from modern Westerners, learning their languages, and trying to understand how they see their world is a very profound experience, because it shatters one's assumptions about what is normal for all human beings.

For example, numbers are a basic part of our lives, and we assume that counting, like language, is part of being human, but the Tauade of Papua New Guinea only had words for one, two, and many, and counting was utterly unimportant to them. So, too, was something else we regard as fundamental, which is time-reckoning, but it was impossible to ask the Tauade how many years ago something had happened because they didn't have weeks, months, or years, and couldn't have counted them even if they had. We live in a society where money is a central fact of life, but traditionally, the tribes I lived with had no money at all, and their lives were organized on an entirely different economic basis from our urban, capitalist society. They had remained much the same in some respects for thousands of years, and ever since I have been fascinated by the question of why some societies should have changed relatively little, while others developed great empires and civilisations, and ultimately modern science and industry.

So if we are going to understand how our kind of society evolved from hunter-gatherers and simple farmers, 'primitive societies', as I shall call them for convenience, in only around ten thousand years, we need to understand what these societies were like, and only anthropologists can answer that question.

Unfortunately, the topic of primitive society is a happy hunting-ground for amateur speculators, and it is hard to open a newspaper without coming across some confident but completely uninformed opinion on what hunter-gatherers or early farmers must have been like. They would have been repressed by magical taboos, or been wild and undisciplined; naturally peaceful, or blood-thirsty cannibals; altruistic and self-sacrificing, or individualistic and competitive, and always trying to maximise their own material interests in the struggle for survival. They were naturally curious, gazing up into the vastness of space and wondering if there was life on other worlds, or only interested in the next meal; and they obviously invented religion as psychological protection against a threatening environment. These fantasies, however, simply reflect the biases of our own culture, and produce something rather like the Flintstones – basically ourselves, modern people with Western attitudes, but dressed up in skins and using stone tools.

Some of the worst offenders in this respect are evolutionary theorists moulded in the Darwinian and materialist traditions, who also take it as an article of faith that man is really only an animal, and must be explained as such. While this approach claims to be rigorously scientific, it is actually profoundly ethnocentric, a Western myth-structure that developed in the nineteenth century. So this book is intended as an anthropological corrective to all this, written by someone who has actually lived in primitive societies.

This perspective is also of great value when we come to the ancient literate civilisations later in the book, because it protects us from the same temptation to explain them from our own point of view, instead of making the imaginative effort to understand them from theirs. Indeed, we shall see that even early modern scientists were not nearly as like us as we suppose. Anthropology is also inherently comparative, and it is only against a comparative background that we can assess our own cultural tradition more objectively, and not assume that Western society is typical of the human race.

This book is about social evolution, then, but it is not about *progress*. Readers are perfectly entitled to say, when they have finished it, that it would have been better if we had never abandoned the simplicity and equality of hunter-gatherer life, or at least had never had the industrial revolution. I am simply trying to explain why things turned out as they did, not deciding if this or that aspect of the process was good or bad, which too easily degenerates into third-rate moralising.

The discussion of many difficult and profound issues has had to be much briefer than they deserve, but I have provided extensive endnotes to explore some of them in more detail, and also references to my other books where I have dealt with many of these subjects at much greater length.

# Acknowledgements

I would like to thank my family for various helpful suggestions as I was writing this book. I am especially grateful to my daughter Julia, and also to my old friends Dr Martin Brett and Professor Victor Snaith, for reading the whole text and giving me many valuable criticisms and comments. It would have been extremely difficult to write this book without the facilities of Robinson College, Cambridge, and the University Library, and I am most grateful to both these admirable institutions.

**C. R. H.**
Shipton Moyne
Gloucestershire

*July 2008*

# CHAPTER I
## How social evolution works

*1. The human revolution.*

Only about 10,000 years ago the entire human race lived as small groups of hunter-gatherers. The adoption of farming and the domestication of animals set in motion a vast transformation of our lives, with the discovery of metals, the rise of powerful states, writing, mathematics, philosophy, the harnessing of water and wind power, the emergence of modern science, and the technology of industrial society. In a few millennia man's capabilities have been transformed, from a primitive state of ignorance and technological impotence, into being not just the dominant species of the earth, but one whose achievements in thought, art, science, technology and government have taken him entirely beyond comparison with the animal kingdom.

This, in scope and speed, has been quite different from biological evolution because it was made possible by something uniquely human – culture – which was as radical an innovation as the emergence of life from non-living matter. By 'culture' I do not mean simply learning new behaviour and passing it down the generations, which many animal groups can do. I mean the ability to use language to transmit ideas – knowledge, values, customs, and social institutions to other people. The origin of language is one of the most obscure and debated problems in human evolution, but however and whenever it began, once it had developed it allowed human beings to be linked together not just by purely animal relations such as mutual grooming, or sharing the same odours, but by shared ideas.

Human society is therefore a new kind of system altogether because its institutions exist in people's heads as ideas, but which are also public ideas communicated by language: one cannot *see* the Prime Minister, for example, but only a man, and someone who does not know what being a Prime Minister *means* has to be told. This can only be done properly by explaining how his role fits into the British Constitution, which in turn involves explaining cabinet government, the rule of law, democracy, and so on. Our whole society, then – the nation, the government, money and the banking system, trade unions, companies, local councils, and so on – forms a world of ideas, a landscape, within which people have to interact with each other, and which powerfully affects their behaviour. [1]

So while we, like our primate ancestors, are still physical beings, who have to survive in the natural environment, we also inhabit the radically new environment of culture, in which people can behave in ways that, unlike the animal world, may have nothing to do with material needs. In April 1975, for example, the inhabitants of all the cities in Cambodia fled into the countryside. This was not because of some physical emergency, such as an outbreak of plague, or because the food supplies had run out, but because the Communist Khmer Rouge government regarded cities as the root of capitalism and social inequality, and wanted to enforce a completely equal society of peasants, without money or the family.

These extraordinary powers of culture over the individual have, however, led some anthropologists to make the extreme and foolish claim that there is really no such thing as basic human nature at all, so that everything we do is the result of the culture in which we have been brought up – even that all the differences between males and females are culturally conditioned. The fact, however, that in all societies, past and present, males, and young males especially, are responsible for the great majority of physical violence is enough to disprove this kind of theory. There are many other universals in human thought, feelings, and behaviour that clearly have a biological origin: our amazing ability to learn language from infancy, for example, is obviously innate, and the same six basic emotions of anger, fear, surprise, disgust, joy, and sorrow, together with their facial expressions, that occur in all human beings derive primarily from biological rather than cultural roots. [2] Throughout this book I shall refer to a number of other features of human nature, such as our love of body-decoration, or our propensity to exchange gifts, which have also been of great importance in social evolution.

Just as some anthropologists try to deny the existence of human nature, some socio-biologists go to the other extreme, and claim that more or less everything in human society, from our patterns of kinship and marriage to religious belief, can be shown to have a genetic basis. While there are far more universals of human nature than some anthropologists have been willing to admit, what we do, however, is not solely dictated by our genes – it depends on our cultural environment as well. We may have an innate ability to learn to speak, but we must still be brought up in a human society for this ability to be developed into actual speech. One can agree that the capacity for physical aggression is part of human, and especially male, nature, but it does not express itself willy-nilly whatever the circumstances, and the motivations for warfare in tribal society and in modern industrial states have quite different patterns. Killing one's hated neighbours from across the river in revenge for their murder of one's relatives, and thereby gaining the sexual favours of the women of one's own group, has no resemblance to the motives of politicians who order soldiers in modern armies to kill complete strangers in distant countries, or to the motives of the soldiers who obey those orders. Appealing to human nature and to 'genes for aggression' is entirely inadequate to explain these very different patterns of warfare: as one distinguished anthropologist has put it, 'the reasons people fight are not the reasons wars take place'. [3] So while human nature has been an essential part of social evolution, this has obviously involved social and cultural factors as well as biology. [4]

## 2. Darwinism and social evolution.

The most fashionable explanations of social evolution are based on the materialist assumption that man is really an animal, even though a strange one, so that social evolution is basically driven by his physical needs. Closely linked with this materialist philosophy is the belief that social, as well as biological evolution, can be explained by Darwinian theory because this is supposed to be so simple and so powerful that it can be extended far beyond the realm of biology to human society. [5] Darwin's theory explains biological evolution by two simple ideas – random genetic *variation*, and the *selection* of successful variations in the competition for survival.

In the same way, it has been suggested, why can't we treat new variations in doing and thinking, inventions, institutions, customs, ideas, and so on, as appearing randomly, like genetic mutations, so that those novelties that are best adapted are then selected in the vigorous competition of daily life? They will survive and be imitated in large numbers, while the failures will dwindle and die out. So the rise of civilisation over the last few thousand years has been a process of blind trial and error leading to discoveries, new ways of doing and thinking, that were better adapted than others to our material needs, and gave a competitive advantage to individuals and groups who adopted them whereby the strongest and most efficient emerged on top of the heap. By the same process, too, knowledge and science accumulated by trial and error, and gradually triumphed over ignorance and superstition. This version of Darwinism is often called social or cultural selection, and on the face of it seems a perfectly reasonable idea. [6]

When trying to explain the evolution of culture, the Darwinist uses the idea of blind variation as a simple and convenient way of representing change and innovation. But it actually has very little to tell us about social change: we do not find, on the whole, that people try out vast numbers of different ways of doing things, and that the most efficient is ultimately selected by a process of competition, in which inferior variants are eliminated. There is usually only a limited range of choices, and people generally opt for the one that is the easiest or most convenient – it would be very odd if they didn't. [7] So, for example, just about every verbal number system in the world is based on 10. Are we to believe that this was the result of some trial-and-error process in which thousands of societies tried out all sorts of other numbers, and that only those systems based on ten survived? Or, did people almost always go for ten in the first place, without any trial-and-error, simply because we have ten fingers, and that was the easy and obvious choice?

But once we accept that people tend to *do what is easiest*, this contradicts the whole notion of blind variation, because what is easiest – physically, psychologically, technologically, socially and so on – will be very restricted and the complete opposite of what is random. Trying to account for change by talking of random variation is in fact a camouflage for ignorance, that excuses us from investigating how the significant innovations in human history have really occurred. [8] While this is much more intellectually demanding than just assuming random variation, it is only this that will show us how things first appear within any society. Instead of randomness or accident, we more often find that not only may there be a limited range of options, but that there can be easy gateways into new ideas or discoveries, or that some important change has been made easier by a particular set of conditions that can themselves often be explained. Logically, too, it should be obvious that before we discuss why something may have survived, we need to understand how it appeared in the first place. The understanding of *origins*, then, will be our first and fundamental concern.

The other pillar of Darwinian theory is selection, the assumption that competition for survival between different ways of doing and thinking in human society has always been very severe, weeding out the maladaptive and selecting the fittest. In our modern capitalist world of rapid innovation, financial rewards for commercial success, and advanced communication there is obviously a very high level of competition; this is true not only of goods and services, but of the market place of ideas and our notions of how we should live. These conditions, however, are highly unusual. In earlier periods, and especially in small, technologically primitive societies, the rate of innovation is very slow with few alternatives to choose from, and, just as important, a number of different ways of doing things may all be viable, so that the competition between them is actually very weak. The level of competition itself, then, can vary greatly, but if this is so, then widespread customs or institutions, such as magic or the vendetta, may not necessarily have proved themselves in the rigorous struggle for survival – it may be that there are simply frequently recurring features of human nature and society that produce them. And if conditions are undemanding, then it will be easy for the inefficient to survive indefinitely because competitive pressures are low.

Magic, for example, was universal in pre-modern society, but this is not because people all over the world performed experiments and found that certain spells worked very well – the reality was the other way about. People who are still at a certain stage of intellectual development find some ideas and ways of thought inherently convincing – such as 'like affects like', or that the name of something has an inherent power over it – and for these reasons will continue to believe in magic even though it doesn't work. It is actually very difficult to disprove magical beliefs experimentally, because all sorts of excuses can always be found to explain why they failed on particular occasions. So the theory that these beliefs have been selected because of their adaptive value to the individual or the group gets the matter exactly wrong: they are universal merely because they are easy to think at a certain stage of intellectual development. And they have survived not because they have withstood some rigorous selection process, but precisely because, in an intellectually uncompetitive environment, it is so difficult to bring them into a decisive confrontation with the facts, because like so much in early culture they *seem* to work without in fact doing so very well. Indeed, one of the most obvious facts of life in general is that when people encounter evidence that contradicts their beliefs they do not accept it/ignore it/ excuse it/do not understand it/define it away and so on.

We must therefore escape from the mind-set that believes that history has been a continuous struggle between the better and the less adapted, and recognise that things may happen because they are easy to do or because conditions are right for them, not because they are necessarily in people's best interests, or are more efficient solutions to their problems. Rather than the survival of the fittest, we often have the survival of the mediocre. [9]

## 3. Materialism and adaptation.

The first question a Darwinist asks himself in trying to explain why a particular custom, or institution, or belief was selected is 'What *use* does it have?' Here I should emphasise that many of the things people do, like planting crops, building houses, and making pots are certainly useful, but since this is obvious to the people themselves the survival of these practices needs no kind of Darwinian explanation.

The opportunity for this arises in the case of many customs and practices whose material usefulness, or adaptive value, is *not* obvious to the people. Anthropologists constantly encounter food taboos, complex rituals, competitive feasting and exchange, beliefs in witchcraft and the evil-eye, or elaborate rules about whom one can marry that seem burdensome or wasteful, and not useful at all. Yet when asked why they do these things, the people will simply respond that it is their custom, or give an explanation based on supernatural beliefs or cultural values.

It is here, in trying to explain why customs and practices persist even when they seem to have no material use, that the Darwinist steps in. He will argue that they really are useful, even though the people do not understand how, and this explains their survival. Anthropologists therefore look for any kind of material benefit that a ritual, for example, may have, like conserving resources or encouraging the exchange of goods. When they have found one, which is usually not very difficult, they claim that this really explains its survival, regardless of the people's own explanations which are dismissed as superficial and unscientific. This type of adaptive explanation assumes that 'culture is '... part of the means by which animals of the human species maintain themselves in their environments. There should be no conceptual difficulty in treating culture much as one would the behaviour of other animals'. [10]

In this materialist and utilitarian view of man, we are always competing for scarce resources, and driven by a fundamentally animal agenda of mating, parenting, and trying to increase the proportion of our genes in the population. Our physical needs have priority over all others, which is why material factors such as population growth, geography, the modes of production, and the need to harness energy, have really determined the course of history. The objective realities of the natural world, too, are inescapable, and cannot be altered by how we think about them. Nature, then, has a fundamental priority over mind, and the notion that ideas and beliefs and imagination are essential factors of human behaviour can be dismissed as 'mentalism'.

This world-view was the product of nineteenth century notions of the survival of the fittest, the school of hard knocks, the 'weakest go to the wall' philosophy of life of the iron-masters and mill-owners of the Age of Steam, immortalised by Dickens in the imaginary Coketown of his novel *Hard Times*. This begins with the famous words of the schoolmaster, Mr Gradgrind, 'Now what I want is Facts. Teach these boys and girls nothing but Facts. Facts alone are wanted in life. Plant nothing else and root out everything else. You can only form the minds of reasoning animals upon Facts: nothing else will ever be of any service to them.' No room for Fancy here, just hard Facts, buying cheap and selling dear, and the workhouse and the pauper's grave for those without the grit to succeed in life.

The whole of social evolution has been seen as a kind of gigantic Coketown, in which the only really important facts are the material ones of food and shelter, energy maximization, and reproduction, and where life is a grim struggle for resources in which the weak and the dreamers are trampled underfoot, and institutions and customs are selected for their survival value. And it is supposed that in primitive society, due to its rudimentary technology and limited resources, the struggle for existence will be particularly severe and selection will operate most rigorously, especially through warfare. Ideas and beliefs are little more than the reflections of social organization and material needs, mere froth on the surface of reality, while ritual, religion, magic, and other forms of superstition will all be selected for the contribution they make to social cohesion or individual self-confidence. [11]

This has not been my experience at all: the Tauade of Papua with whom I lived [12] were endowed with rich supplies of natural resources, and in particular with vast areas of land, and yet in pre-colonial times had one of the highest levels of violent conflict in the world. Their major interest was in raising large herds of pigs which devastated their gardens, and produced innumerable quarrels and even homicides, and which were then slaughtered in such quantities that some of the meat was thrown to the dogs. Enormous labour was spent in erecting imposing villages for these feasts to honour the bones of their dead, but after only a few months these villages and their great men's houses were abandoned to decay. Large areas of barren grassland had been produced by unnecessary burning for their amusement, and it was their traditional practice to keep the rotting corpses of their important men in their hamlets, to absorb their vitality through the smell even though they found the stench disgusting.

These are definitely not the sorts of things that Mr Gradgrind would have approved of at all. He would have grimly censured the Tauade for their lack of forethought and their improvident use of their resources, their irrational violence and, above all, for regulating their conduct by Fancy instead of by Fact. But the Tauade have nevertheless survived perfectly well, and they have done so because Coketown is not actually a very realistic model of the world. In primitive society the material conditions of life are certainly restrictive – there is only a subsistence level of food production, a limited variety of building materials, a small workforce, and a simple technology, but for these very reasons there are many different ways of organising social life, all of which will work. As long as we satisfy our basic material needs, nature is indifferent to how we do this, and how we spend our spare time. For example, a group of men with stone tools can be organized to cut down trees and make planks out of them by clan, or by age-group, or by rank, or by where they live, or by who their friends are.

The belief that in each primitive society there is a single optimum solution to every 'problem' of survival, that will inevitably be discovered by natural selection, is therefore a complete fallacy. What we find instead is that as social organization becomes more complex, the range of options becomes *increasingly limited*: the ways of organizing the workforce of a large modern saw-mill, for example, are far more restricted than organizing our men with their stone axes. The saw-mill must operate at a profit to pay for the expensive plant, and needs a source of power, a rigorous time-schedule, a high division of labour, good transport links, and so on. In Coketown the same principle applies – the immense technology and the vast proliferation of goods require the factories to be concentrated in the town, the trains of iron ore and coal to run on schedule, the factory hands to report punctually for work, the police to maintain order, and all the elaborate apparatus of finance and government to function smoothly. These complex requirements explain why all modern industrial states tend to look much alike: not because a million bizarre alternatives were tried but were weeded out by natural selection, but because only one model could ever be constructed in the first place.

In conditions which are undemanding, and where a number of options are equally viable, we can therefore expect to find the survival of the mediocre as a matter of course, especially where, as in so many primitive societies, all their neighbours are doing the same thing, so that there are no competing alternatives. Many customs and beliefs and practices are mediocre because people do what is easiest or most attractive for them in the circumstances, and this may not be very efficient or even very sensible. It is easiest for most people to follow the crowd, to go on doing things in the same old way, to be lazy, gluttonous, and revengeful, to go to extremes rather than be moderate, to prefer short-term solutions, to follow instructions blindly, and to be suspicious of strangers. Once people have devised something that works better than nothing, the natural temptation is to rest on one's laurels and go no further, unless there is a compelling reason to do so. This is certainly true of the history of technology which, especially in tribal societies, had a strong tendency to settle down into a conservative rut once people had absorbed some new device. Why, then, did any significant change occur at all?

*4. The place of Fancy in social evolution.*

Once we cease to be obsessed with competition and adaptation, and realize that it is easy, especially in a primitive milieu, for a whole range of mediocre practices and institutions and beliefs like magic to survive, then we are free to look in quite another direction. This is the possibility that a way of doing things, whether it is a kind of technology or a type of social organisation or an idea, may also have the unsuspected potential of doing other things as well, which will disclose themselves later when changed circumstances are favourable. Competition focuses upon the victory of the winner over the loser, but this is often very misleading, because in many cases the loser provides the very basis for the winner, and the means by which it comes into existence. We begin to think, in other words, of *construction* rather than selection, of how new and more complex systems of all kinds are built up.

For example, Sir James Frazer in *The Golden Bough* [13] proposed an evolutionary sequence in which Magic was displaced by Religion, which as an explanation of nature has in turn been displaced by Science, where development occurs by the overthrow of the predecessor. This looks backwards and therefore sees only failure falling by the wayside; but if instead we look *forward* I shall show how magic and religion, particularly in the forms of alchemy, astrology, and the idea of cosmic order were essential foundations for the eventual development of science. This same principle of *evolutionary potential* [14] is found in many other areas of culture, where the first reasons for being interested in something may be very different from what turn out to be its most important possibilities. There are sometimes easy pathways into complex discoveries, so that difficult ideas first appear in simple forms that may have no practical value. This all casts an entirely new light on the whole notion of the survival of the mediocre because what is really significant is not just what use something has here and now, but also its other properties and their evolutionary potential in the future.

If early man had really thought like Mr Gradgrind, basically concerned with practical calculations of profit and loss and what would be immediately useful, with sober matter-of-fact and ignoring the fanciful and imaginative, then I do not think that very much would have happened at all in the way of social and cultural evolution. We must first of all realise that primitive man often had very different ideas from us about how to *be* practical. Of course he wanted health and prosperity just as we do, but from the earliest times of which we have any reliable evidence, we know that man did not see the world only in terms of its immediately obvious physical properties, but also as permeated by supernatural forces and beings on which his survival and prosperity depended.

Here we must remember that the environment can be, to some extent, what people think it is, and this may have no basis in objective reality. Nature does not, however, behave like Mr Gradgrind, and always rap us over the knuckles whenever we get something wrong, like thinking the earth is flat; on the contrary, it can hide behind mask upon mask of ambiguity and deception. Since many different, and false, interpretations of nature will therefore all seem to work, the materialist belief that they will be eliminated by natural selection cannot be right, and mediocre ideas will survive just as well as mediocre institutions. Plants will grow perfectly well without the use of garden magic, the rain will fall without rituals to bring it, and the Aztecs did not need to slaughter thousands of human victims to ensure the survival of the sun. For primitive man, water, air, fire and earth, animals and trees, were not just physical objects but filled with mysterious powers that were akin to man himself, so that his sexual acts and his killing of animals and men resonated with cosmic significance. Play and myths and rituals were all part of this world of the imagination, which clothed the natural world in symbolic forms, and which found expression in art, ritual, and the decoration of the body.

Disconcertingly, however, for Mr Gradgrind, these fanciful preoccupations had great practical consequences, for precisely because they inspired people to do things which, from the perspective of Coketown made no practical sense, they led them to explore the properties of the world around them, and so to discover the easiest pathways into its evolutionary potential. It is essential to realise that some of the most important innovations in history had no obvious practical pay-off in their initial stages, so there had to be some other, non-practical, reason for our ancestors to take an interest in them. Personal decoration, for example, has no practical use, but because early man liked ornaments he first used gold and copper to adorn himself, before he could have had any idea of what else these metals were good for in a practical sense.

But while the earliest use of metals was for trivial purposes of personal adornment, like the shells of South Sea Islanders, unlike shells they had enormous evolutionary potential. Initially, this was in a vast range of tools and weapons, and then machines, and this potential has only been fully realized in the last few centuries by the technologies of steam and electricity. It was only because men were willing literally to play about with metals for the non-practical purpose of decorating their bodies, that they eventually came to understand their practical possibilities. We shall see in the course of this book that honour and status have also been immensely important motivations, especially in economics and warfare, which have stimulated people to efforts that mere physical survival would never have done.

As societies became more complex, the importance of the non-practical as the basis of the practical did not diminish. Religion was the main inspiration for all the monumental architecture of the ancient civilisations, (apart from defensive fortifications). The pyramids, for example, were of no practical use; they were huge because this appealed to the human imagination, and flattered the grandiose claims of the pharaohs, but because they were huge men learnt far more about how to build in stone and to organise great public works than if they had been content with small buildings of mud-brick. Simple farmers all over the world observe the sun, moon, and stars as part of their general calculations about sowing and harvesting. But the really systematic astronomical observations by the ancient civilisations went far beyond these limited practical needs, and were based on the belief that the well-ordered society had to be in tune with the heavens, so that astrology laid the foundations of astronomy and for many of the subsequent advances in mathematics. Magic was an essential ingredient of alchemy which, among the Chinese, Arabs, and Europeans led to intense investigations into the properties of a wide range of substances in the search for the Elixir of Life that would confer immortality, and be the means of turning base metals into gold. But without these vast and deluded researches modern chemistry would not have developed.

War, unlike religion, has always been regarded as an eminently practical activity (at least by evolutionists) in which people are supposed to defend their vital interests, and as the archetypal example of the struggle for survival in which man closely resembles the animals. But animals when they fight are, by comparison with man, very rational and cautious, and much of their fighting is ritualised and designed to avoid serious injury, while their aggression is directed to the practical ends of finding food or a mate, or defending their territory. Primitive human warfare, however, is not much about defence of territory, access to scarce resources, or the selection of the fittest, but is usually about insult and honour, or vengeance or the need by men to impress their womenfolk and one another by killing people, and is a by-product of uncentralized political organization that cannot control the endless cycles of retaliation that are generated. It can also be a form of human sacrifice in which killing is a source of supernatural power. But while primitive warfare was, from the practical point of view, largely a waste of time it was an essential basis of the state, and it is probable that if life in tribal society had been uniformly peaceful the state would never have developed. If there had been no warfare of this dysfunctional type in human society, no constructive warfare of the type that built the civilisations of Egypt, Mesopotamia, and China, Greece and Rome, could have emerged either.

Social evolution has therefore been possible partly because, instead of weeding out everything that is not immediately useful, societies carry a good deal of 'dead wood' that may be of no particular adaptive value at the moment. They operate rather like those people who never throw anything away, because 'you never know when it may be useful'.

## 5. Evolution as construction, not selection.

Darwinism thinks of the environment as weeding out failure, and selecting the fittest, but I think of the environment in a quite different way: as a set of constraints and opportunities for active individuals that make some changes easier than others, in a process of *construction*. Just because some social institution, or invention, or idea has ceased to exist, it does not therefore mean that it was a failure in the struggle for survival, because we must also look at what it made possible while it was around. Instead of wasting time trying to find adaptive explanations for particular customs and beliefs, we need to ask about their origins, about what conditions made it easy for people to do or to think X, and then, what X can lead to, either by itself, or combined with other cultural traits – in other words, what is X's evolutionary potential.

The fact that gold is no longer the standard of value in the modern world does not mean that it was an economic failure by comparison with paper money, the equivalent of the dinosaurs in biological evolution, because this would overlook the enormous effects that gold had on trade, and there are innumerable other examples. Aristotle's physics were completely mistaken, but his false ideas focused the minds of early modern scientists on crucial problems, so that better theories emerged. We no longer have the divine kings of earlier societies, but they were nevertheless in their day essential in the development of the state. The reciprocating steam engines that used to pump water from mines and drive railway locomotives have now passed into history, and been replaced by turbines, electric motors, and internal combustion engines. But it would be absurd to describe them as failures in the struggle for survival: they laid the foundations of the Industrial Revolution and helped create the conditions in which the electric motor and the internal combustion engine could appear.

Conversely, there are some institutions such as slavery that were more or less universal but which had no evolutionary potential and led nowhere, or the nomadic mounted warrior with a powerful bow who enjoyed military success for many centuries, but without a firm agricultural base remained essentially parasitic and destructive, and could therefore make no lasting impact on history.

When we are looking at the evolution of societies the important question, therefore, is not 'How common did this become?', or 'Why did that one die out?' but 'What significant effects did it have on the rest of the society while it was around?' The significance of agriculture, clans, cities, metallurgy, writing, and electricity, for example, lay in what further developments they made possible, not just in how successful they were themselves. This means that in social and cultural evolution there may be an *accumulation of necessary conditions* for further changes in important directions, such as larger settlements, greater political centralization, more favourable conditions for invention, the development of literacy, and so on. Indeed, we very often find that innovations are not single changes in something, like Darwinian mutations, but the result of a *combination* of different factors.

This brings us to the final weakness in the Darwinian model of social evolution, which is its complete lack of interest in how *systems* and *structures* actually work and evolve. It is only concerned with whether this or that individual custom or institution is useful, and therefore thinks of a society atomistically, as simply a bundle of adaptations to the here-and-now, a *population* of bits and pieces, of inventions, customs, institutions, ideas and so on, that have each been separately selected in the competitive process. But when I refer to the accumulation of necessary conditions, I have in mind the gradual construction of increasingly complex systems, (whether these be societies or knowledge structures or technologies), and this implies that key innovations must occur *in a certain order*.

So, then, the adoption of agriculture and the domestication of animals laid the foundations for much larger and more permanently settled groups than could be supported by hunting and gathering. New property relations with the land under agriculture made possible the development of corporate groups such clans and lineages, which were the basis of hereditary authority, and these and other conditions, such the development of an economic surplus through tribute, conquest warfare, and trade, in turn laid the foundations for the emergence of the state. Because the state could control large populations, and extract a correspondingly large economic surplus from them, this in turn created a new set of conditions, including urbanization, that were the basis of literate civilisation and high culture, including the development of philosophy and the world religions, and later the emergence of modern science and industrialism. [15]

Growing size, centralization, hierarchical structures, urbanization, and increasing division of labour resulted in fairly similar developments independently in different parts of the world. When the Spaniards invaded Central America in the sixteenth century, for example, they found literate urban societies ruled by sacred hereditary kings with nobles, bureaucrats, and priests, and with temples and monumental architecture, that in these and other basic respects were similar in structure to the kingdom of Spain. The Jesuits found essentially the same in China, as did the first English ambassador to the court of the Mughal Emperor in India at the beginning of the seventeenth century.

It is often objected, however, that every society is unique, and that since agriculture, or the state, or cities, or literacy developed in very different ways in different times and places, there cannot have been a single evolutionary pathway to the literate civilisations of the ancient world, for example. I am certainly not arguing that there was, and this objection misses the essential point that it is precisely *because* there are many different ways of reaching agriculture, or the state, or cities, or literacy that we can expect them to occur repeatedly in history. If, in a landscape, there are many paths leading to a particular location, it is all the more probable that many people will end up there. But, however they got there, people with states, cities and literacy are now subject to similar constraints, and are more likely to have the potential of developing empires, advanced technology, philosophy, universal religious systems, and so on.

It is obvious, however, that the accumulation of necessary conditions will not always go on indefinitely until every society develops the state and, eventually, industrial civilisation. Most societies did *not* spontaneously develop the state, let alone modern science and industry, and this is because in most cases the necessary conditions for these developments were not present. So stagnation and equilibrium are just as likely as evolutionary advance, and here it is important to stress that the special features of particular societies may be of great importance in how they develop. The evolutionary pathway followed by each society is to some extent determined by how it starts out, by the historical peculiarities of its early stages which may stamp themselves on its subsequent development. Throughout the book I will be drawing attention to why some specific features of a particular society's organization or belief system made it easy or difficult for it to develop hereditary political leadership, or the state, or science, or capitalism. Social evolution is certainly not some unitary, general process that is going to be the same everywhere, regardless of local circumstances.

## 6. The evolution of thought.

Since societies are systems of ideas, the human mind must have been central to this evolutionary process. There is a vast body of evidence from anthropology and psychology that the members of non-literate, small-scale societies with simple technologies, 'primitive societies' as I am calling them for convenience, do not in some ways think like the educated members of modern industrial societies. Many Victorians believed that the brains of people whose societies had developed relatively little science and technology, such as those in Australia, Melanesia, or Africa, were therefore different from those of other peoples, such as the Chinese, Indians, and Europeans, where these had become highly developed. But we now know that this was mistaken because African or Melanesian children from non-literate, tribal societies can, if sent to Western types of schools and universities, learn modern science.

So it is reasonable to assume that the brains of people 10,000 years ago were essentially the same as ours, and that what has actually changed are not our genes and our basic intelligence, but *how we use our brains, and the new intellectual skills that we have learned* This approach throws an entirely fresh light on the old problem of primitive thought, because to understand how learning occurs we can study living people, rather than speculating about our unknowable ancestors. We can actually learn a great deal about how culture has evolved from the studies of how the thinking of children develops that have been made by developmental psychologists, such as Piaget, in very extensive studies of children from all round the world. The reason is simple: the child finds some ideas and ways of thinking much easier than others, whether we are talking of the natural or the social worlds, but the child will only master the more difficult forms of thought if he has to face problems that involve them in daily life, and also if his culture can provide the intellectual tools for solving them, such as books and schooling. If these conditions are lacking, then in some respects the thinking of adults will not develop significantly beyond that of children, but these forms of thought will still be quite adequate for people to get by in ordinary life.

The Tauade of Papua New Guinea, for example, had no words for numbers beyond two, so it is quite understandable that they could have no idea of multiplication and division, and in their culture of counting on fingers and toes Tauade adults did not need to develop any further the simple mathematical skills they had acquired as children. They also thought about time and space in equally elementary ways that have close similarities to those of children. Again, children find it difficult to separate the names of things from the things themselves, which is why they easily accept the possibility of magical spells; and when trying to understand how clouds, and wind, and moon behave it is easiest for children to think of them as having some kind of purpose, and as behaving as they do because it is their job to bring us rain, or come out at night.

If technology is simple and if the intellectual environment generally is undemanding, then such elementary types of thought will be able to survive into adulthood, as in the case of magic, and the evolution of human thought is a classic example of the survival of the mediocre. This does not mean, of course, that members of primitive societies actually *are* children, because in all other respects, such as skills, imagination, knowledge and self-control, they are adults like us. But what it does mean is that they will think in some ways that are different from those of people who have learned to read and write, been to school and university, and live in cities where machines are part of every-day life – all rather obvious when one stops to think about it.

Mr Gradgrind demanded that the boys and girls be taught nothing but facts. Facts, however, are not presented to us on a plate, ready made. We have to apply our minds actively to the world of nature to construct our understanding of it, and how we do this will depend, among other things, on our assumptions about it and the intellectual skills we have developed. Once we go beyond the simplest of observations – that water is wet, or that fire burns – our minds constantly intervene with their own representations of reality: for primitive peoples magic and witchcraft are obvious facts, just as it seems to be a fact that the earth on which we stand is flat and stationary, and that the sun and moon are small bodies that move across the sky, and they will not encounter anything that seems to contradict these apparent facts. Our modern scientific form of thinking about nature, which is very different from the thought of pre-modern societies, has had to be constructed over the last few hundred years, and as a result the natural world now appears entirely different to us, and our powers over it have changed dramatically.

Understanding our own society, and thinking about moral issues, is in some ways as difficult as understanding the natural world. The human mind finds it easiest to think in terms of the here-and-now, the concrete and the local, and as long as an institution seems to work, it is easiest to perpetuate it by custom. Since custom is not based on conscious purpose or planning, but grows up unawares, it will be hard to reflect about it or to explain it to strangers from other societies. Moral and political philosophy, the social sciences, and the whole analytical, self-reflective, logical, and abstract forms of thought which we take for granted in thinking about how our society should be organized, are not normal for all human beings. They have only been reached by a prolonged historical and intellectual struggle, in which we have learnt to think about problems of society and ethics, and about human nature and psychology, about language, and about thinking itself.

These revolutionary developments in the power of human thought were not, then, the result of any changes in our genes or our brains, but of the co-evolution of culture and the intellectual skills of the individuals who transmitted it. [16] They were the result of people having to reflect on new problems and situations that were themselves produced by changes in social organisation and technology: by agriculture, for example, or the state, or money, or the demands of bureaucratic organization, measurement, and so on.

## 7. Conclusions.

To sum up, then, I shall show that the broad process of social evolution over the last 10,000 years was not a Darwinian process of variation and selection, nor driven by material needs and technology alone, but was the result of very different principles. Rather than thinking in terms of people doing random things that are then selected for, it is far more realistic to think of them doing what is easiest in the circumstances of a particular social and intellectual 'landscape'; this may well be mediocre, but will still survive because of the low level of competition that is normal. People do some things for non-practical or non-rational reasons, but without these motivations would never have pursued some activities or investigations that were crucial in the development of the state, or science and technology. What really matters is not how immediately useful something may be, but its future possibilities, its evolutionary potential. Evolutionary potential exists because many pieces of technology, social institutions, and ideas have other properties in addition to those for which they were originally adopted, and in the right circumstances they may prove immensely fruitful. There are often many different pathways to some crucial development such as cities, writing, or the state, and in the course of history there is an accumulation of those necessary conditions that provide the right circumstances for evolutionary development.

Whether or not those conditions appear very much depends on the particular organization and cultural traditions of each society, and in some cases these will prevent evolutionary development, and in others make it much easier. In general terms, the course of social evolution is much better summed up as a process of construction and constraint than as one of variation and selection, or of technological determinism. But, while I deny that material factors were ultimately responsible for social evolution, I am not claiming that ideas were, either. *All* theories of social evolution that are based on single causes, whether biology, or geography, or technology, or social organization, or the mind, are hopeless theories. The process by which the modern world was constructed was fundamentally an interactive one between all these factors.

Some may object, however, that I am merely reviving the long discredited idea of 'historical inevitability'. In a famous passage, the great historian H.A.L Fisher said 'Men. . .have discerned in history a plot, a rhythm, a predetermined pattern. . . I can see only one emergency following upon another as wave follows upon wave. . . there can be no generalizations, only one safe rule for the historian: that he should recognize . . . the play of the contingent and the unforeseen.' [17] Karl Marx, for example, claimed that history follows an inevitable path, and Fisher, who wrote this in 1936, might seem to have been vindicated by the complete failure of Marx's predictions about socialism and capitalism in the rest of the century. [18] 'Historicism', as the theory of historical inevitability is often known, has been justifiably ridiculed, so what is the theory of social evolution advanced in this book if it is not historicism?

In any debate about historicism, one of the first objections to be raised is that there is a conflict between historical inevitability and free will. But 'free will', in the ordinary sense of being able to make the decisions we think appropriate in any situation, does not imply the freedom either to do what we like, or to think what we like. We cannot make choices that are unimaginable within our own culture, or that are intellectually too difficult, and we can only act within the limits of what is socially and physically possible. Instead of thinking of people being compelled to do things by mysterious 'social forces', we imagine them instead as making choices within changing landscapes of *constraints and opportunities*, some environmental, some social, some biological, and some intellectual, in which it is easier to do or to think some things than others. (Popper refers to these 'landscapes' as the 'logic of situations'.) [19] There are, for example, the landscapes of tribal society, of the state, of industrial society, of intuitive thought, and of experimental science.

Unlike Fisher, then, we are not primarily concerned with specific events at all, but with the contexts in which events occur, with how particular social and cultural systems work, and how easy it is for them to change, and in what directions. Given that domesticable plants and animals existed, then agriculture was bound to develop somewhere and so, too were the state, international trade, cities, literate civilisation, and even world religions of some kind, 'bound to' meaning that the probabilities were overwhelmingly large. The whole emphasis will therefore be on different kinds of social organization and systems of ideas, and their potential for change in some directions rather than others, and not on particular people or events. The exact where, when, how, and by whom of all this was, of course, a matter of historical accident, but in spite of all the accidents and unique events and personalities of history, there are also fundamental constraints that can produce basically similar results, such as the structure of the state in China, India, Spain, and the Aztecs in the sixteenth century that I mentioned earlier.

There is here a genuine problem that historians like Fisher have simply ignored. It is rather like the game of Monopoly: the players are all different and the throws of the dice produce a completely different game each time, yet the underlying constraints produce essentially the same result – a single player who owns everything and has driven all the others into bankruptcy. This is a good illustration that unique events, even randomness, and free will, are quite compatible with broadly predictable outcomes.

But whereas every game of Monopoly ends with one player owning everything, and while on the world stage it was (probabilistically) inevitable that the state would emerge, of course not all societies must inevitably develop the state. Far from it: as we shall see, the majority of tribal societies had features that made it difficult or impossible for them to become states. In the same way, it was not possible for modern experimental science to have developed in any of the ancient literate civilisations. In so far as its appearance depended on a combination of unusual conditions in Western Europe, that might well not have occurred, the probabilities of modern science and technology, unlike those of the state, were actually rather low. The 'historical inevitability' of social evolution is simply the result, then, of the probabilities that different social and cultural conditions will come into being, and their potential for making certain further types of change more or less likely. It does not in any way contradict the obvious facts of our free will, or that great men and major events, the accidental and the unforeseen, have played essential parts in the actual history of mankind.

# CHAPTER II
## The Simplest Societies

*1. The life of the hunter--gatherer.*

If we wonder why our cats and dogs behave as they do – leaving scent-markers, for example – it is very illuminating to see how they lived in the wild. This is because their behaviour is based on instincts that evolved over millions of years of adaptation to their ancestral environment. And since man spent most of his history as a hunter-gatherer, and only a relatively short time as a farmer or city-dweller, it is often assumed that modern hunter-gatherers, too, can give us a privileged insight into human nature and abilities, telling us what we are *really* like. But although human nature makes it easier for us to do some sorts of things rather than others, it is not instinctive in the way that much of animal behaviour is. Our pre-agricultural ancestors therefore lived as they did not because they were following their instincts, but simply because they had no other choice in their very restrictive conditions.

It has actually taken the whole course of history to reveal the full potential of human nature and human abilities, that were concealed by the constraints of hunter-gatherer life, and needed advanced civilisation before they could emerge. In 1831, for example, Gilbert-Louis Duprez first sang the high C with chest voice, instead of falsetto, in a performance of *William Tell* in Italy; it had an immense impact and was the birth of the heroic tenor voice in nineteenth century romantic opera. [1] This had never been heard before in human history, and while there must have been many ancient hunters with the genetic endowment to sing like Caruso or Pavarotti, this genetic potential needed all the cultural resources of European opera before it could be realised.

The heroic tenor voice is, of course, a very particular and unusual skill, but no one could ever have guessed, from a knowledge of hunter-gatherer life alone, the much broader but hidden potential of human nature and abilities – mathematics, and science, for example, or the lust for status and power, altruism, ideological hatred, self-analysis and conscience, romantic love and sadism – and what our subsequent history would turn out to be once the restraints of the hunter-gatherer way of life were removed. All those familiar figures from history, the inventors, artists, and scientists, the religious teachers and philosophers, the enlightened rulers, the generals, and the tyrants and slaughterers of millions, the sycophants, and debauchees, the political and religious fanatics, the bureaucrats and torturers, have had to wait until a more complex social order had been constructed before they could find expression for any genes there might be for these dispositions of heart and mind. So the notion that in hunter-gatherers we can glimpse our true selves as human beings before they were obscured by the distorting encrustations of subsequent civilisation is the opposite of the truth.

We must now spend a few pages surveying what these limitations were, to see where we started from and so to appreciate more fully the consequences of agriculture when it came. We know a great deal about hunter-gatherers because many such groups have survived into modern times, but, some might ask, how far can we rely on modern information about hunter-gatherers to deduce how people used to live at the beginning of the agricultural revolution? Just because the basic mode of livelihood has survived it does not mean, after all, that nothing else in these societies has changed in the last 10,000 years. It used to be thought, for example, that the Australian Aborigines were relics of this ancient time, like flies in amber, and that their society, especially its totemism and complex marriage regulations, could give us a picture of how our own ancestors had lived before the coming of agriculture. All societies change, however, and the complicated Australian marriage systems may have evolved quite recently, so that there is no secure reason for believing that Aboriginal society gives us a uniquely true picture of how all hunter-gatherers used to live. In some parts of the world hunter-gatherers have been in contact with farming societies for thousands of years and this will have affected their way of life, while hunter-gatherers such as the Bushmen of the Kalahari Desert in Africa and the Negritos of Malaysia, are relic populations of social groups that used to be much more widespread and were driven out of those areas by farming immigrants.

*2. Ecology.*

Having made all these allowances, however, the foraging way of life imposes some basic limitations on how human beings can organise themselves and think about the world. The most important is that it can usually only support very low population densities, generally between 1 – 25 per 100 square miles, and another is the need to follow game and to forage for different sources of food as they ripen in various locations. Studies of modern hunter-gatherer societies in many parts of the world show that there are basic similarities between them, and we are entitled to assume that these would have applied to ancient hunter-gatherers as well. The very low population densities, the difficulty of making permanent settlements, and the equal availability of resources to all meant that groups were small, private property was limited to a few personal possessions, and any social inequality could only be based on age and gender, not on inherited wealth or power. We also have some very rich information from anthropological studies about just those aspects of culture that cannot be deduced from the archaeological record – language, beliefs, values, modes of thought, and social organisation.

By 10,000 years ago human beings had spread to most parts of the globe. In many cases, techniques of hunting and gathering had become very advanced, with dogs, bows and arrows, nets, traps, and the ability to grind and store seeds, and to leach the poisons out of potential foods by a variety of techniques. People had also learnt to exploit rivers, lakes, and oceans by the development of boats, fishnets and traps, weirs, and fishing spears and hooks, and in a few locations the supply of game, or nuts, or acorns was enough to allow fairly permanent settlements, as among the Ainu of Japan or the Yakuts of California. This was also true for the so-called 'Natufians' of the Middle East, for whom stands of wild wheat and barley made settlements possible as early as 15,000 BC. Fishing, in particular, may allow permanent settlements, and the most famous fishing societies, the North-West Coast Indians of Canada, had elaborate timber dwellings, the famous 'totem poles', and a stratified, clan-based society with chiefs at the top and slaves at the bottom. The problem with fishing as an economy is that it is tied to a particular location, because it is forced to depend on the natural supplies of salmon, candlefish, trout, and so on that are provided by the rivers and the sea in that place. Agriculture, however, can be adopted anywhere with sufficient rainfall or other sources of water, and it could therefore expand over a far larger area of the world. Hunting and gathering, of course, could do this too, but it could not support the level of population necessary for more complex societies.

It used to be thought that hunting and gathering was a life of miserable hardship and deprivation, in which people were forced to spend their lives endlessly roaming in the search for food under the daily threat of starvation. Indeed, the claim that human beings evolved in conditions of scarcity is a basic assumption of Darwinism about all animal populations: 'The resources that humans struggle for, which allow them to survive and prosper, are in short supply. This means that humans are caught up in a struggle for survival and reproduction with their fellow humans. This struggle is inevitable and unceasing.' [2] One only has to reflect, however, that most of human history for at least the last 50,000 years or more has been migratory, to see that this could not have been true because people always had the option of moving. While game had to be hunted, and to that extent may be called a scarce resource, the response was co-operation and sharing, not competition, and the struggle for food was not with one's neighbours, but with nature.

In reality, the image of hunting and gathering as dominated by scarcity is seriously distorted. Studies of modern foraging societies have shown that people in many of these societies can easily survive on less than four hours work a day, for only three or four days in the week. [3] The Bushmen of the Kalahari, for example, have been estimated to spend between 12 to 19 hours a week in getting their food, and in their case a species of nut, the *mongongo*, is found in such abundance that it acts almost as a crop. In tropical and sub-tropical latitudes especially, there is a wide variety of plant foods such as roots, nuts and berries, as well as game; this broad range of plants on which they live is less liable to fail than the smaller selection cultivated by the farmer, and the diet of hunters and gatherers is generally well balanced and healthy, so much so that it has been described as the original affluent society. This was particularly true of those groups that could also exploit fish such as salmon and the other products of river and sea. While foragers experience periodic food shortages, starvation has probably been less of a threat to them than to farmers, whose much narrower range of food plants is more vulnerable to drought, or to plagues like locusts and crop diseases, and also more likely to be the source of dietary deficiencies and malnutrition.

But the foraging way of life did not allow individuals or families to accumulate property. The land and its plants and animals are the collective assets of the whole group – and the bands of which it is composed. Those who killed animals in the hunt had to share the meat, while personal possessions were limited to basic tools and weapons that were also frequently borrowed. Everyone therefore had equal access to the necessities of life, either directly or indirectly through sharing, and no one could store up a surplus of food for private consumption; the only storage was to enable groups to avoid seasonal shortages. So the most striking features of hunting and gathering society is how small they have always been, the simplicity of their social organisation and the remarkable degree of equality that prevails, even taking account of differences of age and gender.

Bands are small groups generally of between 25 to 50 or so; this is partly because of the small number of people who can be supported by a given area of land, but that is not the only reason. Even where the natural resources would allow larger groups, the people seem to prefer to associate in small groups of about a dozen families or less, and even these may split up for part of the year. But such small groups are not viable as self-sufficient breeding populations, and if all the members of the group are to find mates they must get them from other bands. It has been calculated that it requires about 175 breeding couples to produce a more or less equal number of males and females in the long term, and the result is a number of inter-marrying bands, which collectively form a group speaking the same dialect and numbering about 500. People often move from one band to another, especially if they have quarrelled and want to avoid each other, so we should not think of bands as tightly-knit little clusters of people who spend their whole lives together, united against a hostile world of nature and other bands. On the contrary, the general picture of band organisation is one of rather casual groups made up of families related to one another through both their mothers and their fathers, and who regularly intermarry with members of other bands. In addition, since supplies of food fluctuate in different areas, it would always have been advantageous to allow movement between bands to take account of this.

## 3. Kinship and marriage.

Family ties are therefore of great importance, but fathers have no special control over their adult children (except to some extent where, in societies like the Eskimo, they may be deferred to as expert hunters). This is because they have no control over access to food, which only requires the basic tool-kit of weapons and implements. A father and some of his adult sons may sometimes live together in the same band for a time, but we don't find that there are larger groups of kinsfolk like lineages or clans, all descended from a founding ancestor, and all of whom have special obligations to one another and are united against other lineages. This is basically because there is no opportunity for groups like lineages to own land and restrict access to it to members of the lineage.

Nevertheless, in these small groups of hunter-gatherers everyone will be a relative of some kind, some more distant than others, and so for these people kinship has to be a basic feature of their lives. In practice, there are only three ways of tracing descent -- through males only (patrilineal), through females only (matrilineal), and through both (cognatic). It used to be thought that matrilineal descent must be the oldest, because it was believed that originally there were no stable unions between men and women, so that children would only have known who their mothers were, and not their fathers. Then, it was supposed, patrilineal descent developed together with marriage, while cognatic descent was the last of all to evolve. In fact there is no reason why any of these modes of descent should be the earliest, but if any form was, it is more likely to have been the cognatic type. This is the type of descent found in most hunter-gatherer societies today, while lineal descent groups are highly unusual, primarily because of the lack of inheritable property in land or stock. Cognatic descent also fits in well with the open band way of life in which individuals are free to move from one band to another. (There is, however, one striking exception to what I have been saying about the lack of lineal kin groups among hunter-gatherers, and that is the Australian Aborigines. Here, although people lived in bands in the normal way, there were also dispersed lineal descent groups associated with different totemic animals, and which regulated marriage.) [4]

It was once believed that early man had nothing like marriage, and that men and women coupled at random, while perhaps later on groups of women had stable unions with groups of men. In reality, neither such arrangement has been discovered in any living society, and we find everywhere that there is the idea of a stable union between an adult man and woman who have children together and support them and each other. Four different forms of marriage are biologically possible:

1. many men – many women (group marriage)
2. many men – one woman (polyandry) ⎤
3. one man – many women (polygyny) ⎦ polygamy
4. one man – one woman (monogamy)

Since 1 doesn't exist, actually we only find 2, 3, and 4. 2 is very rare, and societies that *only* allow 4, monogamy, have been very much in the minority in the pre-modern world. Most societies have therefore had 3, though in practice only rich or important men can have more than one wife, while ordinary men have to be content with one.

In hunter-gatherer society everyone gets married as a matter of course, for survival; the responsibility for hunting is primarily that of the men while gathering is the main job of the women, though they may kill small game as well. Because romantic love is so important to us we often assume that it must therefore be an essential feature of human nature, and a basic motive for marriage in all societies. Sexual attraction is certainly a universal feature of human nature, but while human beings have no doubt been *capable* of romantic love for a very long time, it is not reported by students of modern hunter-gatherer societies. This is not surprising because romantic love, by definition, is not just between a man and a woman but between a man and a woman who see each other as special, even unique, and this requires a large pool of potential mates even to be a viable possibility, let alone the norm. But since bands are very small there may only be two or three possible spouses available at any one time, and people simply have to choose between these regardless of any romantic feelings. It is only in urban civilisation that romantic love actually appears in a form we would recognise. It may also be true that men prefer women with large breasts and narrow waists, and that women prefer tall men, but in practice our prehistoric ancestors had to take what they could get – the fat one or the thin one, as it were. [5]

Evolutionary biologists like to imagine that men have always attracted 'mates' through power and resources, because women want a secure environment for child-rearing. This sounds like meaty, down-to-earth realism, but unfortunately fails to grasp that in hunter-gatherer societies, not only are power and resources pretty equally distributed but that, even at this level of social organization, women are often not free to choose their husbands, because their father or brother may wish to exchange them in marriage with another man. (The freedom of women to choose their husbands becomes much more restricted in farming societies, as we shall see.) The main motive for marriage is certainly practical, however: because men and women pursue the food quest in different ways they have to co-operate, and therefore the nuclear family is the basic unit of labour. Obviously sex is also a fundamental consideration but is not by itself a sufficient explanation, since many societies allow young people to have sexual relations before marriage, and in others adultery is also common; but the husband and wife who have had children become full adult members of their society as a result.

*4. Leadership and equality.*

Because no one can control more resources than anyone else, and because the extended family is the largest effective kin grouping, there is no means by which any family can acquire more wealth or influence than any other, and for any system of hereditary leadership to evolve. Nor does the eldest son have any special status, as he usually does in farming societies, because among foragers there is no property for him to inherit, or lineage of which he can become the head. In fact, there is a general disposition to cut the superior person down to size. One of Lee's informants among the !Kung Bushmen said 'When a young man kills much meat he comes to think of himself as a chief or a big man, and he thinks of the rest of us as his servants or inferiors. We can't accept this. We refuse one who boasts, for some day his pride will make him kill somebody. So we always speak of his meat as worthless'. [6] In the small groups such leadership that exists is weak and informal, and depends entirely on the personal qualities of the individual.

It might be thought that the strongest man could simply seize control and bully all the rest of the group into submission. But this is quite unrealistic: even the strongest man has to sleep, and even if such a would-be dictator were not murdered, people would simply abandon him and go and live somewhere else. The notion of a leader imposing his will on his followers misses the whole point of leadership in simple societies, which is that the leader has to attract people by having something to offer them, not by threatening them, because individuals are free to live more or less where and with whom they like, and have free access to all the necessities of life.

To have power over people one must control something they want: food, land, personal security, status, wealth, the favour of the gods, knowledge, and so on. In other words, there must be *dependency*, and among hunter-gatherers there is precious little of that. In more complex societies, where people are not self-sufficient in defence, or in access to resources or to the supernatural, they will therefore be willing to accept inequality of power because they obviously get something out of war-leaders, or clan heads, or priests. But it follows that in these early stages power cannot be imposed but only 'negotiated' as a *quid pro quo*; it is something that is only conceded to those who can satisfy the needs of their dependents. (Only after the advent of the state can power be riveted on to people by force whether they like it or not, and when it is too late for them to do anything about it except by violent rebellion.)

In these small and flexible societies there is no official 'slot' of authority, like that of a chief that someone has to fill, and without much in the way of organised warfare and without the possibility of controlling access to scarce resources like land or stock, the only scope for personal dominance lies in the realm of religion. As we shall see, some hunter-gatherer societies have important shamans who claim magical powers, and to be possessed by spirits that give them a variety of supernatural abilities. In Australia the older men sometimes claim a monopoly of sacred knowledge and so have been able to control the initiation of young men, and give themselves privileged access to the young nubile women. But these functions depend on age or on personal abilities, and so can't be converted into hereditary offices.

The qualities most valued in leaders are good sense, good temper, hunting skill, and the wisdom of age. Among the Bushmen, 'Leaders work in subtle ways; they are modest in demeanour and may never command but only suggest a course of action'; [7] and the leader of a Netsilik Eskimo band is called *inhumataq*, 'the one who thinks'. Again, 'There are no chiefs, councils, or any formal governing bodies in a Pygmy camp. There is respect for age and experience... at the same time every man in the band has a full say in any discussion'. [8]

Nor, in fact, is there very much for a leader to do. The adult members of the band are quite able to make their own decisions about daily foraging, and relations with other bands do not require any official spokesman or mediator. When quarrels break out within the band, as they often do, the senior band members can often help to resolve them, by talking to those involved and joking, but even here there are no formal mediators who can resolve disputes in an authoritative way. Fear of public ridicule and shame are probably the most important means of maintaining peaceful relations and social order, but when relations between two individuals are seriously strained the standard means of resolving this is for one or both parties to go and live with other bands, where they will always have relatives through their mother or father. Really dangerous men (including those who are believed to kill fellow band members by witchcraft) may be killed with the consent of their kin, which means that no revenge will be taken.

The basic way in which band solidarity is maintained is by sharing food and exchanging gifts, and lending one another tools and weapons. In the words of one Bushman 'The worst thing is not giving gifts. If people do not like each other, but one gives a gift, the other must accept. This brings a peace between them. We give to one another always, we give what we have. This is the way we live together'. [9] But in all these societies every gift is made in the expectation that the giver will receive one of equal value in return. The idea that 'one good turn deserves another' is a basic and universal human belief, (as is the idea that one bad turn deserves an equally bad one or worse). This sort of gift exchange is not, of course, an *economic* transaction, any more than when a group of friends in a pub in our society take it in turns to stand each other a round of drinks. This is a symbol of harmony, and its purpose, as among hunter-gatherers, is not to get more beer for individuals but to enhance the comradeship of the group.

In band societies objects are also used as payments for services, such as to a shaman or a ritual expert, or for someone's help in making something such as a canoe, or to buy the right to use something such as a sacred song or design or a particular ritual; or as compensation for an insult or an injury; or to wipe out a homicide; or given by a man to the parents of the woman they have allowed him to marry. In all these cases the payments are made because what is paid for – the assistance, the rights, the knowledge, the life of the victim, or the wife – can't be supplied by the purchaser himself and has to be converted into things. To be sure, vengeance is a common alternative to the insult or for homicide, and a man may provide a daughter or a sister to those who gave him a wife, but such forms of repayment may be unwelcome or unavailable and so it is here, in the valuation of non-material things like rights, services, and offences, or human beings, by equivalent amounts of desirable objects such as shells or tools – that we have the beginnings of one aspect of money, convertibility, although at this stage it is confined to a few special situations in life.

Hunter-gatherers are often thought to be able to tell us if human beings are naturally selfish or altruistic. [10] While biologists have had serious problems explaining why members of animal groups co-operate, in human groups, however, the question makes little sense. Individuals who persistently tried to get without giving in return would soon be ostracized, so that to this extent altruism is simply enlightened self-interest. The hunter shares his catch because he knows that others will later share theirs with him. But people understandably believe that the bands of our hunter-gatherer ancestors must have been united by more than this, a spirit of group loyalty and comradeship; as someone recently wrote to *New Scientist*, 'we would have been co-operative, compassionate, and *if necessary self-sacrificing*'. Studies of modern hunter-gatherers do not support this. Although sharing and generosity are certainly basic practises, we should not exaggerate their compassionate and comradely aspects, because all this gift-giving is based, as I have already said, on quite conscious self-interest – if you don't help others they won't help you. But there is also a strong undercurrent of envy, which makes people uneasy if they have more than some other members of the group. This sort of 'balanced reciprocity' is therefore quite compatible with a resentful individualism that is the opposite of true co-operation and altruism.

Marshall says of the Bushmen 'Their security and comfort must be achieved side-by-side with self-interest, and much jealous watchfulness. Altruism, kindness, sympathy, or genuine generosity were not qualities that I observed often in their behaviour'. [11] Nor is there much evidence that band members are specially loyal and supportive of one another. Among the Chewong of Malaysia 'Individuals are expected to, and on the whole do, carry on their activities on their own. It is a rare sight to witness someone asking someone else for assistance. Similarly, offers of assistance are also rare. I have many times watched strong young people lying about all day while old, and sometimes ill, people toil with heavy work without asking for or receiving help'. [12] The Paliyans, too, 'work and live in parallel rather than joint fashion and exhibit little co-operation outside their rather loose nuclear families. They are hesitant to become emotionally involved with others and equally reluctant to unite toward practical goals. There is a very strong expectation for autonomy'. [13]

According to Woodburn, 'The Hadza [of Tanzania] are strikingly uncommitted to each other; what happens to the individual Hadza, even close relatives, does not really matter very much. People are often very affectionate to each other, but the affection is generally not accompanied by much sense of responsibility. If someone becomes ill he is likely to be tended only so long as this is convenient.' [14] He discusses, in this connection, how hunter-gatherers often leave the sick to die, and gives the example of a paralysed boy abandoned by his mother and other close relatives only a few miles from water, to which they could have carried him without too much difficulty. He concludes, from this and other cases, that ecological explanations are much less important than the individualistic quality of social life. [15] Real group loyalty and altruistic self-sacrifice, like romantic love, are not universal features of human nature, but the products of more complex societies.

In the 1960s, hunters and gatherers, in the spirit of the age, were also portrayed as especially peace-loving and unaggressive, and it was fashionable to believe that they represented the real nature of Man before greed, militarism and, of course, capitalism, had corrupted it. This amiable illusion cannot be maintained, and there is a good deal of evidence to show that hunter-gatherers could quite well be aggressive and warlike. Very high death rates from fighting are recorded for some Aboriginal groups, for example, in the nineteenth and early twentieth centuries, although these fatalities occurred over many years of small skirmishes. Defence of territory is recorded for some Eskimo groups in earlier times, as it is for the Indians of Tierra del Fuego, the Pygmies, Bushmen and Hadza, and there is plenty of evidence that in hunter-gatherer societies generally, quarrels sometimes led to homicide. The relatively low-level warfare found among the hunter-gatherers seems to have been more often related to vengeance, however, than to serious competition over resources, vengeance involving quarrels over women, insults, and disputes that got out of hand leading to murder. Fighting over women was a significant cause of violence in these societies, a cause of tension that was aggravated by the custom of female infanticide.

But warfare among the hunter-gatherers is inherently limited by the sheer lack of numbers of adult men, and by the lack of social organisation and leadership to coordinate military actions of any significance. Warfare, then, is not as marked a feature of hunter-gatherers as it later becomes with the much larger and more tightly organised societies of farmers and pastoralists. Even without actual warfare, however, strangers fell entirely outside the range of those to whom any concern was due. Among the Eskimo, for example, 'In traditional times fear, intense suspicion, and potential or actual hostility permeated relations between strangers'. [16] The !Kung Bushman says of people from other ethnic groups 'We call creatures who are different from us *!hohm* [wild animals] because when they speak we cannot understand a word'. [17]

## 5. *Religion and forms of thought.*

Evolutionary theorists are always asking how religion began and why it was adaptive, and assume it is a simple disposition to believe in supernatural beings which has continued essentially unchanged from cave to cathedral. In this sense, however, there is no such 'thing' as religion. I shall show in more detail in Chapter IV that religion is not a simple disposition at all, but covers a wide range of thought and feeling, and that its whole structure has changed radically as societies have become more complex. Religion as it exists in modern Western society is therefore a useless guide to religion in primitive society. Even among hunter-gatherers religion is far wider than supernatural beings and includes a range of beliefs that try to make sense of man's relations to the world around him. At this stage in the development of thought people think most easily about causes as the actions of some kind of being, so there is no doubt that supernatural beings are very important, but notions of *order* are even more fundamental because they provide the meaningful context, the stage, on which these beings perform.

While the religious beliefs of different foraging peoples vary a good deal, they all focus on animals, plants, and the physical landscape, the weather (especially thunderstorms), and the seasons. Man himself is not seen as fundamentally distinct from his environment but as an integral part of it, and shares with animals and plants the same life essence. The natural world is therefore seen as animated with spiritual power: in the beginning it was given its form and order by the acts of a variety of spiritual beings, or 'culture-heroes', who also maintain it, and men can establish relations with these beings and with the animals and plants in a variety of ways, especially by dreams, by the trances of shamans or medicine men, or by performing appropriate rituals. Different species of animals, in particular, are not seen just as the individual birds or animals, as we would think of them: the species itself has a kind of eternal existence, manifested in the actual creatures, but which itself can never die. This means that the species can be thought of as having a mystical, 'totemic', association with individuals or, especially in Australia, with clans or other groups, and rituals to increase or preserve the species may be performed.

Mythology, derived in particular from dreams and trances, is often elaborate and well developed, with stories about the doings of the culture-heroes who existed before the present and modern people, who formed the landscape and gave the ancestors the knowledge of things like fire and cooking, and the proper ceremonies, that were necessary to transform them from animals into real people. In this 'Before' or

'Dream Time' state of things the world was unfinished and confused, so that animals and humans were not altogether distinct, and humans did not know how to behave. These supernatural beings are thought to maintain the cosmic order, especially when it is threatened by the violation of taboos such as laughing at animals or marrying someone too closely related. Myths may be acted out to get rain, or success in hunting, for the well being of plants and animals, to protect people from supernatural danger when taboos have been broken, or at social ceremonies such as initiation and the disposal of the dead.

Rituals may involve human bloodletting, real or simulated sexual acts, making sacred drawings, manipulating sacred objects such as stones or carved boards, the whirling of 'bull-roarers', and acting out the roles of culture heroes or totemic beings, but sacrifice is rare (except in a few cases such as the bear sacrifice of the Ainu), and only becomes a really important ritual act after the domestication of animals. But in the unstructured society of the hunter-gatherer, there can be no hereditary priests, as there are in farming societies, so that rituals have to be performed by non-hereditary shamans or older men.

Shamans are particularly important among hunter-gatherers, and their characteristic trances are usually induced by drumming, dancing, and singing often to the point of exhaustion, and are the means by which they can transform themselves into spirit beings or animals, and so in a way repeat the formative acts of the culture heroes. Their basic work is curing people of sickness, but they can also be asked to help with success in hunting, ending bad weather, finding lost objects, and visions or prophetic dreams are integral parts of their work. Shamans are elusive, transforming themselves between humans, animals, and spirits, and are also often sexually ambiguous, being the main examples of men who assume female roles.

One of the most popular explanations for belief in supernatural beings is that they are a projection of earthly human relations, especially that between children and their fathers, into the supernatural realm. The culture-heroes and spirits of hunter-gatherers are typically of great power, inscrutable, unpredictable, often malevolent, and not much interested in the morality of human actions, but are apparently more concerned with maintaining cosmic order by punishing people with thunderstorms or wasting diseases for the violation of taboos. Laughing at certain types of animal or killing female animals, burning bees-wax, killing cicadas, drinking raw blood or throwing leeches on the fire, and the incorrect use of food names are much more likely to provoke the anger of these deities or spiritual beings than murder, theft, or adultery. In all these respects they are quite different from fathers in band society, who would never treat their children as the gods are supposed to treat men, and they have equally little resemblance to such band leaders as exist.

In spite of the often elaborate mythology and pantheons of spiritual beings and culture-heroes, we generally find that hunter-gatherers only use very simple forms of classification by comparison with those that are found in more complex societies. Verbal counting systems, in particular, are often very poorly developed, sometimes with no words for numbers beyond two or three. We can get a good idea why this should be so from the example of a Cree hunter from eastern Canada: he was asked in a court case involving land how many rivers there were in his hunting territory, and did not know:

> The hunter knew every river in his territory individually and therefore had no need to know how many there were. Indeed, he would know each stretch of each river as an individual thing and therefore had no need to know in numerical terms how long the rivers were. The point of the story is that we count things when we are ignorant of their individual identity – this can arise when we don't have enough experience of the objects, when there are too many of them to know

individually, or when they are all the same, none of which conditions obtain very often for a hunter. If he has several knives they will be known individually by their different sizes, shapes, and specialized uses. If he has several pairs of moccasins they will be worn to different degrees, having been made at different times, and may be of different materials and design. [18]

Again, hunter-gatherers reckon the passing of the year by their seasonal activities and the availability of different food resources, rather than by a calendar based on twelve lunar months. The evidence suggests that lunar calendars with named months only start to develop in farming societies, together with more complex number systems, so it is extremely unlikely that our prehistoric ancestors thirty thousand years ago were keeping track of days of the month by notches on bones, as has been claimed. Nor do hunter-gatherers divide space by the cardinal points of North, South, East and West, but use specific features of the terrain to orientate themselves. Basic colour terms are often restricted to dark/light, while the names of the chromatic colours are based on a wide variety of actual objects, and their classification of plants and animals is more concerned with practical details than with system: 'Not only are their taxonomic systems limited in scope but they have a relative unconcern with systematisation'. [19] The whole emphasis of their thought is on the local, the specific, the concrete, and the individual.

This has been called 'memorate knowledge', that is, knowledge based on the details of daily experience, and it has been noted in a wide range of hunter-gatherer societies, as well as some shifting cultivators. I emphasise this relative lack of developed classification because it extends to social relations as well as to the natural world, and 'Just as [the Paliyans] have problems with natural taxonomy, they manifest difficulty providing models or rules to describe social practices such as residence'. [20] The lack of formal mediation in quarrels gives little opportunity to develop any body of law, or to think about criteria for dispute settlement, and in moral thinking generally the whole emphasis is on the practical requirements of getting along together, not on ethical principles.

The personal qualities most approved are those such as sharing, and generosity, and on avoiding disputes and fighting, all of which are obviously of great practical relevance in daily life, but there is no appeal to abstract principles such the Golden Rule, of doing to others as one would be done to. Someone may say 'If struck on one side of the face, you turn the other side toward the attacker', [21] but on closer examination this is not a condemnation of revenge in the Christian or Buddhist manner, but is simply practical advice to avoid fighting within the group. The basic moral orientation is individualistic, and one helps others not out of any altruistic concern for their well-being, but from motives of self-interest, and people generally behave properly from fear of social pressures to conform, not from any promptings of 'conscience'. This, like romantic love, is not something that springs spontaneously and universally from every human heart, but has to be developed in the right social circumstances. Nor is there any idea, as we have seen, of a moral obligation to strangers. The kind of moral thinking that we find among hunter-gatherers is therefore of an elementary kind: people pursue their own interests, obey the rules not from the promptings of conscience but from group pressures and self-interest, accept that good for good and bad for bad is fair, but justify doing the right thing in terms of practical self-interest, not abstract moral principles. [22]

*6. Humans as animal populations.*

When trying to explain the survival of customs and institutions that seem of no practical use, materialists disregard the explanations of the people themselves, and claim that the real reasons why such practices survive is that they have been selected for their adaptive value. This may be plausible for animal populations, where bees, for example, signal the location and distance of nectar by a standardized dance performed by instinct. But in human society there are no instinctive patterns of behaviour of this type, which could be selected for. For example, the Montagnais-Naskapi of the Labrador Peninsula in Canada locate game by divination: a shoulder-blade of the animal species to be hunted is placed in the fire, and the pattern of cracks and spots that are produced on the bone is believed to tell the hunters where to find the animals they want. It has been suggested, however, that this form of divination has survived because it is really a randomising procedure that often *prevents* the hunters finding their quarry. The adaptive value of doing this is that, by preventing over-hunting, it is therefore a means of *preserving* game-stocks. [23]

But what is being proposed here is theoretically impossible. The Montagnais-Naskapi, like all primitive peoples, don't think in terms of chance and randomness at all, and they continue with their divination, not because it is instinctive, but because they believe it helps them attain their goal of successful hunting. If, in fact, it achieves the opposite effect, which *happens* to confer an adaptive advantage, but which they don't know about, this can have no bearing at all on the survival of the custom. If, for some reason, they stopped believing in divination, they would abandon it, regardless of the consequences which, in any case, they would not understand. It is therefore a complete illusion to believe that customs which, unknown to the people, have a beneficial effect on survival, can be selected for this reason regardless of the people's own beliefs. We have here, then, a classic example of the fallacy of using the model of animal behaviour, based on instinct, to explain aspects of human society that are based on ideas about the world.

Again, institutions such as warfare, Australian marriage systems, and the monopoly of young wives by older men, birth control practices, the shifting membership of bands, and elaborate sharing and exchange are all supposed to have evolved because of their adaptive advantages. So, the monopoly of young women by older men is 'really' to ensure that the most fertile women are married to mature men at the height of their powers as hunters; the flexibility of band membership and warfare redistribute population in relation to resources, and birth control and infanticide keep the population at a safely low level in relation to possible famine years. But how have such useful adaptations come about? We have already seen that, since human behaviour does not rest on instinct, there is no other psychological mechanism by which people could be motivated to believe or do things that have beneficial effects, but which they themselves don't know about.

Indeed, we should not assume that individuals at this level of social organization know or care about the welfare of the larger group or 'the optimal exploitation of its resources by the total population'. Individuals may have no more interest in preserving the environment for the common good than they have in keeping the population stable over the generations or maintaining peace between bands. (Deliberate population control only occurs in more complex societies, where compulsion is possible.) They move from one band territory to another in time of shortages simply to satisfy their own hunger; women keep the numbers of births down by prolonged lactation, post-partum sex taboos, abortion and infanticide because they cannot support a small child and a baby at the same time; people move to other bands to avoid their enemies purely for their own peace of mind and men fight other men because they are angry with them. If these individual preferences and practices also happen to work out for the general good that is quite accidental, and in fact they can sometimes have a detrimental effect on the welfare of the group.

People will quite happily exterminate the animals on which they rely, or pollute their environment, or overpopulate it if that is the easiest thing for them to do as individuals, and explanations of individual behaviour by convoluted academic theories of what is adaptive either for the group or for individuals, without their knowledge, are always wrong. The reason that hunters and gatherers have generally lived in harmony with nature is merely that their very simple technology does not allow them to do too much damage. If some practices are actually adaptive (and it is often harder to prove this than might be supposed) there are only two possible explanations: it is either deliberate, as when people store food to tide themselves over a lean period, or else it is an accident.

# CHAPTER III

## The Agricultural Revolution

*1. The domestication of plants and animals.*

Ever since people began reflecting on material progress, from the ancient world onwards, it was always supposed that the life of the hunter-gatherer was one of miserable hardship that drove man to improve his conditions of life, by adopting agriculture, building houses, living in cities, and so on. Since we now realise that this picture of hunter-gatherer hardship is greatly exaggerated, we have to see why people would have wanted to take up agriculture instead.

Agriculture was the fundamental change in human life-style that made all the other developments possible: towns and cities, the state, literate civilisation, and ultimately the modern world. It did this because planting crops enormously increased the number of people who could live in a particular place, and also allowed them to settle down in permanent settlements. It seems to have begun when people started growing wheat and barley in the Fertile Crescent of the Middle East about 8000 BC, not long after the end of the Ice Age (around 9,000 BC). Over the next few thousand years agriculture, based on other crops, such as rice and millet, but also on tubers like yams and taro, also appeared in a number of other places, including Asia, Central and South America, and Africa. [1]

The reasons for this 'agricultural revolution' have been debated for decades by archaeologists and anthropologists. [2] After so many millions of years of hunting and gathering, the sudden appearance of agriculture all over the world within such a relatively short space of time has seemed very striking. There have been two basic sorts of answer to the problem. First, that it was a brilliant invention (probably

## The Agricultural Revolution

in the Middle East) that then rapidly spread around the world because everyone was only too thankful to be liberated from the burden of gathering their food. Secondly, that it was an adaptive response to world-wide environmental problems, and the two favourite candidates for these are population pressure, and climate change.

All these theories get some of their plausibility from the idea that agriculture took an extraordinarily long time to appear, and therefore that something radically new must have been responsible for it, whether this was an inspiration of genius, or some ecological disturbance of the *status quo*. But we actually get a very misleading impression of the 'delay' of the agricultural revolution if we think of man in terms of *Homo erectus*, who first appeared between 1½ - 2 million years ago, and then compare this enormous period of time with the few thousand years in which agriculture has existed. It is much more realistic to compare the time-span of agriculture with that of modern man, *Homo sapiens sapiens*. While the earliest fossils of these, from Ethiopia, are now dated to approximately 200,000 years ago, there seems to have been a considerable gap before there is any evidence of what we would consider characteristically human behaviour, such as burials, art, or the use of bodily ornaments like necklaces. This type of behaviour seems to have started appearing around 100,000 – 75,000 years ago, and any sophisticated use of language *may* not be much older than this either. So it is possible that modern man had not actually developed the cultural and linguistic abilities to domesticate plants and animals before this time. We should also consider that during the last 50,000 years or so, human beings were migrating all over the world, so that exploring these new environments would in itself have taken a considerable amount of time. Looked at from this perspective, the beginning of agriculture around 10,000 years ago is not, then, the kind of 'delay' that should particularly surprise us.

The 'brilliant invention' theory also rests on two false assumptions: that discovering how to grow crops was a serious intellectual challenge, so that it could only be expected to happen once; and that the foraging way of life was very arduous, so that this discovery would have been generally welcomed around the world as a major liberation. We now know, of course, that hunting and gathering is not the life of grinding hardship it was once supposed to be, so the idea that agriculture would be accepted as a major liberation is clearly false – in many ways it required more labour, not less. (This is particularly true of cereals, which have to be threshed and ground before they can be cooked, whereas tubers can simply be baked in their skins.) Nor was agriculture difficult to discover. Hunter-gatherers have a very extensive knowledge of the plants in their environment, and modern anthropological studies show that they are aware of how to grow them, even though they may not choose to do so. [3]

There are many reasons why people might wish to experiment with growing plants, apart from wanting to eat them, such as maintaining useful plants in convenient locations, or encouraging the growth of species that were particularly desired such as fish poisons or gourds. It is quite possible that sedentary fishing communities in the Amazon basin, for example, gave an important impetus to agriculture in this area by experimenting with a variety of plants in their gardens. One of the earliest domesticated plants, the bottle gourd, *Lagenaria siceraria*, found throughout the humid tropics and which occurs in Mexico at least 7000 BC, is not edible at all, but produces gourds to serve as a variety of containers, and many other things that would have been useful to hunters. Knowing how to grow food plants was simply part of this general botanical knowledge, and the idea that agriculture was an intellectual 'discovery' is therefore quite wrong. It no longer needs genius to explain it, then, and was instead the novel *application* of long

*The Agricultural Revolution*

familiar knowledge about plants in general.

This being so, agriculture is much more likely to have developed independently in many different places, rather than been diffused from a single point of origin. In any case, it is hard to see how any process of diffusion from the Middle East could have operated fast enough, especially since the methods of cultivating cereals, legumes, and root crops are so different. It is not really very likely, for example, that the inhabitants of the remote interior of highland New Guinea began cultivating their taro and bananas around 5000 BC as the result of a stimulus from wheat-growers in the Middle East in 8000 BC. But diffusion was certainly very important in spreading agriculture at a more local level, and there was also a major diffusion of cultivated plants from the primary centres: wheat and barley, for example, were transmitted to China from the Middle East about 3000 BC, while the Asian taro and bananas were transmitted westwards. But none of this explains why so many peoples chose to adopt agriculture within the span of a few thousand years.

One theory is that it was the result of climate change. There is certainly no evidence of agriculture before the end of the Ice Age in any part of the world, and after it ended the world climate would have become moister, average temperatures risen, and sea levels become higher. But this was not the first time in world-history that agriculture became possible, since it had always been possible in the tropics. (Over the last 100,000 years, average temperatures at the equator even during the glacial periods were at most only 8-10° F cooler then they are today.) No doubt, the end of the last Ice Age would have had local effects that favoured farming in certain areas; in the Middle East, for example, the wild grasses that were the basis of the cultivated forms of wheat and barley seem to have become more abundant. But, given the independent adoption of agriculture in many other places, the fact that this first occurred in the Middle East had no general significance. At most, then, a warmer climate could have had some local effects that would have favoured agriculture, but it could not have been responsible for its world-wide adoption, since this had always been possible in large areas of the world.

The alternative theory, that it was a response to population pressure, [4] recognises that the life of hunter-gatherers was relatively easy, and so concludes that they would not have taken on the burden of farming unless they had been forced to do so by excess numbers. This is also the kind of very basic factor needed to explain a world-wide phenomenon, like the adoption of agriculture within a few thousand years.

But modern hunter-gatherer population levels typically remain stable, so it is not at all obvious why, in the period we are considering, there should have been such a general tendency for their populations to increase all over the world. Indeed, even in 1500 AD almost a third of the world's land surface was still occupied by hunter-gatherers, including the whole of Australia, many parts of South and North America, the Circum-Polar regions, and many societies of Africa and South-east Asia, (though by then only about 1% of the world's population were hunter-gatherers). Even where agriculture was practised, it was often in conjunction with foraging. In many parts of Papua New Guinea, for example, there was greater dependence on hunting and gathering than on cultivated crops until the last two or three hundred years, when the sweet-potato was introduced. Why were all these areas not subject to excess population as well?

The theory also assumes that the *only* advantage of agriculture is to support denser populations, so 'it will thus be practised only when necessitated by population pressure'. [5] But this is quite untrue. Agriculture's other obvious advantage is allowing people to settle in fixed locations by making gardens, and this gives us the most important

*The Agricultural Revolution*

clue about its origins. People may choose to live in fixed settlements, during part or all of the year, for all sorts of reasons noted by modern anthropologists. They may initially want to live close to food-stores of nuts, acorns or grains, or to gardens of useful plants, or the location itself may be very favourable for activities such as fishing, or have other advantages that they are unwilling to give up. A whole series of craft activities, such as basket-weaving, pottery, wood-working, the making of ornaments, and more elaborate food-preparation such as pestles-and-mortars, are all more easily carried out in permanent or semi-permanent settlements. Fixed settlements are also very convenient for holding ceremonies, such as initiations or other sorts of assembly. The various attractions of increasingly *permanent settlements*, then, for many groups of hunter-gatherers around the world, are a much more plausible reason for the wide-spread development of agriculture, than population pressure. We can also tell from observations of living foragers and shifting cultivators that the food surplus which can be produced by agriculture can have other uses besides supporting permanent settlements: in systems of exchange and competitive feasting between different groups, or traded with different groups as between forest and fishing societies, or for other desired commodities such as stone for tools and shells and ochre for personal ornaments, or in the domestication of animals – the earliest agricultural sites in Highland New Guinea, for example, are associated with pigs. So 'It is unlikely... that the same combination of factors operated in every case, and most archaeologists today would agree that there can be no universally valid model for the adoption of agriculture.' [6] We also need to remember that agriculture need not have been adopted suddenly. There could actually have been a slow shift, over many generations, from primary dependence on hunting and gathering, to primary dependence on agriculture for subsistence.

Once, however, people took to living in this sedentary way, populations would certainly have started to grow, not least because female fertility increases in permanent settlements. [7] In many cases the population would have eventually reached the point where the group's dependence on agriculture became irreversible. So rather than being a response to population pressure, agriculture could have produced it by a 'positive feed-back loop', in which some dependence on cultivated crops allowed population to increase, which then stimulated even more dependence on crops, and so on.

Some archaeologists believe that agriculture automatically produces a surplus, and that this then creates the need for a leader to manage its distribution. Agriculture, in fact, only produces a surplus if people deliberately plant enough crops to do so. This is hard work, and people will only go to all this trouble if they have some special inducement to do so, such as the competitive feasting we shall come to in Chapter V. And if people do find themselves with bumper crops in some years, they certainly won't need a leader to distribute it for them. Leaders actually emerge in the context of kin groups, and their control of access to land, as we shall see in the next chapter.

## 2. *The domestication of animals.*

Settled villages with a supply of cultivated crops would also have provided very good conditions for the domestication of animals, and fortunately for man there are many regions of the world where suitable species exist. They would have been herbivores, so as to eat the sort of food that humans could provide, that could be tamed and were willing to breed in captivity, and been herd animals with a pattern of dominance that could be transferred to human masters. Goats, sheep, cattle, buffalo, pigs, camels, and horses, to name the most important, have all these traits. (The scavenging dog had been domesticated from the wolf by hunter-gatherers in northern Eurasia thousands of years earlier.)

The process of domestication would have occurred in two stages: attracting animals to human settlements and then, in the second stage, controlling their breeding. Pigs are omnivorous scavengers who would have found human settlements attractive, while goats, sheep, and especially cattle crave salt, and villages of farmers could have provided fodder as well. In the early stages of domestication it would not have been necessary to restrain the movement of animals by penning them since a regular food supply would bring them back to their owners' dwellings. The first planting of taro in Highland New Guinea was closely associated with the domestication of pigs, and this is still the standard manner of keeping pigs in New Guinea, for example. But under these conditions the animals breed in the wild, so that until they can be penned and the males separated from the females, it is not possible to influence such characteristics as docility, milk yield or amount of wool by selective breeding.

For Mr Gradgrind and his ilk the only conceivable motive for domesticating animals would have been to eat them, but the matter is not so simple. All the domesticated animals, apart from the dog, had previously been hunted and it is not obvious that the desire for meat alone would have justified all the trouble of domestication. Food, no doubt, was an important reason for domestication, but animals, much more than plants, have very powerful symbolic and religious significance to hunters, and this attitude would certainly have carried over into the new world of the farmer. For example, aurochs, the wild ancestors of cattle which only died out in the eighteenth century, were ferocious and it is likely that mastering them would have been seen as a challenge by the young men, just as breaking wild horses still offers a similar challenge in our own times. Many kinds of sacrificial cults were associated with the earliest domesticated animals, especially cattle and horses, and livestock would also have formed a new type of wealth that conferred prestige on their owners.

In New Guinea, for example, the pig is certainly valued for its meat, but in many areas in the past people could obtain all the meat they wanted by hunting pigs without going to the trouble of producing food for them. It was the introduction of the sweet-potato in the seventeenth century that allowed people to grow much more food than hitherto, and this in turn was the basis on which large pig herds could be built up. But the primary purpose of these herds was not to provide a regular supply of meat to their owners throughout the year, like prudent cottagers in an old English village smoking their hams in the chimney. It was to allow periodic ceremonies in which vast numbers of pigs were killed in an orgy of slaughter, often to honour the ancestral spirits, with dancing and speeches by the Big Men, to which hundreds of guests were invited to receive their pork and so bring glory to the hosts by their generosity and the number of pigs their women could rear. The prestige of Big Men lay in their ability to manage these systems of exchange, and conduct the ceremonies, and without these factors of competition and prestige the peoples of New Guinea would not bother with domestication. Pigs contribute nothing to farming by way of manure, and the hides are not cured. In fact, their main effect on gardens is to destroy them, which requires elaborate fences to keep the pigs out. In the same way some of the Nagas of Assam keep mithan (a type of buffalo) much as pigs are kept in New Guinea, for prestige purposes and sacrifice.

Since these non-utilitarian motives are important in the domestication of animals by societies that we can actually observe, it would be unreasonable to dismiss such motives for their earliest domestication, just because they have left no trace in the archaeological record. Animals are also surrounded by many taboos and this, too, should caution us against explaining their domestication solely in relation to the human stomach. There are so many cases of perfectly healthy sources of animal food that are rejected because they are considered ritually impure, that it is naïve to think that meat is just another source of food. [8] In East Asia, for example, milk and dairy products are widely rejected, and in many parts of East Africa fish and birds may not be eaten. So the Konso have always kept chickens, but regard eating their meat and especially their eggs with disgust: they are used only for their feathers.

While it seems that animals initially provided only meat, hides and leather, their real evolutionary potential was not so much for food, but in reinforcing the agricultural system through manure and as draught animals for ploughs, in transport and communications, as sources of power, and in warfare as the basis of cavalry.

*3. The spread of agriculture.*

The reason that agriculture and the domestication of animals occurred in some areas of the world before others was obviously because some parts of the world are much less well endowed with domesticable plants and animals than others. Deserts and the very high latitudes of both hemispheres can be ruled out at once, while Australia has no domesticable animals, and few domesticable plants except in the north. Eurasia, on the other hand, as the largest land mass, has the greatest variety of plant and animal species, and those that are domesticable have great evolutionary potential. [9]

Another extremely important factor affecting the spread of domestic plants and animals across the various continents is a simple feature of geography. The main axis of Eurasia is from east to west, and this means that it is possible for species to be transported from one end of this great landmass to the other and yet remain within the same climatic zones. But the main axis of the Americas and Africa is from north to south, which has meant that it was far harder for this process of transmission to occur in these continents because so many different climatic zones had to be crossed. [10]

Within a few thousand years, then, in many different parts of the world, our ancestors began moving from a hunting and gathering subsistence to one based on farming, but the two other forms of economy, fishing and pastoralism, have played lesser roles in history. We have already seen that while fishing can support large local populations, and complex societies, its spread is basically limited by the need for access to water, and this means that fishing, as the *main* form of subsistence, could never have become as widespread as farming. But sedentary communities of fishermen would have provided good opportunities for people to try cultivating some of the plants they gathered, and this could have played an important part in stimulating the agricultural revolution.

It was once believed that hunting and gathering did not develop immediately into farming, but were followed by a stage of nomadic pastoralism. This theory has long been discredited – in the Americas, pastoralism was not even possible – and we know that, well before the domestication of crops and animals, there were strong sedentary tendencies in many hunting and gathering societies. It is very likely that the early domestication of animals, such as the goat in the Zagros mountains, was an important factor in a number of areas linked with the domestication of plants, but pure pastoralism is usually a later specialisation that occurred in agricultural societies that had acquired large herds of animals, often where the younger men acted as herdsmen while their fathers tilled the land. (The best established case of a society moving directly from hunting to a form of pastoralism is that of the

Lapps and kindred reindeer hunters of northern Eurasia, although the horse may also have been the basis of a pastoral economy for hunters of Central Asia.)

Pastoral societies with large animals such as cattle and horses are inherently more mobile than farmers, and because stock are much easier to steal than grain, pastoralists are inevitably involved in fighting to defend their herds. This gives them a military advantage over cultivators, as in East Africa where pastoralists invaded and conquered a number of farming societies. The military superiority of pastoralists is most obvious when it is based on the horse. [11] The great military potential of the horse when used by a mounted warrior derived initially from its speed of attack and retreat, and especially when combined with the bow and arrow, it was of decisive importance in the conquests by the Indo-European peoples, the Scythians, Parthians, and Huns in the Roman period, the various Muslim invasions, and of course the Mongol empire of Central Asia. But these achievements, though spectacular, were ephemeral: pastoral societies can never be self-sufficient and always have to live in a symbiotic relation with farmers – as the Chinese say, one can conquer a country on horseback, but to rule it one must dismount. So while cavalry remained an essential military feature of armies until the twentieth century, pastoral societies as such had no other advantage besides the military one over agricultural societies, and led nowhere in social evolutionary terms.

4. *The technological consequences.*

The shift to agriculture did not necessarily involve any new tools. The wooden digging-sticks and stone axes that had served for foraging could be used for tilling the soil, while the hoe and, of course, the plough were later developments. Root crops can simply be cooked in the ashes of a fire, while legumes such as beans and peas only need to be shelled and boiled. Reaping wild grasses, however, required sickles which were made from sharp microliths mounted in wooden handles, and these cereal crops needed to be hulled in mortars, with grindstones to convert the grains into flour, while pots were very useful but not essential for cooking this. So agriculture and the domestication of animals primarily depended much more on new techniques, rather than on the discovery of new tools, even though in many parts of the world it was associated with the manufacture of polished stone, 'Neolithic', axes and other tools. Archaeologists tend to attribute special significance to tools because they are such a significant proportion of what is left for them to study. But

> ...for the greater part of human history, labour has been more significant than tools, the intelligent efforts of the producer more decisive than his simple equipment...The principle primitive 'revolutions', notably the Neolithic domestication of food resources, were pure triumphs of human technique: new ways of relating to the existing energy sources (plants and animals) rather than new tools or new sources. [12]

While the adoption of agriculture did not generally depend on any significant innovations in tools, it had a profound effect on their subsequent development, however, because it ultimately made possible a vastly greater division of labour and craft specialisation. Hunters and gatherers could each make their own simple tools and weapons, but pottery, weaving, and metal-working are much more technically demanding skills that require a new range of tools and techniques. The specialists who master these may even have to withdraw from food production entirely in order to concentrate on their crafts. This in turn needs a constant demand for the pots, cloth, tools and weapons that are produced so that the artisans have an assured supply of food from their customers or patrons. This economy of scale was itself made possible by the much larger populations produced by the adoption of agriculture, which therefore not only supplied the conditions in which pottery, weaving, and metal working had special practical relevance, but also the necessary level of demand for these products as well. [13]

Not all forms of agriculture are the same, of course, and the most important difference is in their intensiveness. As shifting cultivation is replaced by manuring, fallowing, perhaps the construction of terracing, the use of the plough, and especially by irrigation, land becomes an increasingly valuable asset in which a great deal of labour has been invested. Under shifting cultivation it is relatively easy for those who have quarrelled with their neighbours to move away and create new gardens elsewhere, but much harder to move when this involves bringing virgin forest under the more intensive forms of agriculture, especially where irrigation is involved. So the intensification of agriculture makes farmland more valuable, so that people will be more reluctant to give up their land and move somewhere new, and this brake on mobility is one factor enabling political leaders to enforce their authority. Intensive agriculture also increases the population density, and this further enhances the value of land, so that inequalities in inheritance may have very important effects on social inequality in general.

Agriculture was therefore the essential basis of an enormous increase in general social complexity, and ultimately made literate civilisation possible, because only an agricultural economy could support populations of a size in which cities and the state could develop. [14] But before we get to this stage of social evolution we have to understand the tribal society out of which literate civilisation developed.

# CHAPTER IV
## The New World of Tribal Society

Farming populations are not only far bigger than those of hunter-gatherers, but they are able to live in permanent settlements because they can be close to food supplies densely planted in one place, instead of having to scour the landscape to find them. Even in the case of shifting, or 'slash-and-burn', farming the settlements typically last for several years at least. Although population levels greatly increased, in traditional societies they were normally well below the carrying capacity of the land, as we shall see in the next chapter. The Darwinian idea that animal populations constantly have to compete for resources, is therefore a thoroughly unreliable guide to human society. Where population pressure might threaten to exceed food resources, as on Polynesian islands, greater social organization makes it easier to stabilise population levels. Apart from the traditional means of infanticide and *coitus interruptus*, they can enforce compulsory celibacy for younger brothers, sea voyaging by young men, and, in extreme cases, the expulsion of a particular group. [1]

Societies also have various means of redistributing land for those who need it, and the real struggle is to win those resources from nature, which needs *co-operation,* not competition. People now have much more to co-operate about than hunter-gatherers, such as clearing the forests and vegetation by cutting and burning, and preparing the soil, and possibly additional tasks as well such as terracing or fencing, building communal houses, planting trees and domesticating wild animals, defence, and ceremonial life. It seems to be a universally accepted rule that to mingle one's labour with something, such as clearing land and cultivating it, or planting a tree, establishes a good claim to own it.

This means that farmers are now involved in a new relationship with the land and animals, that of ownership and property rights. Being first, as in the first occupation of land, is another universal rule of ownership, and so is the idea of inheriting something from one's forebears. It is very easy, then, for hereditary groups to form around the ownership of land: this ownership is not the private property with which we are familiar, but group ownership, in which individuals merely have the right to *use* the land when they need to. The existence of property in land (and also in domestic animals), together with higher population densities, can intensify the possibility of disputes and provocations over property rights and animals. This in turn would make some degree of group solidarity in self-defence an obviously appealing option, and not surprisingly there is a higher level of group conflict in early farming societies than among hunter-gatherers. (As we shall see, conflicts involving property typically involve provocations rather than scarcity.) In a situation of warfare it is a basic advantage to have larger groups, which can also co-operate in other ways such as ceremonial activities and producing a food surplus for feasts. The domestication of animals provides a convenient source of sacrificial victims and, as we shall see, the development of more complex social organization and hereditary offices made more complex ritual possible.

Rights to property and access to resources generally, defence and warfare, ceremonial and ritual life, and all sorts of community activities provide a social landscape in which it is very easy for a variety of groups to form that are based not only on residence, but also on kinship and descent, work, voluntary associations, age, and gender. This creates a very different sort of milieu from the atomistic individualism of band societies.

## 1. The new kinds of group.

The most important new types of group are those based on kinship and descent. They originally seem to form through their claims to own particular tracts of land, and we can see this beginning among the Tauade, where each tribe was made up of a number of clans each descended from ancestors who had migrated there in the past, and been given some forest to clear for gardens by the group that was living there first. Clearing the forest gave these ancestors the right to claim that land as their own, and they were welcomed because they gave more strength to the group for pig rearing and warfare. Rules had to be developed about how these rights of ownership and group membership were to be inherited, but we need to be clear that the rights of individuals to use some land also involved belonging to a *group*. The implications of this group membership extend far beyond property to one's whole social status and religious well-being. Even the weak and dispersed Tauade clans each had their own name, the boundaries of the ancestral land were remembered, they did not kill one another or intermarry, and the bones of their dead were deposited in special clan caves.

One of the commonest rules of inheriting membership of a descent group is to base it on that of one's father, patrilineal descent. Since in all societies it is men who control the descent group's land (and other assets) the patrilineal rule combines inheritance and control very conveniently. Over time such groups may become very large and are often referred to as clans, while the smaller divisions of clans are sub-clans or lineages.

A much smaller number of farming societies, however, (about 15%) have matrilineal descent which is traced through the mother, and this is rather more complicated. [2] It should be pointed out, first of all, that 'matrilineal' does not mean 'matriarchal', and there is no known society in which the roles of the sexes are reversed, and where women go to war and dominate politics, and own the property, while men look after the babies and do the household chores. In matrilineal society the men are still the dominant sex, and control the descent group's property, but their heirs are not their sons but their sisters' sons. To ensure this, brothers have to maintain more control over their sisters than in patrilineal systems, where a man's legal control over his wife is much more complete. As a result of this conflict between men's roles as husbands and brothers, the marriage bond is weaker so that women do have a rather higher status than in patrilineal societies.

But in many cases – over 30% – lineal descent groups do not develop and descent is instead traced through both males and females ('cognatic' or 'bilateral' descent), all of whom in principle have the right to live on the group's land. While in theory it might look as if cognatic descent *groups* would be unworkable because everyone in a small primitive society is related to everyone else, they are perfectly practical as long as people's choices of which groups to live with are limited, and the groups are linked to land. For example, if a man has the right to become a member of his mother's kin group, but only if he goes to live with them when he grows up and becomes a full supporter, or if, when people marry they have to choose to live with the man's group or the woman's, then the range of options is drastically reduced. And if men in successive generations usually choose to stay where they have been born on their fathers' land, the cognatic descent group *on the ground* may look very much like a patrilineal one. In some ways cognatic descent groups have an advantage over lineal ones, since whereas patrilineal groups may either become too large for their land, or fade away, it is easy for people to move between cognatic groups in response to these imbalances. It is also possible for different rules of descent to operate at different levels of a large descent group, as in Polynesia where chiefs and those of high rank inherit patrilineally, whereas local groups of commoners can be recruited cognatically.

The founder of a descent group (whether patrilineal, matrilineal, or cognatic) has a special status, not only because of the prestige of being first but because all his descendants are seen as an expression of his fertility, 'the seed of Abraham', as it were, a great tree springing up from a small shoot. The ghosts of the lineage founder and all the other ancestors are widely believed to have great power to punish or reward their descendants, so that ancestor worship is found all over the world, and the head of the clan or lineage may have the responsibility of representing his kin in rituals of ancestor-worship. But how does someone become the head of his clan or lineage?

This brings us to the vital issue of seniority. In every family the eldest child is able to dominate his or her younger siblings, at an age when the three or four years' age difference that often results from birth-spacing and infant mortality, will mean a great deal. Among hunter-gatherers, eldest sons had no advantage because there was no property for them to inherit, and no authority for them to exercise, but all that changed with the agricultural revolution, and the development of property-owning descent groups. In all traditional societies males and not females are the leaders, so that the superiority of eldest sons that is generated by this very simple family dynamic has profound significance for social organisation as a whole. (The first-born child is also of special importance to its parents, because its birth transforms them from juveniles to adults.)

For all these reasons, birth-order is a very important basis of social status that is widespread throughout the world, and over many generations it can produce major inequalities within descent groups, because some people are descended from the founder through a long line of eldest sons, whereas at the other extreme some men's ancestry is entirely though junior lines. This has powerful social consequences, because it can produce an aristocratic class of men who are the lineal heads of their descent groups. Even without actual ancestor-worship, the spiritual potency of the founder in some cases may be concentrated in the head of the lineage, who typically has the power to bless its members, their fields, crops, and animals, settle their disputes, and may also have considerable control over their land as well.

None of this was the result of variation and selection; seniority of birth was neither adaptive nor maladaptive, but simply the natural result of a normal feature of family relations, while descent groups themselves were the easiest and immediate route for early farmers to take in organizing access to land and resources. But seniority and descent were essential foundations for the subsequent development of the state, and in the development of social inequality. This is because clans and lineages are all different from each other through their descent from different founding ancestors and, because these founding ancestors were often themselves in unequal relationships, it means that the clans can very easily become components in a system of radical inequality.

For example, among the particular Tauade tribe with whom I lived, one clan, the Karuai, are considered superior to all the others because their ancestors were the original inhabitants of the group territory, and had allowed the ancestors of the later clans that arrived to live there. Among the Konso of Ethiopia (with whom I also lived [3]), four of the clans are considered superior to the other five because the four senior clans are believed to have begotten the five junior clans:

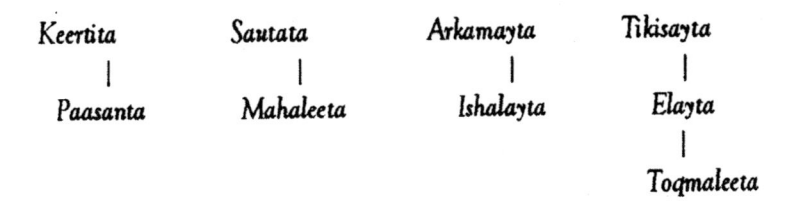

| *Keertita* | *Sautata* | *Arkamayta* | *Tikisayta* |
|---|---|---|---|
| \| | \| | \| | \| |
| *Paasanta* | *Mahaleeta* | *Ishalayta* | *Elayta* |
| | | | \| |
| | | | *Toqmaleeta* |

In other societies, some clans are regarded as the descendants of the conquering group, with a royal clan at the head, and the other clans as the descendants of the losers, or even as slave clans. So inequality of some sort between descent groups can be taken as the norm in tribal society, and this is often given a religious basis as well, in which different clans are associated with different species of animals or other natural phenomena, which we will come to shortly. In some societies certain clans may have special functions, such as priests or war leaders. It is also a very common rule that members of the same clan may not marry one another, even when the clans have thousands of members.

In any clan or lineage there will obviously be different generations, of young unmarried men, fathers, and then the most senior generation who may supply a number of elders to form a clan council. But it is also possible to group people on the basis of relative age *alone*, regardless of their descent groups, and systems of age-grouping often grow out of initiation ceremonies for boys. Initiation ceremonies are very common in tribal societies, often because the men want to take the control of the boys away from their mothers and incorporate them into male society. To do this they may make them undergo months or years of seclusion under the supervision of older men, or there may simply be a rite of passage to mark their entry into the adult world, often accompanied by ordeals such as circumcision or other bodily mutilations in which the boys are tested for their courage and endurance of pain. Boys are usually initiated in groups every few years, and this has the potential of creating a hierarchy of groups based on age which can each be given their own name.

In several parts of the world there are quite elaborate systems of age grouping, in which there is a hierarchy of uninitiated boys at the bottom, then warriors, with elders at the top to settle disputes and take political decisions. It is a fact of nature that all men start as boys, and, barring accidents, all can at least hope to grow old, so that systems of age-grouping are inherently egalitarian, and also inherently cut across groups based on birth – the clans and lineages. Age systems often appear in societies where the herding of stock is important because this is a task well suited to young men, leaving their fathers to cultivate their land, and the frequency of pastoralism in East Africa is probably why age-systems are so common there. But they are also perfectly viable in purely agricultural societies, because their structure is so simple, and because initiation is so frequent in tribal society.

There is therefore a basic conflict between the principles of age and descent: descent is inherently unequal and in a sense divisive because we all have different ancestors, whereas age is inherently egalitarian because we all grow older. And because it cuts across descent groups it can also be seen as uniting society, instead of dividing it. Correspondingly, the authority of councils of elders is basically different from the kind of authority wielded by those who inherit their office in the senior line of their descent group. It is not surprising, then, that in societies which have age-grouping systems the significance of the descent group is reduced, and there is a somewhat egalitarian and 'democratic' ethos. Whereas lineal descent groups are the standard basis for inherited political office, age-systems are inherently unable to provide this, and the most they can do in the way of individual leadership is to elect temporary war-leaders; political authority and decision-making is the responsibility of groups of elders who, like war leaders, are transitory. [4]

The distinction between male and female roles, especially in relation to ritual, may often become much more institutionalised as well in tribal society, sometimes to the extent that the men form associations of their own, or sleep in special men's houses, while in some societies there are elaborate rituals expressing male resentment of female fertility.

*feeling of displeasure*

In a milieu of chronic warfare, with a highly masculine ethos, women may be represented as threats to male courage and strength. The male domination of warfare in agricultural societies intensifies because men are involved in the defence of settlements, and property such as land, crops, and stock from attack, and they also decide who is to join the tribe and have access to its resources. Men also dominate tribal politics because these are concerned with the problems of war leadership, alliance formation and peace negotiations with neighbouring groups. Male tasks, by contrast with those of women, also typically involve longer and more specialist training, greater mobility, and more physical strength and danger, warfare being the archetypal example.

Women on the other hand have a more restricted role confined to the home, child rearing, and food-producing. Women's tasks are those which do not require long apprenticeship, or travel to any great distance from home, or much physical strength and, as we might expect, are closely bound up with preparing food and child-rearing. These, together with techniques of food-preservation and the making of female tools and appliances, and in some cases household medicines, may of course take many years to acquire, but girls typically begin their apprenticeship in these skills much earlier than boys, by learning from their mothers in the kitchen and the garden, while their brothers are playing.

The role of women is also affected in the area of marriage, because groups often like to exchange women to create and maintain alliances, and this may often be achieved by complicated marriage rules that oblige men to obtain their wives from certain kin groups, and give their sisters and daughters to others. Kin groups can also exercise considerable control over their young men where they supply the bridewealth for them to acquire wives. In these societies, however, unlike our own, women do not look for husbands who will be good providers; in the first place, women normally have to accept the husbands chosen for them by their fathers and kin; and in any case it is mainly the women, not the men, who do most of the providing both in the fields and at home. To be sure, the husband needs access to sufficient land, but this is normally not a problem because in most tribal societies land is controlled by the kin group. [5]

*2. Coping with large numbers.*

One of the unplanned results of all this complexity is that it makes it possible to co-ordinate the much larger populations that result from agriculture. The ability to hold large numbers of people together is fundamental to social evolution: it is the basis of specialisation and the division of labour, particularly in crafts and technology, and of the development of hierarchical structures, while only large social groupings can eventually provide the economic surplus to support the state, and the literate civilisations of antiquity.

To see how complexity allows the co-ordination of large populations, we need to understand an important mathematical property of groups. This is that whereas group membership increases arithmetically, the number of relations between group members increases exponentially. If the number of members in a group is $n$, then the number of potential relations between them is $\frac{1}{2}(n^2 - n)$. This may seem difficult but really is not: if there are, say, 4 members in a group, then $4^2 = 16$, minus $4 = 12$, and divided by $2 = 6$. So the number of possible relationships between all members of the group will be 6, as we can see in this illustration of two parents and their two children:

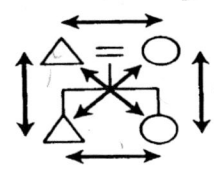

If we add one more member to the group, making it 5, the number of possible relations is not 7 but 10; and if we add 5 more members to produce a group of 10, the number of possible relations is 45, and so on, so that a group of 50 has 1225 possible relations (PRs):

| n | Prs |
|---|---|
| 5 | 10 |
| 10 | 45 |
| 20 | 190 |
| 30 | 435 |
| 50 | 1225 |

The population of a typical hunter-gatherer band is roughly 25, but the average size of a local group of slash-and-burn farmers in Papua New Guinea is about 250, and the larger villages of the Konso in Ethiopia can be 2500:

| n | PRs |
|---|---|
| 25 | 300 |
| 250 | 31,125 |
| 2500 | 3,123,750 |

The result of such enormous increases in possible relationships is that individuals start having major mental difficulties in keeping track of all the possible relationships involved, and the number of possible quarrels increases exponentially as well. One way of dealing with this problem is by subdivision into residential and kin groups. When I first went to live with the Konso they numbered about 60,000 people, who were divided into 9 patrilineal clans dispersed among about 36 walled villages. Each clan was divided into many lineages, and all the members of a lineage lived in the same village under their lineage head, the *poqalla*, and there might be twenty or thirty *poqalla* in each village. It was quite

## The New World of Tribal Society

easy for people to remember the *poqalla* of their neighbouring villages, and their clans, and so if strangers met, at a local market for example, simply finding out their village and their clan at once narrowed down their social identity to a handful of lineages.

Co-ordinating a large number of people is also much easier if there is a hierarchical structure of authority and information-flow. Suppose, for example, that we have a group of 8 members, and an item of information has to be communicated to them all. If there is no established sequence of individuals for this, then the total number of interactions must be 28, $\frac{1}{2}(8^2 - 8)$, and in this case many people will be telling others what they already know. This is very slow and cumbersome and can be made much more efficient by a hierarchy:

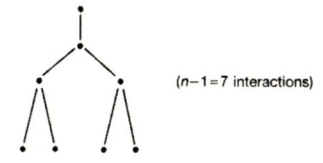

$(n-1 = 7$ interactions$)$

The number of interactions is now only 7, and hierarchies, in fact, are the key to organizing social groups larger than a couple of dozen because they also make authority, ranking, and specialization very easy to arrange. So if we have, for example, two groups of 250 members each, as among the Tauade, and if these groups want to negotiate something, such as terms of peace or compensation, then if all members (total 500) want to talk to one another the total number of possible relationships will be 124,750 whereas if each group were represented by a Big Man to whom each member of the group gave his opinion, the total number of possible relations will only be 498, a 250-fold reduction. Again, a large Konso village of, say, 2500 may be subdivided into 6 wards each with a population of about 400, or about 80 male heads of household in each ward, and they typically choose about 6 councillors for each ward. These in turn send a couple of representatives to the council of the whole village, so that about 12 men can take the decisions for a population of 2500. Finally, organizing not just individuals but whole groups into hierarchies, such as provinces, districts, and local villages, enables even larger numbers of people to be co-ordinated, and this is the particular achievement of states, as we shall see.

Hierarchical structures also mean that there is a small minority of men at the apex of each hierarchy who constitute an elite – Big Men, lineage heads, or councillors, for example – and these men all know each other, and typically marry one another's sisters and daughters. They form the 'nodal points' of a network through whom a disproportionate quantity of important information passes and in this way, too, hierarchies are vital aspects of social co-ordination and increase in importance as societies become bigger and more complex.

Of course, the people involved have no idea of the mathematical properties of social networks and hierarchies in co-ordinating large numbers of people, and descent and residential groups are formed because of the practical requirements of agriculture, leadership, and all the other aspects of daily life. These mathematical properties were not selected for: they are inherent in the nature of things, but they allow the control and coordination of the far larger populations that the agricultural way of life made possible and are a good example of how order can arise spontaneously. This is just another example people creating institutions for limited but perfectly understandable purposes, without realising their further implications.

3. *The religious significance of tribal institutions.*

In the world of Coketown we are used to setting up factories, trade unions, banks, local councils, clubs, insurance companies and all sorts of other institutions which are designed to do something clear and practical, and people join these groups from individual choice, not because they have been born into them. The various types of group in tribal societies, however, are obviously very different because they are bound up with the biology of human nature – gender, descent, and age, in particular. Men's societies are based on gender, while clans and lineages are based on descent from particular ancestors in which male ancestors are typically given more emphasis than female ones. Age groups are based on relative age at the time of initiation, and on the biological fact that we all grow older, and even people who simply live in the same place can be seen as having a biological link with one another, because they have all eaten and shared the food produced by its soil.

People have little choice about joining these groups, and while they regulate access to resources, marriage, and status, this biological basis of the institutions of primitive society makes it very easy for them to be seen as embedded in nature, bound up with the working of the physical universe. The very idea that our social institutions could be linked with the cosmos is unimaginable to us, but this is because we think of our little planet as lost in the vastness of space, and our own species as the result of millions of years of evolution. But the Nuer of the Sudan believe that the very tree under which the first human being was created was still alive within living memory, and it takes a great effort of imagination to think ourselves back into the truly tiny world of primitive society. They have no idea of the vastness of space and the distance of the stars and the true size of the sun and moon, and they have never gazed up into the night sky wondering if there might be alien life forms in the depths of the universe. Until people start making serious astronomical calculations there is just no way of telling how far away the stars are, and the general assumption of primitive peoples is that they are no more distant than the twinkling lights of neighbouring fires across the valley at night. (I was asked, quite reasonably, by the

Konso if aeroplanes can fly among the stars.) They do not think of the sun as a hundred times larger than the earth and millions of miles away, but as a small object that travels across the dome of the sky quite close at hand, and burrows under the ground at night, and of course there is no idea of the earth as a vast globe: or indeed any idea of the earth at all, but only of the little area in which the tribe spends its life, which is effectively the limit of their existence.

It is a universal assumption, then, that social organisation is intimately linked with the cosmos, and if society goes awry then men's relations with the natural world will also suffer. For the Konso, men are associated with the sky whereas women are associated with the earth, and these differences were expressed in many details of sacrifice and other religious ceremonies. The male rain god Waqa and the sky are so closely associated that they are almost one and the same, and the rain is seen as a reward for those people who live together in harmony. For example, the Konso had an extremely elaborate age-grading system, which imposed all sorts of burdensome rules on them. These could prevent a man getting married until he was thirty-five, while those who married before attaining the grade for marriage were required to abort any children conceived. When I asked them why they had such a system, their reply was that it made the crops grow. By this they meant that because it established harmony between the generations, Waqa gave them rain. Harmony between the generations meant for them not only the absence of quarrelling, but that only one generation at a time should beget children, so that young people and the grandparental generation were forbidden to do so. It also meant that the young men, the warriors, performed the basic male function which was killing enemies and game animals, while the elders blessed them and the rest of the community, and there were many elaborate ceremonies for this. So it is somewhat misleading to say that primitive peoples believe in 'supernatural' beings, because for them what we think of as the supernatural can't be clearly separated from the natural world or from society.

Clans, too, in tribal society are not seen merely as social institutions but are usually thought of as having a close, 'totemic', relationship with species of birds, animals, and plants, and their founders and heads are seen as endowed with sacred powers. The Konso, for example, believe their nine clans were founded by the children of the first human beings, and so they also believe that all the other peoples in their area of Ethiopia belong to these clans too, even though they may have different names. Each clan is closely linked with the natural world: Arkamayta, for example, with Waqa and the sun; Ishalayta with the pygmy antelope; Keertita with the Sagan River, and so on. And members of each clan are also believed to have their own distinctive characters and behaviour. So Arkamayta clan appeal to Waqa and the sun to bring peace; Ishalayta, like the pygmy antelope, are innocent, happy, kind-hearted and harmless; Sautata cause harm to the property of others, so their totems are the animals which destroy crops – elephant, mouse, baboon, and locust; and the Paasanta are believed to have much intercourse with spirits and use a lot of witchcraft, so their totem is the kidney where witchcraft ability is located.

The head of a Konso lineage, the *poqalla*, inherits his position in the senior male line from the founding ancestor, and the name *poqalla* means 'sacrificer'. Unlike the Tauade Big Man, he is not concerned with the management of exchange and making speeches, but with mediation and performing annual ceremonies to bless his kinsmen, their fields, and their livestock. He also inherits a much larger share of his father's property than his younger brothers, and can use this to buy more land, so that the *poqalla* is also the richest member of his lineage. Members of the lineage also give him tribute of meat and other valuables from time to time, but he is also obliged to give them food and seed grain in times of famine, while those whose land is insufficient will be provided with more in exchange for their labour. Religious and social authority are therefore very closely related in tribal society.

This relationship can be seen even in the case of Tauade Big Men. In the old days, when a Tauade Big Man died his body would be put into a specially built enclosure which women were not allowed to enter. Pigs were then slaughtered inside the enclosure and the sacred bull-roarer was whirled, away from the gaze of the women. If enough boys were available they would be kept inside the enclosure in a little hut for several months where they could imbibe the vitality of the dead chief and were taught to be tough and aggressive. The Big Man's corpse, meanwhile, had been put on a special platform in his hamlet where it was allowed to rot, and it was thought that people absorbed the powers of the Big Man in the smell. Big Men also had a special association with certain birds of prey and sacred oaks, and were believed to be essential for the general health and well being of the group. But they did not actually inherit their rank, because they had to have the strength of personality and the leadership abilities to act as Big Men, and sons often did not inherit these qualities from their fathers. (In Polynesia the idea of sacred power, *mana*, being inherited by men of rank is much more highly developed than it is in Papua New Guinea.)

So in tribal societies, religion and social institutions are bound up with one another in a way we cannot hope to understand if we think of religion as it is today in modern Western society. We take it for granted that religion and churches are about God, and quite different and separate from banks, and hospitals, and farming, and political parties, and all the other activities and institutions of daily life. We think of religion as something special, and essentially otherworldly, and almost the opposite of the 'practical'; and we assume that religion is about 'faith' – as opposed in some way to what can be proved – but all this is our own peculiar view of things, and unless we discard it completely we cannot begin to understand our ancestors in tribal society and early antiquity. Primitive ritual is basically public, and not about personal spiritual experience or the salvation of the soul, but about the pursuit of Life, prosperity, and social well-being. While belief in some sort of after-life is universal, people have little interest in its details and the general attitude is that prosperity in this world is what matters. Supernatural beings may punish sins such as lying, oath-breaking, theft and murder, but they do it in this life and not the next.

Trying to define religion as belief in supernatural beings is therefore quite inadequate and simplistic. Supernatural beings always operate in a wider context of order, of how things began, of how men became human, and of schemes of cosmic order, and symbolic classification. 'Religion' covers not only beliefs in a variety of supernatural beings such as gods, nature-spirits, ancestor-spirits, ghosts, and souls, but the impersonal forces of ritual, magic, divination, luck, and fate. These have very different psychological origins: the human ability to find pattern and order; difficulty in distinguishing between the social and natural worlds; dreams of the dead, ghosts, and out-of-body experiences; trances and other altered states of consciousness; the emotional satisfaction of ritual; and the metaphor and imaginative story-telling of myth, to name some of the most important. Religion has therefore been constructed in the course of history out of these very different psychological components, and given the fundamental changes it has undergone, there is about as much sense in looking for a 'God gene' as in looking for a gene for economics. If economics is about the production and distribution of commodities, then religion is about *the ways in which people find meaning and order in the world around them*, and this can be done in many different ways, and at many intellectual levels. People try to apply what they believe about this meaning and order to obtain Life and prosperity through a variety of rituals, which may have little or nothing to do with any supernatural beings. No doubt, because it is very easy to explain the causes of things by something like a human purpose, supernatural beings are very common in primitive thought, but order is more fundamental in the search for meaning, and it is to ritual order that we now turn.

### 4. Ritual order.

Many rituals of tribal societies are what we call 'rites of passage', notably birth, initiation, marriage, and death, but some are particularly concerned with obtaining Life – health, fertility, prosperity, many children, good crops, flourishing herds of animals, and killing many enemies. They are public events, often involving sacrifice, and not occasions for private worship or personal mystical experience. I have attended many sacrifices, but I never detected any of the religious awe and heightened emotions that some theorists of primitive religion have imagined. While there is temporary excitement at the moment of slaughter itself, by far the most important thing about ritual is not that people should have some special kind of emotional or spiritual experience – that is more a feature of the world religions in literate civilisation – but that it should be correctly performed, since following the correct procedure is fundamental to manipulating the cosmic order.

Ritual has always had a close association with play and games (many of which actually originated in ritual), with ancient theatre, and with board games, some of which began in divination. They are formal, highly ordered activities which have no *immediate* practical aims, like building a house; and they create an artificial, separate world outside ordinary time and space, according to precise rules of their own:

More striking even than the limitation as to time is the limitation as to space. All play moves and has its being within a play-ground marked off beforehand either materially or ideally, deliberately or as a matter of course. Just as there is no formal difference between play and ritual, so the 'consecrated spot' cannot be formally distinguished from the play-ground. The arena, the card-table, the magic circle, the temple, the stage, the screen, the tennis court, the court of justice, etc., are all in form and function play-grounds, i.e. forbidden spots, isolated, hedged round, hallowed, within which special rules obtain. All are temporary worlds within the ordinary world, dedicated to the performance of an act apart.

Inside the play-ground an absolute and peculiar order reigns. Here we come across another, very positive feature of play: it creates order, is order. Into an imperfect world and into the confusion of life it brings a temporary, a limited perfection. Play demands order absolute and supreme. The least deviation from it 'spoils the game', robs it of its character and makes it worthless. [6]

With farming, the opportunities for rituals greatly increased, and there were far more things to do in rituals, notably the sacrifice of animals which now assumed great religious importance, as did the sacred places where these ceremonies were held. Hunters do not have much opportunity to sacrifice animals because normally they do not capture them before they kill them. But when people own herds of cattle, sheep, and goats it has seemed appropriate to many groups that some should be offered in sacrifice, which became a widespread ritual of farmers and pastoralists. The Konso invoked the Sky God, Waqa, the god of rain, and also Earth, when they sacrificed, but sacrificial blood itself has great power, and they referred to a place of sacrifice as 'the navel of the land'. Agricultural societies have a new relationship with the Earth, which may take on the guise of Woman in whom the seed is sown (an idea that would have made no sense for hunter-gatherers). We also find that, opposed to the Earth, is the Sky, the source of the rain that is so essential for the crops, and the Sky is often represented as male, impregnating the Earth. The changing seasons and activities of the agricultural year, especially sowing and harvest, also have an obvious and powerful resonance for religion that is devoted to securing Life, and are the basis of the ritual cycle.

Materialists automatically assume that farmers introduced calendars based on lunar months so that they would know when to sow and harvest their crops. But crops grow in relation to the weather and the seasons, which may vary each year by weeks, and plenty of farming societies, especially the shifting cultivators of Papua New Guinea, can get by without any formal calendar by relying on their knowledge of the seasons and observing natural signs, like the rising of the Pleiades. The primary use of named months is actually for co-ordinating social events, and as a guide to when religious ceremonies should be performed. Since religious ceremonies are fundamentally concerned with good harvests and the health and fertility of men and beasts, this is why there is a need for these calendars to be linked to the seasons and to agricultural activities.

The settled life of the village, too, acquires great symbolic importance as the basis of the civilised and the tame, and the opposite of the wild bush and forest, the abode of non-human forces. Those 'rites of passage', which are celebrated to some extent in foraging society, often become much more important in tribal society because groups are larger and more formalized. These also provide, as we have seen, new opportunities for hereditary authority which very easily acquires ritual functions in these ceremonies. While the religion of tribal society was therefore greatly influenced by the new features of agricultural life, it was not so much 'caused' by them, as an imaginative response to the opportunities they offered.

We can see this imaginative, creative thought at work by looking in more detail at the various ways in which social groups in primitive society are organized around a few basic principles of order. These are not the principles of management efficiency we find in modern business and government, like specialised committee structures, hierarchical tables of administration, budgetary controls, and schemes of delegated authority. Instead we find all over the world a range of symbolic forms of order, and some of the most significant are complex cycles of ceremonial exchange; ritual boundaries and transition across them; pairs, opposition and alternation; colours, especially black, white, and red; sacred numbers; centre and periphery; right and left, and other aspects of the human body; and purity and impurity. [7]

One of the most important is that of 'moieties' (two halves of a group) and pairs of groups generally, especially involving the exchange of women in marriage, and the ceremonial exchange of gifts and ritual services, as when two groups bury each other's dead, for example. Another form of pairs involves opposition, as between men and women, or between eldest brothers and younger brothers, or between the fathers' and sons' generations in an age system, when the two generations are seen as alternating in the competition for power, or between hosts and guests as in New Guinea feasts, or between sacred and secular authority. The symbolism of Left and Right [8] typically expresses the opposition between men and women, superiors and inferiors, life and death, sacred and secular, order and disorder, the civilised and the wild, and so on. This is only one example of the significance of the human body in a great deal of ritual, such as the head and hair-cutting, purification, and blood-letting.

The notion of purity and impurity itself is derived from the human body, since the basic image of impurity or dirt is faeces, and more generally involves the intrusion [Einbruch] of the biological or, more accurately, the animal side of man, into social life. [9] Sexual relations, childbirth, menstruation, eating, defecation, and death all, in various ways need to be controlled, either literally or figuratively, because they remind us all too clearly of our animal nature, and disrupt the ideal order which religion tries to construct. Cannibalism, incest, bestiality, eating food raw, and nudity all have powerful resonances of animality, and hairiness often symbolises the wild or animal state, so that cutting or shaving the hair may be a rite of incorporation or re-incorporation into society. [10]

Ritual boundaries, too, are extremely important in a world of rigidly demarcated [to separate] categories, so that the gateways to settlements as well as to ceremonial centres may have a sacred character, and there is also a ritual opposition between centre and periphery, with the centre as sacred. Boundaries are closely linked with notions of purity and impurity, [unclean foul] so that certain sorts of people and animals can be thought to defile a sacred space by crossing into it, and those who have left the group territory may have to be purified when they return, as by having their head shaved. Passing from one ritual state to another in the 'rites of passage' is analogous to transition in space, and very much involves boundaries and states of purity and impurity, so that birth, initiation, marriage, and death often involve rituals of purification appropriate to boundary crossing. Initiation ceremonies are part of this complex as, of course, are marriage and funeral ceremonies. These are not just the private concern of individuals, but of groups, who lose a valuable asset when a woman's fertility and child-bearing abilities are transferred to her husband's group, or when a member of the group dies and it has to adjust its internal structure as a result.

The basic symbolic colours of black, white and red occur often in ritual, and typically represent dirt or night, purity or milk, and blood,

## The New World of Tribal Society

and sacred numbers may also be important in ritual. For the Konso, for example, 2 represents woman, 3 man, 5 marriage, 6 the death of a woman, and 9 the death of a man: 3 and 5 are good numbers, whereas 6 and especially 9 are very bad numbers that are thought to bring misfortune.

An opposition between the life of the village, the tame or social, and that of the bush or forest, the wild, often occurs in tribal society. A good example comes from the Umeda of the Sepik River in New Guinea, who live in villages that are made up of a number of hamlets. In the centre of each hamlet are their coconut palms, which provide shade as well as nuts, and here the older men congregate during the day where they play with their infant sons. As these boys grow older, however, they are ejected from the centre of the hamlet to its boundary area between the hamlets and the wild bush. Here the Umeda plant areca (betel-nut) palms, *pul*, which also means 'fence'; while out in the bush are sago palms, *naimo*. These are usually only a source of grubs, but the women can pound up their pith and cook it. The three sorts of tree also have a profound cultural meaning for the Umeda because they respectively symbolise the ordered social life of the hamlet, with the adult men at its core; the boys in the marginal border between bush and settlement, and the wild bush, associated with the women. The trees are also closely associated with the mythological culture-heroes from whom their whole social order originated, and with the two moieties, Edtodna and Agwatodna, into which each village is divided.

Edtodna is the male moiety, whose founding ancestor was the culture-hero Toag-tod, coconut man; members of the moiety are also associated with the coconut palms that symbolise the centre of the hamlet and the original founding ancestors of the hamlets, together with its adult men. Toag-tod also had a younger brother, Pul-tod, areca-nut man, whom Toag-tod later made his adopted son. In the

89

legends Pul-tod, areca nut man, is a marginal, semi-human figure: he murders his father and marries his mother, is symbolically castrated and wanders around in the bush shunning society, especially women, and is 'not a real man'. The physical appearance of coconut and areca palms is also very appropriate to that of the mature men and boys: 'Coconuts are characterised relative to most palms by their greater height, and slow growth. Their long life-cycle corresponds roughly to that of a man. When old they acquire a gnarled, immemorial appearance. Areca palms are in this everything that coconut palms are not: slender, flexible, rapidly springing up, green-stemmed and swiftly maturing'. [11]

The male moiety of Edtodna is opposed to the female moiety Agwatodna, whose first ancestor was Naimo-tod. He emerged from *naimo*, the sago palm, with two daughters, who provided the first wives of the Edtodna moiety. The female, Agwatodna moiety, is associated with the wild bush, as opposed to the life of the village, which is associated with the coconut. Taking the name of the sago palm, *naimo*, itself, *nai* means skirt, and there is a close resemblance between the dangling fruit stem of the sago palm and the skirts of the women, which are actually made from sago palm leaves, *na*. *Mo* is associated with *mov*, 'fruit', and 'vulva': 'the *naimo* palm is mythologically the origin of daughters (*mo-tod*), i.e. marriageable girls. Girls (*mol*) are to be identified with fruit (*mov*) and... *naimo* is characterised by a truly incalculable number of individual fruit (*movwi*)'. [12] This is a good illustration of how supernatural beings – in this case, the culture-heroes – can only be understood as part of a wider order that embraces what we would call the social, the natural, and the supernatural. [13] But while this sort of symbolic order imposes meaning on the human and natural worlds, it is not what we would call 'functional' in the way that a modern business or local government organization is functional.

## 5. Lack of functional requirements.

Because tribal societies have very simple technologies, people have supposed that they must be almost entirely at the mercy of the natural environment so that everything they do, all their customs and beliefs, must somehow be dictated by the need to survive. It was therefore long assumed by anthropologists that every custom or institution in every primitive society was there because it performed some vital adaptive function. It is certainly true that primitive technology is relatively powerless to do much to the natural environment, both for good and ill, compared with the dams and bulldozers and factories of modern society. But it is just because primitive technology is so simple that tribal society is free to organise itself in a much wider variety of ways than we are, because its productive units, mainly the nuclear family and small groups of neighbours, are small, unspecialised, and self-sufficient.

The design of a settlement, for example, need have nothing at all to do with how the people make their living, and so they can be of very different shapes and sizes. Members of the same local group can live in scattered homesteads, or in separate hamlets, or in a single large village, and there can be all sorts of rules about who may live where depending on their ancestry and social status. Age-grading systems can occur in the very different economies of cattle-pastoralists, shifting cultivators, and intensive farming societies, while patrilineal descent, the senior status of the eldest son, and polygamy can occur anywhere.

The symbolic and religious significance of so many tribal institutions means that while they may be elaborate this may have little to do with any practical purpose (as Mr Gradgrind would think of it). The symbolic ordering of the Umeda hamlet, the age-grading system of the Konso, and the elaborate dances of the Tauade make sense as part of a cultural system of meaning, but they are also, in a sense, 'knobs that turn nothing', unrelated to the efficiency of their society. Each is what one anthropologist has called 'a dramatised philosophy, or a way of acting out a folk faith, rather than an instrumental organisation'. [14] Once we realise that, within certain limits, a wide variety of customs and institutions will all be viable in primitive societies, it becomes clear that this search for adaptive explanations for them is the pursuit of a non-problem. Anthropologists who have spent so much time trying to give functional or adaptive explanations for tribal institutions, to show that each is beautifully adapted to the special needs of a particular society, have therefore been the victims of an illusion: that because societies like ours have many institutions that are genuinely functional, therefore this same degree of 'functionality' occurs in all societies, and that if only we were clever enough we should be able to detect it in primitive societies too.

Functional efficiency of organization only becomes of major significance with political centralization and the emergence of the state. Once it becomes necessary to maintain a central government, to allocate official tasks, to prevent rebellion, to raise and distribute the necessary revenues, to organize truly effective military forces, to organize large-scale public works such as irrigation, and so on, we are in a very different world from that of tribal society, and one that we can genuinely call 'functionally organized'.

If, then, tribal societies are free to develop such a variety of institutions and customs, why don't they just become increasingly different over time, like languages, whereas in reality we find that in many cases chiefdoms and states develop independently in different parts of the world? Why should some societies apparently have this innate tendency towards political centralization? We have seen some of the ways in which this can develop: the superiority of certain descent groups; the special status of the founder of the descent group; ancestor worship; the seniority of the eldest son; and the emergence of the hereditary authority of some clan and lineage heads, which is always associated with special religious status as well. In the next chapter we will look at some further factors that can strengthen political authority, which are trade, tribute, and warfare.

# CHAPTER V
## Economics, War, and Politics in Tribal Society

*1. Primitive economics.*

Many economists think it is part of human nature for the individual as rational, 'economic man', to try to maximize his material possessions, to trade, and to expect payment for his labour. So our notions of economics, like those of religion, have to be turned completely upside down when we try to apply them to primitive society, and poor Mr Gradgrind would be much at a loss to understand what was going on. In the modern industrial world we are not self-sufficient: we cannot, as individuals, even produce our own food or heat our own homes, let alone make our own cars and all the other things we need or would like to own. We have to buy these in the market-place with the money that we earn from our employer by selling our labour to him, just as people in the service industries sell their labour to us. The supply of these goods and services is itself regulated by the demand for them in the market-place. So the market economy needs an all-purpose form of *money* by which everything – labour, resources, and goods – can be converted into a single measure. The amount of money one controls is therefore the basis of one's material well-being, and of one's social status, and the more of it one has the better. [1]

The economic system of primitive society is the exact opposite of this because, although the simple household of a man and his wife or wives, and their children, is part of a wider system of kin-groups and neighbourhood, it is nevertheless the primary unit of labour, and can basically produce all the food, fuel and shelter that it needs. To be sure, there are some occasions when a number of families must combine, as in collecting materials for house-building, clearing the bush and putting up fences, threshing grain, making dams to collect water for stock, and so on, but these are all very simple activities in which each family retains its freedom of action. The division of labour is also far simpler than ours, based mainly on the differences between men's and women's work, with age as a subsidiary factor. Labour, in particular, is not for sale: it is instead part and parcel of social life, so that when people co-operate in production and the other activities I referred to, it is not for payment but on the basis of their social obligations to one another, as kin, age-mates, neighbours, or friends.

In this situation, then, there would be no point in money because, until markets appear, there is nothing to buy with it. Nor, obviously, is there any point in producing more food than one can eat, or building a larger house than one needs. Once people can satisfy their basic material needs of food, fuel, shelter, and protection from human and animal predators, they naturally prefer leisure to work because there is nothing they could get by working harder. So typically the land controlled by groups of such families produces far less food, and supports a considerably lower population, than it could if it were farmed at maximum intensity. It is also possible for people, especially young men, not to begin productive work for many years after puberty, and for older men to cease work well before they are physically obliged to do so. [2] In subsistence economies, then, there has to be something else to do with food, besides eating it, before people will be willing to produce a surplus, and ceremonial exchange is one of the commonest motives for this.

Competitive feasting and exchange of high-status food, such as pork, is particularly common in Papua New Guinea, and while they have no economic purpose, and are solely for prestige, these ceremonies do require major increases in production. Immense labour is typically expended on enlarging the pig herds and the gardens, and constructing the elaborate dance village. Among the Tauade [3] these ceremonies were organized by the leading Big Men, who alone had the authority to negotiate peace with their neighbours so that they could attend the feast, and could organize the co-operation of the households in the tribe. While these households participated so that they could repay their own debts of pork, they still worked hard to produce a large surplus for the prestige of their tribe. The idea was to prove their superiority to their guests by their generosity and lavish hospitality, and guests might stay for two or three months of feasting and dancing. But once the festivities were over, the great dance village was soon abandoned to rot, because the Tauade did not have the sort of social control that would have enabled them to live in such large villages on a permanent basis. [4]

While the Big Men would be able to contribute more pigs than ordinary men, because they had two or three wives to raise the pigs, their prestige did not come from their pigs, but from being able to organize the feast, and from their oratorical skill in making boastful speeches. Many economists have not understood that in primitive society status does not directly depend on wealth at all, but on what a man does for his group in terms of political and war leadership, spokesmanship, dispute settlement, his status in the kin group, organizing ceremonies, or performing religious functions, and many of these roles may be inherited. It is true that men of high status are typically richer in land and stock than other members of their group, but they are richer by reason of their status: the Konso *poqalla* inherits more land than other men, and the Tauade Big Man can attract more wives to tend his pigs because he is a Big Man. In addition it is a universal rule that such men are required to be generous, and to be mean and avaricious is the most serious blot of all on the reputation of a leader.

Whereas in capitalist society wealth is the basis of status, in primitive society, therefore, status is the basis of wealth, and one does not acquire more status simply by accumulating more wealth. The reason is that in our society wealth can buy the labour of others, and can also be invested in ways that produce profits, but in primitive society profit does not exist: there are no mills or factories to be invested in, and one cannot invest in land because there is no labour force that can be hired to work it, and there are no opportunities for middlemen to buy cheap and sell dear. So just as there is no point in money, there is no point in hoarding wealth. The only thing one can do with wealth, then, is to use it in social relations, and create *obligations* by giving it away and so enhance one's status.

This misconception that, in primitive society as in our own, material wealth confers power over others has made the appearance of private property in history seem very important, because it is assumed that this must be the basis of inequality and class differences, with individual families competing for material assets like modern suburbanites. [5] But that does not follow. The Konso, for example, are unusual among tribal societies in having private ownership of land, and a man can buy more than he inherited, especially if he works hard. He will earn respect by this, but it does not in itself give him social influence and authority, because these come from one's status in the kinship system, the age system, and election to the ward and town councils. The private property of the Konso does relatively little by itself, therefore, to produce economic inequality, which as we saw is the result of the kinship and inheritance system, and it remains embedded in the traditional social order.

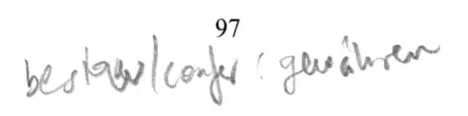

In subsistence societies, then, the production and exchange of goods is dominated by obligations to family and other kin, by ceremonial and religious duties, and by the political organization. A whole series of customary payments are made to satisfy social obligations, such as compensation for homicide (blood money), for a wife in the form of bride wealth to her kin, as compensation for injuries and insults, and as presents to be ceremonially exchanged with regular trading partners, or to be used in competitive feasting. These payments have to be made in high-status goods that are not used in ordinary trade for useful things, and one of the commonest forms of payment is livestock, such as pigs or cattle. 'Despite the relative abundance of cattle and the scarcity of vegetable foods, the Masai did not seek to trade off cattle for food. Such was unthinkable because cattle were the essence of wealth – indeed, from the Masai point of view, there could be no such thing as a "surplus" of cattle.' [6] Traditionally the Konso, too, regarded it as shameful to sell animals in the market; their meat should ideally be shared in a communal meal after sacrifice.

Other kinds of things, too, are highly valued and exchanged not because they are useful but because they are beautiful and rare, as distinct from food, tools, and utensils. Love of personal adornment is a very ancient human trait, and sea-shells, in particular, such as cowries, mother-of-pearl, balers, and conus shells have long been prized for head-dresses, necklaces, armlets, and bracelets, and brilliant feathers, dogs' teeth, boars' tusks, and colourful stones are also prestigious personal ornaments. These honorific wealth objects do not function as money in our sense because, like cattle or pigs, they can only be used for certain limited and honourable sorts of exchange, and are deeply embedded in social relations. Lineage heads, and Big Men can gain a great deal of influence simply by the control they can exert on the flow of prestige goods for exchange:

> In Central Africa, power relationships are established, consolidated and maintained through the control of prestige articles – products which are not necessary for material subsistence, but which are absolutely indispensable for the maintenance of social relations. An

individual needs prestige articles at a number of critical occasions during his life – at puberty rites, for bride-price, as payment for religious or medical services, to pay fines, etc. [7]

But exchange can also take the form of tribute given to a lineage head such as the Konso *poqalla* in return for the benefits which he gives to his kin, and in societies where inherited political office has developed still further the surplus takes the form of tribute that is paid to chiefs in the form of foodstuff such as grain and cattle. With Bantu chiefs, for example,

> The chief's household is the largest and most elaborate in the tribe. He usually has many more wives than other men (some chiefs have over fifty), and many public officers and domestic servants are directly attached to him. . . Most of the resources that he uses to maintain his household, reward his assistants, and fulfil his other tribal obligations, come directly from his subjects. They not only work for him [in various ways] but also pay him tribute in kind. Almost everywhere one or more very large fields are specially cultivated for him by individual districts or other local segments. . . He provides the seeds and takes the crops, but the people do all the work under the leadership of their respective sub-chiefs or headmen. . . In most groups every woman also gives him a basketful of her own grain; its presentation is a ceremonial affair organized by the local rulers, and some of the grain is made into beer which the people assemble to drink as part of the harvest thanksgiving. [8]

Chiefs in all societies are also required to be generous, and redistribute this wealth to their people:

> The prestige of a [Maori] chief was bound up with his free use of wealth, particularly food. This in turn tended to secure for him a larger revenue from which to display his hospitality, since his followers and relatives brought him choice gifts. . .Apart from lavish

entertainment of strangers and visitors, the chief also disbursed wealth freely as presents among his followers. By this means their allegiance was secured and he repaid them for the gifts and personal serviced rendered to him. . . There was thus a continual reciprocity between chief and people. [9]

The other potential source of an economic surplus, besides prestige articles and tribute, that can be controlled by chiefs and enhance their power, is from trade. Adam Smith, like many economists and evolutionary theorists, believed that it is a human instinct 'to barter, truck and exchange one thing for another', but trade is not an instinct and is purely opportunistic. [10] Some local trade had always existed among hunter-gatherers between groups in different ecological niches – between coastal or riverine and inland groups, or mountain and lowland dwellers, or forest hunter-gatherers and those outside the forest. But these differences between regions were increased by agriculture and the domestication of their special plants and animals, and the growth of crafts, that led to increased trade between them. Trade relations of this sort did not necessarily produce markets, however, but could be carried on by partnerships between two individuals from each group, in which there was no haggling or anything else that resembled market conditions.

Where specialist craftsmen exist they do not necessarily need markets either, but in small societies can produce their pots, or stone adzes, or woven mats directly for those who want them as part of the general system of exchange. In Polynesia, for example, 'In the crafts relationship, the patron initiates the exchange, giving one kind of goods and performing certain ritual acts to honour the workers; the *tufuga* [craftsmen] reciprocate, delivering products that elevate the patron', and the crafts were hereditary in specific families. [11] Craftsmen can also work directly for a clan or village in return for payment in kind. But in large populations, and with an increasing demand for craft-products, the specialisation and expansion of crafts such as metal-working, weaving, and pottery were the major reasons for markets in tribal economies. Even so, these markets might be on the margins of social life, [12] primarily for local craftsmen and long-distance traders, with minimal effects on the pattern of local production.

The greater the variety of goods sold in markets, the more difficult it will be to barter them directly for one another. If even, say, 10 different sorts of goods are involved then they would have $10^2 - 10 = 90$ different prices in terms of each other, and matters are greatly simplified if all these prices can be expressed in terms of a single commodity. This will be something useful, such as grain, cocoa beans, bars of salt, axe-heads, or brass rods, and can function quite satisfactorily as primitive money in the market place. But these sorts of money are basically of low status, just run-of-the-mill arrangements to cater for daily needs, and quite different from prestigious wealth objects.

Prestige goods, however, are suited to long-distance trade because of their high value-to-weight ratio – unlike pots or grain, for example – and also because if one is encountering a foreign community for the first time one does not know what they will find useful, whereas one can be sure that they will always appreciate prestige objects.[13] Such goods can then become converted into something more like money, like the clam-shell wampum of the North American Indians, for example, which was originally restricted to ceremonial exchange and religious functions, but became increasingly used by Europeans trading with the Indians for furs as a general form of money. Cowries and other forms of shell money, and of course gold and silver, are further examples of the same development of prestige goods into money produced by foreign trade.

Trade routes had to pass through different territories, and it was very much to the advantage of any local chiefs if they could establish control over the salt, gold, slaves, silk, incense and so on that passed through their trade routes and markets, and claim a share. (Even Tauade Big Men demanded presents from strangers who used the paths that they claimed to control.) The buyers and sellers would be thankful for a chief who could keep order at markets, hear disputes, and guarantee the safe passage of traders, so the chief could claim tribute and taxes in return as well as using the market as a prime source of local information. Access to a regular source of prestige articles would obviously increase the chief's influence and authority within his own society, and we shall see that control of trade routes is often associated with early states.

While trade and tribute were therefore to become very important bases of political power, the extension of the use of money also threatens the traditional social order of tribal society, but in a different way, because it erodes that moral network of ties and obligations by which kin and neighbours are bound together, by treating them all as commercial transactions. We can get an idea of how this happens from the case of Konso craftsmen (blacksmiths, potters, and weavers in particular) and their relations with the farmers. These crafts were originally brought into Konso by outsiders who did not own land and were not allowed to acquire it. While the farmers lived off their own fields and co-operated with each other in ways that were based on their status in ward, lineage, age-group, and so on, the craftsmen could only survive in the market situation by selling to people regardless of their status and the groups to which they belonged, for the highest price they could get. So the relationship between the craftsmen and the farmers was based on individualistic, commercial considerations and not on those of status, and was therefore seen as destructive of the traditional social order. The craftsmen were regarded as greedy and selfish, with no loyalty to their villages or regard for the wider community.

As trade with other parts of Ethiopia increased, however, and a range of desirable goods from outside Konso became available, many farmers took up weaving to earn the money to buy these goods in trade stores set up by immigrant merchants. Unlike food, knives, and pots, and other traditional necessities, for which the demand is inherently limited, the appetite for non-essential luxury goods and 'consumer durables' is potentially insatiable. It is this limitless demand that has motivated people to produce a surplus for trade rather than for communal activities such as feasts, and has been one of the major factors in transforming subsistence economies.

So private ownership of land is not itself one of the turning points of social evolution; it is the wider development of individual economic initiative through markets, trade, and money that erodes the collective, tribal type of social order, where production and exchange are subordinated to the demands of co-operation and status. [14] What Marx called 'the cash nexus', monetary transactions between individuals regardless of their other social ties, becomes increasingly prevalent, and also socially disruptive, as we shall see in Chapter X. The market, too, developed into one of those cases of true competition that we shall encounter again and again in the course of this book. But money was not a sudden invention, and one must remember that 'Barter merges into primitive money and primitive money into modern money through barely perceptible shades of distinction'. [15]

## 2. Primitive warfare.

Primitive warfare is also driven by the same non-material motives as much of primitive economics. Insults, revenge, honour and the desire for glory have always been far more potent causes of conflict in primitive society than competition for scarce resources. Nor does primitive warfare produce much social change, either. History has taught us to think of warfare as the arbiter of national destiny, a series of decisive battles which determine the civilisations that will survive or perish. At the level of warfare between the professional armies of advanced states and empires there is a good deal of truth in this, but many anthropologists think that primitive warfare, too, must be an example of natural selection but in an even more rigorous form, weeding out unfit societies with a vengeance. But while a high level of violence may eliminate some feeble and timid individuals, primitive warfare is generally inconclusive: only a few people get killed at any one time, and some property is destroyed or looted, but there are usually no permanent winners or losers. Of course, humans have always migrated, and throughout history some groups have moved into new areas and have intermarried with, driven out or killed the original inhabitants, but what we do not find at the tribal level of social organization is the military conquest of one group by another, which then imposes political rule on the losers, in the way that the Romans conquered the ancient Britons, for example.

In general, primitive societies are not sufficiently well organized, either politically or militarily, to subjugate their neighbours and this kind of competitive success, as a rule, is only obtained by societies with some degree of political centralization and military discipline, troop specialization, [16] and the ability to conduct extended campaigns in enemy territory. The man who fights in primitive society is essentially a warrior rather than a soldier, that is, he is an armed civilian, not under proper military discipline, who fights basically as an individual, pursuing his own aims or those of his kin. But good discipline and effective systems of command are hard to come by in tribal societies. The warrior culture, dominated by the search for personal glory, goes together with indiscipline and the lack of clear command structures:

there are usually no proper formations that fight as units on the battlefield, but unorganised groups, with a tendency to individual heroism and self-reliance: 'The warrior, in contrast with the soldier, almost always yields to the temptation to accomplish useless little victories, the slaughter of one man or the crushing of one small party. This fritters away the force's strength so that more often than not no real advantage is acquired by the victor nor permanent injury done to the defeated.' [17] But true warfare, that is, warfare that can accomplish political conquest, requires a unified command structure, disciplined formations, the ability to conduct campaigns, as opposed to mere raiding, the organization of adequate supply to maintain a campaigning force on foreign territory, and the clear aim of subjugating the enemy. [18] These characteristics, however, only appear during the emergence of the state.

In any case, in tribal society neighbouring groups are often very much alike in organization and culture, so that primitive warfare does nothing as such to change the actual organisation of society, and tends simply to perpetuate the existing social system. Among the Tauade people might flee from the tribal territory after particularly intense fighting and take refuge with relatives and friends in neighbouring tribes. But after a while they would return, and the normal pattern of life and social organization would be resumed. No tribe could conquer the other, and the same was true of the Konso where the battles between the villages never produced any political or social change, and merely maintained the *status quo*.

It is worth pointing out here that being good at fighting is not *by itself* a basis for political leadership in primitive society. In states with professional armies there are many examples throughout history of victorious generals usurping the throne because the troops preferred to obey their general's orders rather than those of the king. But in primitive society a fight leader is not a professional general, and does not control a specialised group of warriors: he only has his kin and neighbours to support him in a raid, and they have no particular reason to obey him in the ordinary course of daily life. In any case, political leadership in primitive society takes more than being good at fighting. It needs social skills such as tact, diplomacy, the ability to speak well in councils and meetings, and the guile to make advantageous alliances with other important men. This is why men with political authority in primitive society are supported by councils of senior men and do not act alone, and where authority is inherited it is always legitimated by descent and religious status.

The aim of warfare in tribal societies is predominantly that of vengeance, manly honour and displays of martial ferocity by which men gain the respect of their fellows (and possibly the sexual favours of women), and killing for its own sake which, as I have said, can become a kind of human sacrifice. But these rather frivolous motives are repugnant to the disciples of Mr Gradgrind because, they believe, people would never engage in an activity that risked death, injury, and destruction of property unless very serious and practical issues of survival and control of natural resources were at stake.

So, it is claimed, despite people's conscious reasons for fighting, all warfare must really be motivated by economic needs, and the competition for animal protein and other resources; or the need to preserve the solidarity of the group; or the need to space out the population to prevent the degradation of the environment. [19] Unfortunately, no one has ever explained how people could think they were fighting for one reason, such as revenge, but really be fighting for some other reason, such as spacing out the population, which they would hardly understand, let alone care about. No psychological evidence has ever been produced that people could, unbeknown to themselves, be subconsciously motivated by the ecological or social needs of their group, and it has never been explained just how it was that their warfare practices could be so neatly adapted to their environment without their knowledge. [20] And since tribal societies typically operate well below the carrying capacity of their land, the whole argument that warfare confers an unrecognised ecological benefit is inherently feeble.

New Guinea provides the classic refutation of the belief that primitive warfare was primarily inspired by materialist aims such as acquiring more land or other resources, because it had one of the world's highest levels of violent conflict but also very low population densities, and ample natural resources. Among the Tauade, for example, there were only about 20 people per square mile, and they had far more land for their gardens than they could have used. Yet before the Australians established order, the death rate from violence was about 1/200 per year, or 500 per 100,000, by comparison with the 2–5 per 100,000 of modern Western society, and the Tauade pattern of warfare and killing was typical of Papua New Guinea as a whole.

Tribes were small – around 200 on average – and were split up into small hamlets because of the quarrels and vendettas within the group; each hamlet was dominated by one or two Big Men, and while tribes fought constantly with their neighbours there was almost as much fighting between the hamlets of the same tribe. But warfare among the Tauade and in Papua New Guinea generally was not widespread because it was adaptive, but simply because there were some general conditions that produced it. We can begin with the universal characteristic of males, especially young males, to learn aggressive behaviour very easily, especially in a culture that stressed the ethos of vengeance and competitive feasting. We then add to that the mutual suspicions and hatreds between the different tribes in this milieu, which were heightened by the gardens and fixed settlements, valuable stands of pandanus-nut trees, and the rearing of large pig herds for competitive feasts. These all provided many opportunities for quarrels over theft and the destruction of gardens. But these conditions by themselves were not enough to explain the traditional levels of violence.

Throughout most of Papua New Guinea there is a general absence of men with the authority to mediate in disputes, [21] and while Tauade Big Men had some control over their immediate followers, they could not actually settle disputes among them, let alone within the tribe as a whole, nor could they exercise effective control over individuals. This lack of any system of chiefly authority allowed individuals or small groups to start fights or take revenge on their own initiative, which produced a self-perpetuating system of endless cycles of retaliation. Vengeance could also be taken upon *any* member of the killer's tribe, not only his close kin, and taken by anyone who felt sufficiently upset to do so, not just by the family of the victim. Again, men who committed acts of violence within one tribe could easily escape and live with friends and relatives elsewhere, without having to face the consequences of their acts. Compensation was readily accepted both for personal injury and for homicide, and while this could settle individual disputes, it created a social milieu in which people knew that they could literally get away with murder. This lack of political authority and the high mobility between groups are, of course, familiar features of hunter-gatherer society, and it seems that many of the farming societies of Papua New Guinea have retained these features because, in particular, the sweet potato was only adopted in the last two or three hundred years, and before then there was much more reliance on hunting and gathering. [22] The sweet potato allowed population levels to increase dramatically, but the role of Big Man was unsuited to the political control of these larger groups, and the greater productivity of the sweet potato also supported larger herds of pigs. The competitive atmosphere surrounding the pig feasts in turn increased the antagonistic relations between tribes, and men who had killed had high status which they advertised by the shell ornaments they wore on their foreheads at dances, and were highly attractive to women. The social organization, therefore, maximized the opportunities for conflict.

Competition for scarce resources was not, however, a significant factor in Tauade warfare, and men did not steal pigs or pandanus nuts because they had no food. They had plenty of land, and even when they temporarily drove a tribe off its land so that they had to take refuge with friends and relatives in neighbouring groups, the losers always returned to their old homes when tempers had cooled, and those who were victorious did not take the land of those they had defeated. Nor did they steal their resources, but usually destroyed them, killing their pigs, burning their houses, and cutting down their pandanus trees.

It was obvious from all this destruction, slaughter, mutilation of bodies, and cannibalism that Tauade warfare was motivated by vengeance and love of power rather than by economic considerations. In similar vein, Genghis Khan (who knew what he was talking about) is said to have remarked: 'The greatest joy is to conquer one's enemies, to pursue them, to seize their belongings, to see their families in tears, to ride their horses and to possess their wives and daughters'. *Obviously*, other factors were involved as well, particularly the form of social organization, but we must recognise from examples all over the world that these forms of behaviour emerge very easily in the appropriate circumstances, and cannot be explained by Darwinian and materialist theories.

Cannibalism, for example, was a typical feature of New Guinea warfare, [23] and a popular theory of cannibalism maintains that it was caused by a need for high-quality animal protein in areas where the supply of meat was inadequate, but this is not really very plausible. [24]

The need for animal protein can easily be exaggerated, since vegetable protein from legumes in addition to cereals is an adequate substitute. We might also reflect that for more than two thousand years, vegetarianism has been a defining requirement for high caste Hindus, who are nevertheless not known for their cannibalistic appetites. Cannibalism, like head-hunting and the castration of enemies, should rather be seen as part of a broader motivation for war than the need for protein: the desire to express virility and power by killing and destruction for their own sake.

It was very easy, in fact, for warfare in tribal societies to become a kind of religious cult, and in Fiji cannibalistic symbolism permeated their whole way of life; it was expressed in

> the specific drumbeats announcing the taking of *bakola* [cannibal victims]; the pennants flying from the masts of victorious canoes signifying *bakola* on board; the ovens reserved for cannibal feasts; the special stones near the temple on which *bakola* were carved up; the sacred trees on which their genitals were hung; the (natural) bamboo splints used to carve human flesh and the elaborately fashioned forks used to eat it; the distinctive dances, songs and unrestrained joy with which young women, dressed in finery, greeted the return of successful warriors; the sexual orgies while the bodies were cooking; the ritual consecration of warriors who had killed and the enshrinement of their war clubs in the temples; the miserable afterlife of unsuccessful warriors, pounding a pile of shit through all eternity; the gourmet debates about body parts; the taboos on human flesh for certain persons; the cures effected by pressing cooked *bakola* flesh to the lips of afflicted children; the sail needles made from the bones of notable *bakola* and the poetry from their fate. [25]

Although they abhorred the idea of cannibalism, the Konso are another good example of the sort of cult that can develop around the slaughter of enemies, focused in this case on castration and the penis. Let us imagine we are visiting a sacred place of the Konso, looking out from their mountain heights over the lowlands. Every village has a number of the sacred places that are called *moora*. This word originally meant a cattle enclosure, and comes from the time long ago when sacrifices took place in them. Each *moora* has a border of shade trees, under which are stone seats, and there is a special gate that marks the ceremonial entrance. Inside the *moora* are some standing stones, called stones of manhood, and besides being obviously phallic in appearance they commemorate past battles when the warriors of the village killed some of their enemies. There are also one or two very heavy boulders lying on the ground in the middle of the *moora*. These are lifted by the young men to show off their strength, and who also sharpen their spears on them to bring themselves luck before they go hunting lion or leopard.

Around the edge of the *moora* are some striking carved wooden statues, some still with their red ochre paint and others weathered [affected by exposure to weather] to a silvery grey. These commemorate distinguished warriors: the hero himself is in the middle, and besides having a phallic ornament on his forehead, the *xallasha,* he is represented nude and is carved with a large penis. On either side are several female figures, also nude, who are his wives, and then a number of sexless figures. These represent the victims he killed in battle; in traditional times the Konso castrated their victims and wore their penises on their wrists as a kind of bracelet, and so the statues have no sexual parts. (Warriors of a neighbouring people are described as returning from battle 'carrying in their hands the severed genitals of their slaughtered foes'.) The statues were painted with red ochre because red is a 'good' colour, standing for meat and prosperity as well as for blood.

At one end of the *moora* is an *ulahita*: this is a bundle of several tall poles made from the stripped trunks of young juniper trees standing on a stone platform, and these represent the young warriors of the village, the members of Xrela age-grade. The *ulahita* and the stone platform on which it stands are also called *miskata*, because no fertile woman is allowed to sit on it by the *ulahita*, for fear that her femininity would weaken the young men. In a similar way, if a bull were to climb onto the platform it would also challenge their virility and have to be sacrificed. The juniper trees are bound together with a vine called *xalala*: this was also the name of the long plaited hair which the warriors used to throw over their shoulders when they were raiding their enemies. The *xalala* is very sacred, and is also used as a garland around the neck of an animal about to be sacrificed. When a man has killed a lion or leopard he will hold a triumph ceremony for them in the *moora*, and drape himself in *xalala*, which is also put around the stones of manhood. Also in the *moora* is a tree, in whose branches are put the skulls and other bones of animals that have been killed in the hunt. Finally, beside the *moora* is a large hut, with a clay pot on the top of the roof that is embellished with model phalluses. This hut is used by the local married men to sleep in, because they believe that their virility is drained by too much sexual intercourse with their wives. Everything in the *moora* seems to focus on bloodshed, the killing of humans and animals, and the penis and castration, but despite its association with slaughter the *moora* was described to me as 'the navel of the land', a place through which the people, the animals and their land and crops receive Life. Headhunting in South East Asia, and the torture and sacrifice of prisoners in North and Central America, are further examples of ritual practices for agricultural fertility that were essential parts of primitive warfare.

Primitive warfare could therefore be an aspect of the search for Life and in some ways be a prolonged form of human sacrifice, rather than a means to some utilitarian end like loot, or access to markets or the conquest of territory. Being often an aspect of religion, it raised daily life above the humdrum level of grubbing for food, and like religion it sometimes drove people into extraordinary exertions that they would never otherwise have bothered with. On the other hand, in many societies including the Konso there was an ambivalent attitude to warfare because of its obvious destructive consequences, and the shedding of blood in battle might conflict with the sacred status of a chief or lineage head and his peacemaking functions. [26] We will come back to this issue in the final section.

Anthropologists, well aware that primitive warfare is not about conquest, have struggled as we saw to find all sorts of subtle ways in which primitive warfare is adaptive, because they have believed that if something is common it *must* be adaptive. I am suggesting, however, that they have been trying to solve a non-problem. Primitive warfare is widespread simply because there are a number of common factors that lead to it: the ease with which people learn to hate other groups and with which young males learn physical violence, the lack of any centralized authority to make peace in tribal societies, the need of warriors and their leaders to prove themselves, the self-maintaining properties of revenge-cycles, and associations between vitality and the killing of enemies. [27] Primitive warfare could continue indefinitely in many areas precisely because it was inconclusive and insufficiently lethal. A number of similar local groups, very inefficient by more advanced standards of political and military organisation, could stagger along for century after century, not really being able to do each other much harm, in a state of minimal social change.

The real evolutionary potential of warfare in tribal society is not that it readjusted population imbalances, or that it was a source of protein, or that it worked as a form of natural selection by eliminating unfit individuals or groups or customs, or had any other pseudo-adaptive advantages, but rather that it stimulated higher levels of co-ordination within the local group than would have been needed in a situation of peace. In the first place, fear over personal safety in a milieu of chronic warfare leads people to live closer to one another than they would otherwise do, in defensive settlements, and the usual response to an increase in warfare is an increase in settlement size and group co-operation in building stockades and other defences. The war party inevitably required the co-operation of a number of adult males of every household, and fighting was, together with ceremonial exchange, one of the basic means by which wider social bonds were forged. When freedom to migrate was restricted in densely populated areas (unlike New Guinea) dependency on one's local group and loyalty to it would obviously be intensified. Mutual dependency of this sort meant that fighting was also the basis of politics in tribal societies where alliances and enmities with other groups were crucial issues, and so it was a powerful reason for co-operation and for some men to accept the leadership of others, not only as fight leaders, but also as spokesmen for the group in peace negotiations, which were very important.

Being a fight leader *by itself* could not be a basis of political authority, as we have seen, but when chiefly authority existed, legitimated by descent and religious status as it always was, if it could also be combined with war leadership this was a very effective means of enhancing the chief's political power. The willingness to accept this would of course increase when there was a significant external threat, and so competing chiefdoms are one of the basic scenarios in which the centralization of political authority occurs. When stable inherited political authority emerged, it was the long established traditions of warfare that could be developed into a means of conquering one's neighbours by armed force, which was a principal route into the emergence of early states.

None of these *consequences* of warfare were, of course, *why* people actually fought. This they did for quite other reasons that we have already examined. They were simply the consequences that were built, as it were, into the structure of tribal society.

### 3. Core principles and the roots of political authority.

So if we have sacred, hereditary leaders, responsible for blessing their dependents, who control the allocation of land and are wealthy from tribute and the control of trade routes and markets, and are also war leaders or at least patrons of warfare, these are favourable conditions for the development of chiefly, political authority that extends to a number of different descent groups. Because descent groups are inherently of unequal status it is possible for the head of one clan to be accepted as the head or paramount chief of the whole society, especially if he has high religious status as well: 'Religion provides a wider basis of kin and community than utilitarian motives. Everywhere among kinship societies common ritual or religious interests draw people together far beyond their ordinary interests…'. [28]

Many believe that the human race naturally prefers equality, so that inequality of the kind that we are discussing has to be imposed against people's will. If the human race naturally prefers anything it is reciprocity, a fair return, and if circumstances create dependency, and therefore require subordination and inequality, so be it. People are quite prepared to give respect and deference, their labour and their goods, to a leader if they think they get something in return, but what they do resent is the failure of reciprocity – the superior who takes all and gives nothing, the person who abuses the powers of his rank, the incompetent and the losers and the self-indulgent in positions of authority.

Obviously, many tribal societies do not, however, develop into chiefdoms with political authority because the necessary conditions are not present. We need to remember that originally these tribal societies would have been founded in particular historical circumstances, by very small groups of people whose idiosyncrasies and personal decisions would have had a profound effect on subsequent generations. This often referred to as 'the founder principle', and it operates when a few individuals have a disproportionate effect on the traditions of institutions or societies in their initial stages and this must often have occurred in primitive society. The result is a distinctive world-view, a set of 'core-principles' about social organization, values, and beliefs in terms of which subsequent generations continue to interpret the world despite migrations and new environments. [29]

For example, it has been possible to reconstruct ancient Polynesian society as it would have been in Tonga, Fiji and Samoa around 1000 BC, [30] and some leading Pacific scholars consider that

> . . . Polynesians carried about with them a set of principles for interpreting the world and organizing their social lives. From this standpoint Polynesian social formations are expressions, under a variety of historical and ecological conditions, of a basic world view that includes specific notions about kinship, relationships between human beings and ancestral gods, and a host of related beliefs. [31]

Polynesian society since then has been based on a number of patrilineal 'clans' (*hapu, kainanga*), made up of the descendants of a common ancestor, whose sub-groups lived together in large houses with associated land, but which at this local level were open to cognatic recruitment as well. The chief of the clan was the *ariki,* with both social and religious functions, like the Konso *poqalla,* and his status was based on the belief that he had inherited a specially large amount of *mana* in the male line from the founder. This was a supernatural force that was inherited much more by men than by women, and by eldest sons in particular, because seniority of birth was a principle of great importance, that linked the eldest son especially closely to the ancestors. The ancestors had the power of growth and were the source of *mana*, so the *ariki* presided over rituals honouring the ancestors that were very important in Polynesian religion. They were held at the *marae*, the sacred place of the clan, with food offerings and the drinking of *kava*, a drink made from an hallucinogenic root thought to bring the worshippers into contact with the ancestors and the gods. But although the *ariki* inherited their *mana*, this was not enough, and they had to demonstrate their possession of *mana* repeatedly by organizing successful feasts, exchanges, and increasingly, warfare. Status rivalry between chiefs was therefore a basic component of Polynesian society and in a few cases, such as Tonga and Hawaii, led to the development of monarchy. (Polynesian islands, however, were usually too small to allow the state to develop or, as in the case of New Zealand, of too recent habitation.)

In New Guinea, on the other hand, it is likely that the extreme importance of ceremonial exchange and its management has been a major factor in preventing the development of hereditary chieftainship, because Big Men needed personal abilities that could not be inherited. [32] When they became too old to maintain their dominance their followers simply drifted away, and so it was not possible for this type of leadership to develop into a system of hereditary authority of the kind that is very common elsewhere in tribal society. In the same way, Big Men could not mediate in disputes because they were simply the leaders of factions within the tribe, and this limitation on their leadership was a fundamental reason why local organisation remained of a very simple type and could not expand very far.

It was also hard for centralized political authority to develop among the Iroquois of North America. They had hereditary chiefs who were the heads of matrilineal descent groups, controlled trade routes and their clan's treasury of prestige goods, were responsible for organizing public ceremonial and religious rituals, and providing generous hospitality, and acted as spokesmen for their community in diplomacy with other groups. Yet they were not paramount chiefs with significant political power because of a number of other features of Iroquoian society. In the first place, the Iroquois, like many peoples of eastern North America, made a very important distinction between peace chiefs, such as the ones I have been describing, and war chiefs who, as the name implies, were responsible for the conduct of raids, and the torture and killing of prisoners and witches. This distinction 'had to do with the idea that violence and maintaining order were incompatible from the cosmological point of view'. [33] The peace chiefs were elected from the chiefly line by the older women of the clan, and there was a general ethos of equality: 'at all levels of Iroquoian society, care was taken to avoid the appearance of coercion or of one person being given orders by another'. [34] It seems that the women of the clan tried to avoid electing chiefs who might become too ambitious, and could depose them if they did. In addition, while peace chiefs may have organized rituals of increase, they had no personal monopoly of sacred powers and many other people officiated at these rituals, while shamans were also very important sources of supernatural power.

Again, the core principles of the societies that speak East Cushitic languages, such as the Konso, have some distinctive features that made it very hard for them to develop chiefdoms. [35] The Konso live in about three dozen large villages which are divided into three regions, and at the head of each region is a leading *poqalla*, who combines priestly and chiefly functions. They performed essential rituals for all their region, blessing its members and the pair of sacred drums that in each region were symbols of peace, and had to observe a number of taboos including the requirement to live in isolation outside the villages. They were particularly rich because they inherited especially large estates, and were given tribute by the villages of their region, and also collected tax from local markets. They could intervene in battles between the villages, and also act as judges in disputes between individuals of different villages but could not take any part in warfare, either its conduct or its planning, and had no armed supporters to enforce their decisions, because this conflicted with their sacred status as bringers of peace. They were not the leaders of any of the villages, nor were they the heads of any of the nine clans of the Konso: these clans did not own land and the members were dispersed among all the villages, so they were not effective corporate groups. The fact that the regional *poqalla* could not be war leaders, did not have a potential power base in any one village, and were not the owners or controllers of clan lands were important factors preventing them developing into chiefs with real political authority.

The Konso villages were the real centres of power, but their decision-makers were councils of elders elected on merit, not the lineage heads, and the councils and the warriors of each village were also organized on the basis of the age system. Certain legends suggest that the Konso age system was adopted, at least partly, as a means of restraining what might become the arbitrary authority of the *poqalla*, and we have seen that age-systems are inherently egalitarian. The fact that political authority, then, was elective, non-hereditary, and based on councils and the age system, were further reasons why its centralization could go no further.

It is therefore very significant that whereas all the East Cushitic speaking neighbours of the Konso had age-systems, and no kings, (with one exception that emerged in strange circumstances), the West Cushitic societies all had kings and no age-systems. [36] The clans in

West Cushitic society also had much greater political importance: there was a royal clan, below which were commoner clans, with some slave clans at the bottom. Not only were West Cushitic clans of unequal status, but far from being scattered like those of the Konso, their members were all located in specific territories, with local hereditary clan leaders who controlled the allocation of land. The West Cushitic king was not only the religious head of his people, but also their war leader; in some cases it was believed that the royal dynasty had been there from the beginning of time, and in others that it had achieved its position by leading the conquest of the original inhabitants. In many of the societies speaking West Cushitic languages, then, the state did develop, notably in Kaffa and Janjero, and this is because in these societies the descent principle was paramount, residence was dependent on clan membership, the chief and subordinate clan heads controlled the land, there was no form of age-system, and the chief was not only the supreme religious figure but could also combine his religious status with that of being a war leader as well.

## 4. Conclusions.

Tribal societies have a natural tendency to expand simply through population growth, while disputes of many kinds and warfare are further reasons why migrations are such a typical feature of tribal society. As lineages expand over many generations they accumulate more and more branches, while the living members become more distantly related. There is therefore a natural tendency for the heads of the different branches and their kin to want to establish their independence and set up on their own, and the hereditary principle means that there are likely to be challenges by rival claimants for the chieftaincy, who either try to overthrow him or to secede.

There are, then, inherent limits to the number of people that can be held together by tribal social organization, or even by chiefdoms. So it is very significant when something prevents this endless splitting up and spreading out, by creating the conditions in which people find it more attractive and advantageous to stay put. This may be because there is no free land to expand into without warfare, or because of a general increase in warfare for other reasons, so that expansion is prevented by conflict with neighbouring societies. An external threat also makes people more willing to accept the need for solidarity in the face of a common enemy. Greater willingness to stay put can also be the result of being in a very favourable environment, like the Nile Valley, where conditions beyond its borders are threatening and inhospitable. These are all conditions favouring the expansion of chiefly authority, and the emergence of centralized political control, in the form of the state, will be the subject of Chapter VII. But first we must deal with the nature of primitive thought, which has been in the background of much of the discussion so far.

# CHAPTER VI
## Primitive Thought

In primitive society, beliefs in witchcraft, magic, divination, evil spirits, the ghosts of ancestors, gods, and so on are obviously very common. For Mr Gradgrind and his successors, of course, this is all superstitious nonsense, but if they are just illusions, and if there is really no connection between the social and the cosmic order, why, then, have such beliefs been more or less universal until modern times? The standard answer has been that they were selected for because, despite being illusions, they had useful results in reinforcing social solidarity (as when people perform rituals together, or believe that supernatural beings punish wrongdoers), or, as in the case of divination and magic, give people confidence when faced with a threatening environment.

There was never any attempt to produce serious, systematic evidence in support of this theory of selection: no one could give examples of tribes disintegrating because they did not believe in supernatural punishments, or villages paralysed with fear because they had abandoned magic. It was simply assumed that every form of belief must be adaptive because, if it were not, then it would have been eliminated by natural selection. The basic weakness of the selectionist theory (besides lack of evidence), is that the comfort-value of beliefs, or their contribution to social solidarity, can't explain why people should actually find them credible in the first place: rituals may certainly promote social solidarity, but how does this explain why those who perform them believe that the social order is part of the cosmic order? People may gain confidence from burning an enemy's hair-clippings, but why do people in so many societies think that this sort of procedure will actually harm the victim? No doubt a Tauade, walking along a path, who recites a spell to reach his destination more quickly (without going faster) does so because it makes him feel more buoyant and optimistic, but this does not explain why he thinks that words in themselves have power over events, and why he does not understand the relation between time, speed, and distance. [1] (Many supernatural beliefs, such as witchcraft and evil spirits, are not comforting at all, but I discuss the theory of 'the supernatural as wishful thinking' in more detail in note 2.)

It is these assumptions about reality that are fundamental, and we can only hope to explain them by looking more closely at how our understanding of the world develops, not by speculating on the adaptive value of optimism and general self-confidence in the struggle for survival. The obvious conclusion is that these, and other primitive beliefs are universal, not because they provide some selective advantage, but simply because human beings find it very easy to think in this way, because it is difficult to refute them, and because they are not so destructive or so self-defeating that they are abandoned, or cause the extinction of the groups which hold them. So in this chapter I shall try to explain those features of primitive mentality that make it easy for the symbolic ordering of the world and certain kinds of supernatural beliefs to be accepted. Much more broadly, I shall show how it is possible for 'first ways of looking at the world' in general to survive very easily in the technologically undemanding and pre-literate environment of primitive society. [3]

*1. First ways of looking at the world.*

The human mind is not like an empty bucket that is gradually filled with information by adults, or by passively observing the world around one; each individual has *actively* to construct his understanding of the world, of things and of people, by interacting and experimenting with it himself. While we are born into a particular culture, which we did not make, our culture can only be transmitted by individuals, including ourselves, so what the majority of individuals can understand must have a fundamental effect on the kinds of ideas and beliefs that can develop over time in any culture.

Some ways of thinking are more elementary than others, and provide the foundation on which the more advanced and complex types of thought can be constructed, when the social conditions are right. But, as we shall see, it is quite possible for simple modes of thought to survive perfectly well in simple societies. We can begin by looking at how children's knowledge of the physical world develops. They do not learn merely by copying adults but by *their own activity and experiments* on the objects around them, and by assimilating the results of these into their own modes of thought. For example, infants below about 18 months (who are at what is called the sensori-motor stage of mental development) have to construct a stable view of the world around them entirely without the assistance of adults; they have to learn that toys do not really vanish when they are hidden behind a cushion but continue to exist and can be found again; that objects do not really get bigger when they get closer, and smaller when they get farther away; and they have to learn how to co-ordinate what they see with what they hear and with what they can touch. They achieve this understanding of the physical world by actions, not by words, and by patterns of actions that are repeated with an increasing range of objects, and with increasing discrimination as they learn what different objects are good for. They also come to realise that they, too, are objects in a stable world of other permanent objects, and this general process of development is the same for all infants whether they grow up in a hunter-gatherer society or in a modern city.

At about 18 months they start to form mental images of different aspects of the world. These images are not just visual but involve imitating the actions and events that the child is trying to understand,

and he can communicate these images to other people not only by imitation but, increasingly, by language and also, if the means are available, by drawing and models. This is the beginning of what is often called the 'intuitive' [4] stage of thought, the point at which, intellectually, the child starts to break out of its private world and to become a social being who can participate in culture. From this point on, the child's society and general environment start to have a major influence on his intellectual development. Imagery allows the child to perform actions internally in its own mind, and is therefore a major advance on pure physical activity. But the fact that the child can now represent the world to himself in imagery, and refer to it in words, means that while it is possible for him to think that he understands what is happening, in fact there are major problems that prevent these image-based representations from corresponding with reality in some respects.

So, for example, if a child of three or four is asked to draw the stages by which a curved piece of wire or stick is straightened he will do it something like this:

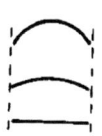

This is obviously wrong, because in reality the ends would get farther apart, but children at this stage are unable to analyse processes and to co-ordinate different elements of a process, such as the relations between the curvature of the wire and the distance between the ends. They tend to concentrate, to 'centre' their minds, on *one* aspect or dimension of the event and ignore the others, and are generally unable to hold two related but different ideas in the mind at once. If a child at the intuitive stage is asked, for example, to arrange a series of ten blocks in order of size he will be unable to do so: while he can say that this one is bigger than that one, he cannot grasp that the same block may be *both* bigger than one and smaller than another. Without understanding this, it is impossible to construct a series of blocks from smallest to largest.

Again, if these children are shown water being poured from a short, fat glass into a tall, thin glass they will say that there is more water in the tall thin glass because they are concentrating on one dimension alone, the height, and not taking the different diameters of the two glasses into account as well. Even though some may also agree that the amount of water remains the same because nothing has been added to it, they still cannot 'conserve' quantity – realise that it remains the same – because they cannot operate with two different dimensions simultaneously, and understand that an increase in one is *compensated* by an equal decrease in the other. While the child may agree that if the water were poured back into the short fat glass there would be the same quantity as before, he still believes that there is more water in the tall thin glass. It is as if pouring back the water from the second glass into the first were completely unrelated to pouring from the first glass into the second, and without understanding reversibility, process presents major problems. Compensation between height and diameter has first to be grasped before the pouring back, the *reversibility*, can be understood as a crucial test that something – in this case quantity – has remained the same, or been conserved, despite the changes in appearance.

Whereas intuitive thought remains tied to the appearance of things, to static and unco-ordinated imagery, development beyond this involves 'operational' thinking, building up systems of thought that can cope with transformations in the appearances of things, and realise how some properties remain the same throughout these transformations. Conservation, based on *compensation* and *reversibility*, is an essential element of these stable systems of relations; but operations also involve *decentration*, the ability to think of things as having more than one relation at a time – like the block that is both bigger than one and smaller than another. Operational, unlike intuitive thinking, is therefore a much more powerful means of understanding the world; it is 'systems thinking', in which the parts and the whole can be thought about together, in mobile and dynamic relationships, without the contradictions inherent in static, *image-based* thought.

In our society the ability to conserve quantity typically starts developing around the age of six, but the conservation of other properties such as length, weight, area and volume takes several more years, and the same is true of an operational grasp of number, space, time, and classification, which I shall discuss in more detail later. Concrete operational thinking is therefore bound up with the sorts of practical problems in weighing and measuring that we encounter in daily life on, say, a construction site, or when we are doing some DIY in our home. It also involves a basic mechanical understanding of how things work, of why the pedals on our bicycle won't go round, for example, or how a steam engine is different from the engine in our car, and so we would expect it to be developed by experience of technology, in particular.

The development of what is called 'formal operational' thought goes beyond concrete thinking, and begins to occur in our sort of society during adolescence, if children have the appropriate schooling and the opportunity to experiment, and is basic to scientific and philosophical thinking. Whereas concrete-operational thought is limited to thinking about relations between actual objects and events, the formal operational thinker can consider systems of relations in a much more abstract way, without the need for concrete examples. He will start thinking in abstract terms of the forces of physics, and the structures of chemical molecules, and be able to express these in mathematical formulas and equations. He can also consider hypothetical situations in the form of statements about them, and so can mentally review all the possible (as distinct from the actual) combinations of factors and variables that could be involved. This allows the person to understand probability, to produce experimental hypotheses and work out methods of testing them, to be able to carry out purely logical deductions based on verbal statements alone, and to think about his society as a whole and to speculate about ideal political arrangements. I must emphasise, however, that not only do people not leap from one stage of intellectual development to the next, but that they are often at different stages in relation to different sorts of problem, depending on the circumstances in which they have grown up. Even in our own type of modern society, not more than about 50% even of adults are capable of formal operational thought, and there will also be many people whose thinking is at the formal level, for example, in the understanding of political and ethical issues, but still at the level of concrete operations in mechanical problems.

This development from intuitive to formal operational thinking also involves profound changes in two other major areas of our relations with the physical world. One is in our awareness of our own minds, and the other is the distinction between the social and natural worlds, but I shall deal with these later. I first want to draw the obvious parallel between the intuitive thinking that psychologists have described among children, and the type of thinking so often found in primitive society.

When tests of conservation have been given to adolescents and adults in non-literate, tribal societies it has been found that a high percentage of them cannot conserve quantity, or area, or volume either. They are nevertheless perfectly capable of getting through the day without this ability because they live in what, technologically, is a very simple world in which they do not need to pour water from short fat glasses into tall thin ones. But it is not just a matter of the lack of jugs and glasses in primitive society. There is also a general lack of measurement, with no standard units, no rulers, no scales and weights, and no clocks or other ways of measuring time apart from looking at the position of the sun. [5] The lack of measurement means that it is very difficult to think about things in terms of their different dimensions so that, for example, there will be no idea of measuring the area of a piece of land by multiplying its length by its width. While there are always words for big and small, heavy and light, long and short, or near and far, there are no words for size, weight, length, or distance, while big and small, or heavy and light, are seen as different and opposite, not as different points on the same scale or dimension.

So in this kind of society people's awareness of physical objects is not based on quantitative, dimensional analysis, but on how they look and feel, and on what it is like to work with them, bound up in the context of their daily use. In the same way, language is only experienced as speech, similarly bound up in practical situations and dialogue between real people, rather than as something encountered in writing that can be analysed on the page. This means that there is a vast range of problems that can never even occur to the members of primitive society, but which our children will encounter in their daily lives and at school. Our children are also provided with the means to solve these problems by their schooling, and social experience generally, which are completely lacking in primitive society.

Let me give a simple example. When I was living with the Tauade I wanted to ask them for how many years various pieces of land had been lying fallow. (They lived mainly on sweet potatoes, and their gardens only produced for about three years, after which they had to be abandoned for a time.) You might think that 'How long has this piece of land been fallow?' is a very ordinary question indeed, the kind that one early farmer, for example, might well ask another, but they were unable to tell me. The reason is that they had no word for year (or for month or week), and would anyway have had great difficulty in keeping a record of them, because they could only count on their fingers and toes, and their words for numbers did not really extend beyond two. This is not because they were stupid, or because their brains were different from modern brains, but because in their very simple environment they had been able to get by without a calendar or a number system. Again, they had no idea of measuring the area of their gardens. They could see how large they were, and they knew from experience how much new land they had to bring into cultivation each year, but this was judged without measurement of any kind.

So it is not really very surprising that even adult members of technologically primitive societies should have the same difficulties as children of six or seven in our society in solving such problems, simply because they have never had to think about them, and puzzle themselves about problems of area and volume, or the relations between speed, time and distance. It will help if we now look in more detail at some parallels between intuitive notions of number, space, time, and classification and those found in hunter-gatherer and tribal societies.

## 2. Intuitive and operational thinking.

A good example of applying operational thinking to the physical world is this problem about a house in a garden. In the first picture, A, there is a garden with a house in the middle, while in B the house is in the corner of the garden:

Put houses in as shown. Which garden has more ground outside the house, or do they have the same?

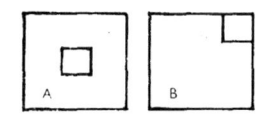

[6]

# Primitive Thought

The normal reaction, whether of adults or children, at the intuitive stage, dominated by imagery, is to say that there is more ground in B. To go beyond imagery, and realise that the amount of ground in the garden stays the same (is conserved), one has to be able to think in terms of area as the product of length times breadth. Then it is obvious that the area of the garden minus that of the house has to be the same in both cases.

This sort of problem is so normal for us that we are no longer aware of the thinking involved, but in fact it requires units of measurement, the ability not just to count, but to multiply and subtract them, and the idea of a space as defined by the two dimensions of length and breadth. We have seen, however, that dimensional thinking and measurement scarcely occurs at all in primitive society, and this means that ideas of number also remain very undeveloped. Here we need to remember the emphasis in hunter-gatherer culture on the *individuality* of things as opposed to their general qualities that we noted in Chapter II. If any counting occasionally needs to be done, they can use fingers (and toes), or stones, or notches in sticks or bones, or knots in a piece of string, to record the totals of things, without any words for numbers at all, and where number words exist they are often no higher than two or three. [7] This relative lack of number words is found throughout Australia, much of Papua New Guinea, the Bushmen of the Kalahari, and some of the hunter-gatherers of South America. It is a good illustration that something we regard as absolutely necessary, and unthinkable to be without, has really had rather a short history in human culture, and that people with very simple technologies can get along quite well without counting in words.

Not surprisingly, when systems of number words develop they are almost always based on the five, ten, or twenty suggested by the fingers and toes, with decimal systems being by far the commonest. But the early use of numbers is thoroughly concrete, bound to the physical objects that are counted. When the Kpelle of Liberia, for example, put objects together, take them away, or share them among sets of people, they never work with pure numbers but always with objects as well – '2 chickens and 3 chickens make 5 chickens', but the expression '2 and 3 are 5' would not be understood. [8] In Papua New Guinea, 'Systems of counting seem to be used entirely for counting numbers of concrete objects, such as wives, children, houses, pigs, and in the case of the larger counting systems [such as 'round the body' systems] the numbers of shells to make a bride price'. [9]

Primitive counting, then, does not treat numbers as abstract classes that are arranged in a series, but simply as groups of concrete objects each with a rigid and static association with a finger or other body part, or a name. It is only when complex market transactions developed, or people in the ancient civilisations started having to *measure* things, such as length, area, volume, weight, and so on and then do *calculations* with these figures, such as the number of mud-bricks needed for a wall of a certain size, that they could develop a more advanced understanding of number. Think, for example, of the tiresome calculations we have to make when we are ordering wall-paper for a room, when we have to work out the area of the walls, minus the openings, and then relate this to the width and length of the rolls of paper to calculate how many rolls we want, and the cost per square foot. To do this we need to understand how addition, subtraction, multiplication, and division are related, so that we can see, for example, that 4 x 3 is carried out by 3 successive additions of a group of 4, and produces the same result as 4 successive additions of a group of 3, and that division, in the same way, is a sequence of subtractions. The result is that people will be able to grasp numbers in an operational way as mobile *hierarchies* of classes, instead of static collections of objects. For example, 12 can be seen as composed of 3 groups of 4, or 4 groups of 3, 6 groups of 2, or 2 groups of 6, and so on. They will also be able to understand the relation between cardinal numbers, which express quantity, and ordinal numbers, which express order, so that the cardinal number 4 also occupies the $4^{th}$ position in the order 1, 2, 3, 4, 5. . . , but the $2^{nd}$ position in the series 1, 4, 8. . .

We have seen that in primitive society area is not thought of as length times breadth, and space in general is not thought of in terms of the straight lines, dimensions, and angles that we take for granted, but intuitively in terms of what is known as 'topological' space. This takes no account of angles, or straight lines, or distances, or co-ordinates, and is often compared to a rubber sheet. If, say, a face is drawn on this sheet, however it is pulled, stretched, and distorted, all the basic features of eyes, nose, mouth, and ears will all stay in the same relative positions to each other, however much the angles and distances between them may change. Topologically, the eyes are each *separated* from the nose but are *next* to it; they are in a certain *order* – eye/nose/eye – and the nose is between or *included* by the two eyes. If we think about a landscape in a similar way it is easy to see that its various features can be thought about as those of separation, 'next-to' or proximity, order, and between-ness or inclusion, and in a primitive environment understanding these will be enough to represent the physical world.

Village layouts and sacred places with their centre, boundary, and gates are like faces: we remember here the three trees of the Umeda – the coconut at the centre of the hamlet, the areca palm on the boundary, and the sago palm in the bush. Because topological space is closely related to actual physical features it easily acquires symbolic values. East, for example, is very widely associated with the rising sun and is therefore life-giving and auspicious, and people may be buried facing that direction when they die, whereas the West is bad because it is associated with the setting sun and with death. The body and the house are the focus of basic topological concepts like right and left, high and low, centre/periphery/and outside.

|          |          |          |
|----------|----------|----------|
| Right    |          | Left     |
| male     |          | female   |
| strong   |          | weak     |
| sacred   |          | secular  |
| superior |          | inferior |
| noble    |          | plebeian |

|               |                |
|---------------|----------------|
| High          | Low            |
| sky           | earth          |
| male          | female         |
| head          | feet/buttocks  |
| sacred        | secular        |
| noble         | plebeian       |

| Centre  | Periphery   | Outside |
|---------|-------------|---------|
| men     | young men   | women   |
| hosts   |             | guests  |
| village | garden      | bush    |
| culture |             | nature  |

[It should be noted that items in these columns have to be read across, not downwards. For example, female is symbolised by left, male by right, just as secular is symbolised by left, and sacred by right. But this does not mean that female also equals secular or plebeian.]

So topological space is essentially *static*, concrete and image-based, unlike our geometrical and projective space, which is designed to deal with every possible *movement* and *transformation* of position and shape, and is the same everywhere. Among the Temne of Africa, for example, the cardinal points are not used to co-ordinate movement and position in space, but have a static, symbolic meaning:

> For us the cardinal points are co-ordinates for establishing location. The Temne never use them in this way, though should the necessity arise they will use one of them to indicate the general direction in which a place lies. Their cardinal points contain [symbolic] meanings which qualify activities and events in various ways. [10]

Topological space is therefore quite inadequate when we want to represent actual tracks, positions, and bearings of different locations on some kind of map, or even to measure the area of a piece of land, as we saw in the problem about the two gardens. An operational, geometrical understanding of space, based on dimensions and measurements, only develops, however, where land has to be measured with some exactness, as in the ancient civilisations, or where long-distance navigation becomes highly specialised, as in Polynesia.

The intuitive understanding of time also has severe limitations. Everyone, including little children, can understand process as a succession of events, because our whole lives consist of processes of one sort or another, from getting up in the morning to going to sleep at night, the rising and setting of the sun, the course of the seasons, the growth of plants and animals, and so on. In primitive society, like our own, the day will often be divided into conventional periods, such as 'breakfast', 'taking out the cattle to graze', 'noon', and so on. There may be weeks made up of market days, and a sequence of named months in which different agricultural activities are carried out, the whole of which form a year. [11] So calendars treat events as if they were points or stretches along a path; just as in topological space all we need to know to find point G, for example, is where it is located in the sequence E, F, G, H, I, so in 'topological' time all we need to know is where 'bringing the cattle home' comes in relation to 'midday meal' and dusk'. In this sense, intuitive time is a kind of static, spatialized time like the static features of a landscape, and in the same way as places in a landscape, some events are more distant from another in time than others. [12] In intuitive time, then, one only needs to know the order in which things come (succession), and whether they last a relatively long or a short time (duration).

The Tauade had no words for week, month, or year, and yet they had a word that can be translated as 'time', *lova*: *oilova*, 'this time', 'now'; *telova*? 'What time, when?'; *opolovan*, 'olden time'. But what they really mean by *lova* is merely 'sequence of events'. So *oilova* would more accurately be translated as 'this sequence of events', i.e. the sequence of events in which the speakers are presently involved in; and '*telova* will you go to Port Moresby?', 'in what sequence of events will you go to Port Moresby?', 'after the plane has brought my letters to the mission'. But sequences of standard events do not form measurable series like the hours of a 24-hour clock: one cannot say that three twilights equal one morning, or that from first light to going out into the fields is the same amount of time as from the coming back of the cattle until supper time. The same is true of divisions of the year, even if they are expressed in named months. Among the Nuer

> It is true that the year is divided into twelve units, but Nuer do not reckon in them as fractions of a unit. They may be able to state in what month occurred, but it is with great difficulty that they reckon the relations between events in abstract numerical symbols. They think much more easily in terms of activities and in terms of successions of activities and in terms of social structure and of structural differences than in pure units of time. [13]

So because time is represented in terms of different sequences of unquantifiable events, and of social structures such as generations or age-sets, time can't therefore understood as a universal standard of measurement that is the same everywhere, but instead is broken up into a sequence of local events. These can never be compared with one another, and without a set of objectively defined, quantifiable and commensurable units, these qualitative sequences of events may move at different speeds in different places. So there is no reason to think that a day here is the same as a day anywhere else, and time may seem to speed up or slow down, or even have gaps. [14]

While people realise that some periods of time last longer than others – walking to a near village as opposed to a distant one, for example – we do not find that the journeys of the two walkers are ever directly compared, as in a race. This means that the relations between speed, time, and distance are never explored in intuitive thought, whereas operational time depends on a grasp of the relations between them. In primitive society the relative speeds of moving people or objects are not important issues in daily life, and in any case there are normally no means of measuring either time or distance. But let us suppose, for example, that two cars, 1 and 2, start from two points $A_1$ and $A_2$ simultaneously, and travel on two parallel tracks towards $B_1$ and $B_2$ and then stop at the same instant:

1   $A_1$.....................................................................$B_1$
2   $A_2$.........................................................................$B_2$

To understand this problem involves, first of all, a grasp of simultaneity: of course, we can all see if two things happen together, if it starts to rain just as we set off on a walk, or if the sun is setting when we come home, but the simultaneity of operational time involves two processes and judging when they both start *and stop* in relation to each other, which is essential for understanding the relation between time, speed, and distance. If car 1 goes faster than car 2 it will travel farther than 2 in a given period of time, but someone who is still at the intuitive level in the understanding of time will think that they could not have stopped simultaneously. This is because to such a person the car that travels the longer distance must have taken a longer time to do so, while travelling a shorter distance must take less time, and they think this because they cannot take account of the different speeds of the two cars, and understand how greater speed will cover a greater distance *in the same time*. A concrete operational grasp of time understands that the amount of time taken for the journey has to be calculated by dividing the distance covered by the speed at which someone travels. Without grasping this, the notion of time is therefore tied to distance and remains spatialized. This elementary relationship between speed, time, and distance is a good example of operational thinking, because someone who understands it will also realise that distance covered is the product of the speed and the time, and that speed is the distance divided by the time – a mobile system of transformations.

Classification is an absolutely basic mental function, but some sorts of classification are easier than others. When intuitive thinkers are asked to classify a variety of things, they do not put them in categories such as 'utensil', 'furniture', or 'food', but in clusters of things that 'go' or 'belong' together, so that knives, forks, and spoons will be included with tables and chairs, and with food and drink, because they all go naturally together for a meal in what we can call a 'complex'. In the same way, models of cow, pig, house, and man, 'go' together because he is a farmer. Utensil, furniture, and food, on the other hand, are what we call taxonomic classes, because they are based on some common characteristic, regardless of their associations in ordinary life, and are characteristic of operational thought. We have already encountered a very good example of the complex in the symbolic classification of Umeda society, based on associations between men, boys, and women with the centre and periphery of the village, and the bush, and three types of palm tree. These associations produce the complexes of: adult man/centre/coconut; boy/periphery/areca palm; women/bush/sago palm. Another good example is from Dahomey, where the goddess Mawu represents the female principle, and therefore fertility, motherhood, gentleness, and forgiveness; while the god Lisa represents power (warlike or otherwise), strength and toughness. They are also associated with day and night: Mawu is the night, the moon, freshness, rest, and joy; while Lisa is the day, the sun, heat, labour, and all hard things. [15]

These are extremely practical classifications of the world, based on the associations between features of daily life, but this sort of classification can't form the basis of taxonomic classes. Someone might ask, however, why we should care about taxonomic classes in the first place. While they are unnecessary in primitive society, they are a very powerful way of ordering experience, and advanced legal and administrative systems, logic, and all scientific thought, would be impossible without them, as we shall see.

They are formed by abstracting certain criteria from all their associations in real life and making them the defining criteria of a particular class. Take, for example, our taxonomic concept of 'vehicle'; we are not concerned with how expensive it is, or if it is kept in a garage, or who drives it, or what makes it go, but only with the minimal defining features: 'A means of conveyance provided with wheels or runners and used for the carriage of persons or goods'. So one of the defining criteria of a vehicle is that it runs along the ground (whether on wheels, runners, or tracks), and as such it is in a different class from boats or aeroplanes. We can further distinguish between different sub-classes of vehicles as 'cars' or 'trucks', and sub-classes of, say, cars by their makes. These are all 'taxonomic' or logical classes and they are not only defined by clear criteria but also form a hierarchy from the general to the increasingly particular:

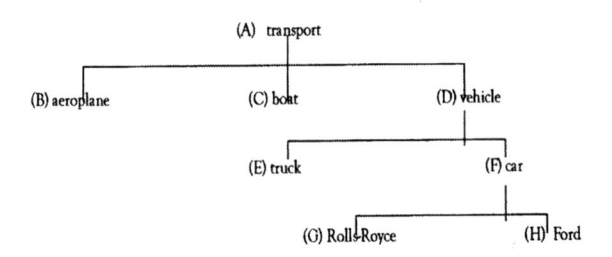

```
                        (A) transport
          ┌─────────────────┼─────────────────┐
    (B) aeroplane        (C) boat         (D) vehicle
                                       ┌──────┴──────┐
                                    (E) truck      (F) car
                                               ┌──────┴──────┐
                                        (G) Rolls-Royce    (H) Ford
```

They are also fundamental for the development of formal-operational thought: for logic, philosophy and, ultimately, for science. For example, in this hierarchy the class A is divided into sub-classes B, C, and D, and D in turn is divided into sub-sub-classes E and F, and so on. In other words, we can understand a hierarchy of logical classes as a system of class inclusions, so that class A>D>F>H, and therefore that A>H. We express these inclusions by use of the concepts of 'all', 'some', and 'none' (e.g. *all* members of F are only *some* of D, *no* E is F', *some* F is *all* H and so on).

But someone who does not understand the logic of class inclusion and logical 'all', 'some', and 'none', will not be able to understand certain forms of reasoning involving logical proof. For example, if someone tells us that *all* animals in Australia are marsupials, and that the koala bear is an Australian animal, we accept as a matter of *logic* that therefore the koala must also be a marsupial, even if we have never been to Australia and know nothing about the animals that live there. We can understand the idea of logical proof as distinct from factual truth, but this is a sophisticated, formal-operational distinction that would be quite unfamiliar to anyone in a pre-literate society. They would reject the suggestion that the koala must be a marsupial precisely because they had not been to Australia, and therefore could not be expected to know anything about the animals there. The whole idea that statements are not just about facts, or expressions of feelings, but can also be taken as having a logical form and so can be proved, regardless of whether they are actually true in the real world, originally occurred to people in Greece, India, and China less than 2500 years ago. These were groups of specialist thinkers who spent a great deal of time arguing about philosophy and religion, and how society should be organised, and they all discovered the need to be able to analyse the force of each other's arguments simply as arguments, and to find some objective way of disproving them. But for most of history, and for most cultures, logic in this sense has certainly not been available as a mode of thought.

*3. The mind and the world.*

Initially, at the intuitive stage of thought, children have no understanding of the mind, or of how it relates to the body, and a good example is how they come to understand the significance of speech and words. At the first stage, when the child can even understand the meaning of the question (at about 6), he supposes that we think with the mouth when we speak, and also identifies thoughts with breath, air, and smoke, or else equates thinking with hearing and so regards this as something we do with our ears. Words, and especially names, are regarded as a part of the things they refer to, and the function of the ears and mouth is therefore limited to collaborating with things – receiving words and sending them out. Words themselves therefore have strength, or weight, or speed, or any other property of the thing referred to. At the first stage, therefore, there are two related confusions – between thinking and the body, and between the sign or the word and what it signifies.

The development of the understanding of names is particularly revealing, for 'name' is a much clearer concept for the child than 'word'. At the first stage, when the child learns the name of something, he supposes that he is thereby reaching to the essence of that thing and discovering some real kind of explanation of what it is like. Things did not exist before they had names or, if a name exists there must be something in the real world to which it corresponds. Because names are properties of things, they are discoverable and we can come to know the names of things just by looking at them.

At the stage of concrete operations, names are supposed to have been given to things by God, or by the first men, but may still be thought in a sense to be 'in the things' or else as being 'everywhere and nowhere'. Even if we can't recognise a thing's name when we see it, the child still supposes that there is an inherent 'rightness' about names – the word 'sun' involves 'shining, round, etc'. The child can now understand that there is a problem about the relation between words and things, but still fails to solve it. Only at the third, formal, stage does he come to realise that names are just conventional labels, handed down by tradition. They are progressively understood to be located in the voice, in the head, and then in thought itself, which has now come to be located in the head as an unobservable and immaterial process.

It is also at the stage of formal operations that the child finally begins to understand the adult concept of the mind in general, and that thought is a set of representations conveyed in language. Our notion of the mind is therefore not obvious to all normal human beings: it has been constructed very laboriously by thinkers and scientists only in the last few thousand or even hundreds of years of literate civilisation, and we can only grasp it because we live in a complex society with an advanced technology, and have had many years of schooling and literacy. Schooling in our society requires children to explain their reasons for making particular choices in test situations, demands that they give reasons and justifications for beliefs, and challenges to the individual's own point of view in debate and argument ('That's just your opinion – how do you know?'), so that schooling in particular develops the awareness of one's own mental processes and the ability to talk about them. [16]

But in primitive society children do not have the opportunity to be challenged about the nature of names and language in these ways because there is no formal schooling. Whereas children in our society learn in the artificial environment of the school where they have to solve problems that are outside their ordinary experience, and engage in debate, in primitive society the child is gradually introduced into the full life of an adult, 'and is almost never told what to do in an explicit, verbal, or abstract manner. He is expected to watch, learning by imitation and repetition [in the context of ordinary life so that] education is concrete and nonverbal, concerned with practical activity, not abstract generalization. There are never lectures on farming, house-building, or weaving. the child spends all his days watching until at some point he is told to join in the activity.' [17] The object of education is not cleverness, or to question or experiment or to think for oneself, but good sense, wisdom, and the ability to perform as a good neighbour and kinsman in work and social relations. The child is highly motivated to conform, and his basic learning commitment is not to things or ideas, but to people, especially those closest to him socially.

As a result, our notion of the mind does not develop in primitive society: For example,

> The Dinka [of the Sudan] have no conception which at all closely corresponds to our popular modern conception of the 'mind', as mediating and, as it were, storing up the experiences of the self. There is for them no such interior entity to appear, on reflection, to

stand between the experiencing self at any given moment and what is or has been an exterior influence upon the self. So it seems that what we should call in some cases 'the memories' of experiences, and regard therefore as in some way intrinsic and interior to the remembering person and modified in their effect upon him by that interiority, appear to the Dinka as exteriorly acting upon him, as were the sources from which they derived. Hence it would be impossible to suggest to Dinka that a powerful dream was 'only' a dream, and might for that reason be dismissed as relatively unimportant in the light of day, or that a state of possession was grounded 'merely' in the psychology of the person possessed. They do not make the kind of distinction between the psyche and the world which would make such interpretations significant for them. [18]

The early Greeks identified thought with breath, and speech and thinking with the lungs, [19] and for the Trobriand Islanders,

> The mind, *nanola*, by which term intelligence, power of discrimination, capacity for learning magical formulae and all forms of non-manual skill are described, as well as moral qualities resides somewhere in the larynx. The natives will always point to the organs of speech, where the *nanola* resides. The man who cannot speak through any defect of his organs, is identified in name and treatment with all those mentally deficient. [20]

And among the Gahuku-Gama, 'cognitive processes are associated with the organs of hearing. To 'know' or to 'think' is to 'hear'; 'I don't know' or 'I don't understand' is 'I do not hear' or 'I have not heard'. [21] The lack of a distinct notion of the mind is also related to a lack of a distinct notion of the body, and leads to a combination of the mental and the physical in what Read calls a physiological psychology:

> The biological, physiological and psychic aspect of man's nature cannot be clearly separated. They exist in the closest inter-dependence,

being, as it were, fused together to form the human personality. To an extent to which it is perhaps difficult for us to appreciate or understand, the various parts of the body, limbs, eyes, nose, hair, the internal organs and bodily excretions are essential constituents of the human personality, incorporating and expressing the whole in each of their several functions. It follows that an injury to any part of the body is also comparable to damage to the personality of the individual sustaining the injury. Similarly, the loss of any of the bodily substances through excretion is, in a rather obscure sense, the loss of something that is an essential part or element of the whole, a loss to the personality itself. [22]

They will still be 'connected' to their owner, which is why burning a man's hair clippings, for example, will seem a reasonable method of inflicting real physical harm on him.

Once we realise that primitive peoples do not have our idea of the mind we can also understand why they will inevitably think of all the symbols used in their rituals, the water, the garlands, the sacred gates, and so on as having real supernatural power. When we talk of 'symbolic meaning' we can use our notion of 'mind' to make a clear distinction between the symbol and what it stands for. So when we see an object that has symbolic power, such as our national flag, we regard our feelings about it as existing in our minds and not in the flag itself. But our notion of the mind is not available to primitive man, so for him the power of the symbol can only be located in the object itself.

When we see a picture of someone we know, we think of the associations conjured up by seeing the familiar face as located in our mind. But suppose we had no idea of the mind, of the processing centre that brings together visual impressions and the memories of this person? Then what else could we do but locate those feelings in the picture itself, and think of them as existing outside us, and of the picture as having a mysterious life of its own? The same is true of symbols used in rituals, and also of speech, when, for example, spells are uttered, and the words associated with the things named in the spell are thought to be under the control of the words that are spoken. So ritual, and the words used in magical spells, curses, and blessings are not just *expressing* peoples' hopes and fears: they are actually creating a new reality, as when a Tauade believes that saying the name of a strong local wind will make him as swift as that wind in battle, or that another spell will make his journey shorter. Against this background, therefore, magic and witchcraft will have far more credibility than they can possess in our kind of society.

Intuitive thought also makes no clear distinction between natural and social laws:

> . . .until the age of 7–8 there does not exist for the child a single purely mechanical law of nature. If clouds move swiftly when the wind is blowing, this is not only because of a necessary connection between the movement of the wind and that of the clouds; it is also and primarily because the clouds 'must' hurry along to bring us rain, or night, and so on…If boats remain afloat on the water while stones sink to the bottom, this does not happen merely for reasons relating to their weight; it is because things have to be so in virtue of the World-Order. In short, the universe is permeated with moral rules; physical regularity is not dissociated from moral obligation and social rule…What, then, do intentions matter? The problem of responsibility is simply to know whether a law has been respected or violated. Just as if we trip, independently of any carelessness, we fall to the ground by virtue of the law of gravity, so tampering with the truth, even unwittingly, will be called a lie and incur punishment. If the fault remains unnoticed, things themselves will take charge of punishing us. [23]

A boy, for example, has been stealing apples and on the way home falls into a stream because the bridge breaks; he will believe that the bridge broke *because* he had been stealing apples. [24]

Since the social and natural worlds are not clearly distinguished, the most elementary way to explain physical phenomena is based, naturally enough, on how people behave. People act for a purpose, and because of their inner, essential nature, and so intuitive thinkers suppose that the movement of an object expresses the force or vitality inside it, and also that this vitality is purposeful and goal directed. 'Every substance is endowed with a *sui generis* force, unacquired and untransmissible, constituting the very essence of its activity…If we try to find out exactly what a child means when he says that a force sets an object in motion, we always discover the idea of mutual excitation.' That is, the movement of a body is regarded as due both to an external will and to an internal will, to command and to obedience. There is no *transmission* of force: 'the external force simply calls forth the internal force which belongs to the moving object', as when a child says 'The road makes the bicycles go.' [25] The movement of things such as clouds and streams, for example, is seen as inherent in them, and called forth by what they have to do in the scheme of things. The child does not think, then, as we do, of force being transmitted from body to body but as belonging to all bodies, not transmitted but awakened – the weight of a stone, for example, is regarded as a force that actively opposes the efforts of a person to lift it.

Intuitive thought therefore thinks of cause and effect much more easily in terms of essences than as the result of relations between objects. The shadow, for example, is not understood as the result of a beam of light from a light-source that is obstructed by an object, and casts a projection of this on to the ground, but as something that emerges from inside the person or object that casts it, a thing that has its own special properties. The fact that a shadow can't be seen in cloudy conditions or at night does not conflict with this belief, because it is supposed that the shadow is still there but is just temporarily invisible. Since a shadow is therefore a special kind of thing, the shadows of people, in particular, are often supposed to be related to the soul, or at least vulnerable to magic or witchcraft. For the same reason, in primitive thought darkness itself is not regarded simply as the absence of light, but as something existing as a separate kind of thing in its own right.

The properties of objects, including people, may also be seen as something that can be transmitted directly between people and objects, and such beliefs are universal in primitive society. So if travellers are very tired on a long journey and fanning themselves with leaves, they may throw the leaves away in the belief that that their tiredness will leave them and pass into the leaves. A mother may not let her children eat the flesh of a species of white-bearded monkey because she thinks that they will catch old age from it. When a tree does not bear fruit, a gardener may ask a pregnant woman to fasten a stone to one of its branches, so that her fertility will pass into the tree, and so on.

The intuitive idea of causes, therefore, is a universe of independent and spontaneous substances that behave as they do because of their essential natures, but it is also a universe in which these substances are linked by their concrete associations in the experiences of everyday life. These 'realms of experience', such as the bush, the forest, the village, the lowlands, and so on are complexes that may come to be regarded as having some kind of inner vitality of their own. Among the Dinka of the Sudan, for example, there are three divinities who are thought of as a family: Garang the father, Abuk the wife, and Deng as son or husband of Abuk. Deng represents the phenomena of the sky associated with rain, and hence also rain-clouds, thunder, lighting, and sudden death and also, by association, coolness, pastures, cattle, milk, procreation, abundance, light, and life. Abuk is a female divinity who presides over women's affairs: gardens, crops and food, and the earth generally, while Garang represents the heat of the sun and certain heated conditions of the human body.

Since the primitive world is filled with purpose and meaning, there is no room for our notions of probability and accident in explaining why things happen For example, if a tree falls on a man and kills him, people will obviously understand, physically speaking, what caused his death but they will also want to know why the tree fell on him, in particular, and not on someone else, or why it killed anyone at all. Because the world has meaning, any event with human significance must have an explanation, and it is only in the case of an insignificant event, such as a tree simply falling down without doing any damage, that they will say 'It just happened'. Such beliefs are the easier to hold because people also have no ability to think statistically, that is, to think about events in numerical terms and understand that the larger the number of cases, the more likely it is that accidents or deaths will occur.

For example, I used to treat the members of the Konso villages where I lived for a variety of complaints such as conjunctivitis, dysentery, cuts and burns, in which I was very successful. This was because the complaints were not serious and the people reacted well to modern antibiotics, but a number of people would never go to the Mission clinic and insisted on coming to me, because they said the clinic was a bad place where people died, whereas they had never known me to fail. They simply compared the *absolute* success rates of myself and the clinic instead of our *relative* success rates, relative, that is, to the far greater number of patients treated by the clinic, some of whom were also much more seriously ill than any I saw, and more likely to die in any case. The assumption that significant events must have a meaning in the larger scheme of things, and the inability to think statistically, form, of course, the basis of the universal belief in omens and divination. (Piaget found that, in the same way, children find it natural to think of events as inherently linked together in patterns and only develop an understanding of probability with great difficulty at the stage of formal operations.) [26]

From what we have now seen of the basic features of intuitive thought, it is obvious that it provides the underlying basis for all those beliefs in ritual, magic, witchcraft, divination, spirits and nature-gods and goddesses with which we began this chapter.

### 4. Social and moral ideas.

Primitive peoples find it just as difficult to think analytically about their own societies as they do about the physical world or about themselves. Of course they know in a practical sense how their own society works, but this knowledge can't be expressed in an articulate and coherent way. In small groups, which continue for generation after generation in the same place, with a simple technology, it is very easy to develop rules (such as those for kinship and marriage) based on principles that do not have to be made explicit. Small changes can be made in each generation which will fit in with the general pattern of life, but this 'fitting in' need not require a conscious awareness of the whole pattern. The symbolic links between the natural world and the clans and lineages, age groups, men's societies, and the rituals associated with these in tribal society also make conscious analysis of one's own social order very difficult.

The organization of Tauade society, for example, can be summarized in a page, but none of my informants could provide me with a description of their society which in any way approximated the one I gave earlier. In one sense the Tauade knew perfectly well how their society was organized, but this knowledge was based on a great mass of concrete personal information about individuals whom they knew and their relationships with them. It would therefore have been impossible for a Tauade to have put this type of knowledge into a connected set of general statements about their social organization and the relations between tribes, clans, lineages, hamlets, Big Men, and so on.

In these small face-to-face societies people therefore have a detailed concrete knowledge of all the individuals with whom they come in contact, and extremely good memories for the events and details of daily life. So, just as the technology of primitive society is cognitively very undemanding, it is also possible for people to function quite adequately in tribal society, without developing the more advanced forms of thought that are required in modern industrial society. Many psychologists have studied how social and moral understanding develops in the individual and, very briefly, their general conclusions are that children first understand society in an intuitive manner as the concrete relations between individuals, in an atomistic, unsystematic fashion, yet with a rigid notion of rules and conventions. Obeying the rules is good in itself, and there is little awareness of the mental states of others, or of one's own mental processes. The self is defined by such physical features as size, gender, and so on, rather than by inner, psychological, features while, in considering other people's responsibility for their actions, it is thought that punishment should be based on what they actually do, and their motives and intentions are not taken into account. The idea of justice is limited to the notion of a fair exchange, a deal.

An understanding of social order gradually develops which involves greater awareness of social roles; in this order what is right involves living up to the expectations of others, and having good motives. The social order is seen as authoritative, and social and natural law are not clearly distinguished. Then gradually society comes to be understood in a more concrete-operational and systematic manner, and hierarchical structures and role differences are understood more clearly. An example of this would be an understanding of the structure and working of a political system or an administrative organization. What is right consists in maintaining the society as a whole, and more distinction is made between intentions and actions when awarding punishment. There is the idea that one should help others because in this way one can expect help oneself.

Finally, at a stage corresponding to formal operations, a society can be explicitly thought of as a total system to which its institutions stand in a part-whole relationship; conventions are understood as arbitrary rules that might have been different, adopted for the general good of society, while moral principles are distinguished from custom and law. The individual is distinguished from society, and it becomes possible to think of moral obligations to all human beings, regardless of the society to which they belong. The idea of justice has advanced to the principle of the Golden Rule: doing to others as one would like them to do to us. One's own society can become the subject of criticism, and hypothetical social orders discussed. The idea of the self becomes predominantly defined by psychological and spiritual attributes, and is much more differentiated and integrated; the cognitive functions of the mind are realised, and the self can be the judge of the self, so that the idea of conscience is fully developed. While this type of thinking is standard for educated people in literate civilisations, many people even in our own industrialized societies do not attain this level of thought, and it certainly does not occur in tribal society.

Historically, the basic ways in which social and moral ideas have evolved are broadly parallel to their psychological development in the individual. The band societies of hunter-gatherers have been called 'atomistic' societies because the lack of clear group structures such as descent groups, age-systems, or residential groups means that the social order is little more than the network of actual relations between individuals. While there is a great emphasis on reciprocity and a constant exchange of gifts between individuals and the sharing of game, we saw that many have commented on the individualistic quality of these societies, the emphasis on personal autonomy and independence, and low levels of inter-personal concern and mutual assistance. [27] The lack of judicial authority, or of formal modes of adjudicating disputes also means that it is hard to articulate social norms in any precise way that is accepted by all, so moral judgements in individual cases tend to be relative to the personal relations of those involved in the disputes.

The shift to intensive agriculture inherently allows larger and more permanent groups to form, so that disputes eventually come to be settled by elders or hereditary leaders in accordance with generally accepted norms. There is punishment by duly constituted authority, and it is now possible for crimes against society as a whole to be distinguished from private wrongs. What people should and should not do in such a traditional social order is prescribed in terms of status, and this provides a firm basis by which individuals can decide what they should do in a variety of situations. Such common expressions in these societies as 'the way to behave', or 'how you should live', are essentially appeals to people's knowledge of this conventional order in which 'the way to behave' does not need abstract ethical concepts like 'duty', 'justice', or 'virtue', or notions such as the Golden Rule. Discussions of moral issues are not concerned with the individual who is searching for a general rule of how to behave to others, but on the realities of social life, and the need for mutual assistance in a small rural community. Morality is essentially prudential in spirit – if you don't help others they won't help you, or you will be unpopular, and so on.

While they can, of course, give lists of what we would call 'virtues'

– good temper, bravery, generosity, and so on – there is no attempt to organize these into a coherent conception of excellence as a human being, such as the cardinal virtues of wisdom, prudence, self-control, and courage. Nor is there any psychological exploration of the inner self and its complexities. Although people obviously have some awareness of the motives and intentions of others, and distinguish, for example, between accidental and deliberate homicide, they still think of inner states as basically inaccessible: vengeance and punishment are based on what a person has actually done, not on any investigation of why he did it. For the same reasons, people are motivated to behave properly not by the promptings of the inner voice of conscience, but from fear of social condemnation and losing their good name and reputation.

In these societies the moral system is therefore indistinguishable from the social order itself. And, as we have seen, the social order is believed to be an integral part of the cosmos and the natural world around it. 'The traditional way of life is hence taken for granted and there is no critical assessment of the validity or usefulness of customary practices and beliefs'. [28] Within the society itself there is, then, no independent intellectual ground on which to base any abstract critique of that order as unjust or otherwise morally deficient. This has to await complex, literate civilisation, and the development of philosophical thought, when formal operational thinking about society and ethics starts to appear.

## 5. Conclusions.

Primitive thought is a classic demonstration that people think in ways that are easiest for them, and that in the simple and intellectually undemanding environments of hunter-gatherers and tribal society, first ways of looking at the world will be perfectly adequate. It is for these reasons that we find intuitive or at most, some concrete-operational thought in these sorts of society. Magic, divination, witchcraft, symbolic classification, and so on are not universal because they have any selective advantage, but simply because they are easy.

The idea that our mental abilities, our 'cognitive skills', will increase when we are faced with more complex technology, social organization, and more advanced education, should be more or less self-evident. Those who have never been to school or university, and have never even learned to read and write, lived in cities, or encountered machines, will obviously think in more elementary ways than people who have had these experiences. So it is not surprising that formal operational thinking, in the form of logic, philosophy, and mathematics, only first appeared among the intellectual elites in some of the ancient literate civilisations as the result of more challenging social conditions.

More recently, in the course of the last hundred years or so of the industrial revolution, the populations of modern Western society have been exposed in an unprecedented degree to more advanced technology, education, and scientific thinking generally. It is also well known to psychologists that during this period IQ scores have risen dramatically in the industrialised nations, as distinct from the rest of the world, where this process has been much less marked. IQ tests measure a whole range of different cognitive skills, but these include the extent to which formal operational reasoning replaces concrete thought, and it is in these areas of thinking in particular that cognitive development has been so marked. [29] This development from concrete to formal operational thinking is just what one might have expected if my view of the relation between thinking and social evolution is correct. Such a dramatic increase in some aspects of IQ was obviously not the result of any changes in our genes, but occurred during a period in which the members of our type of society were exposed to those types of experience that stimulate scientific and logical thinking.

# CHAPTER VII
## THE STATE AND CIVILISATION

*1. The state: a new type of society.*

Between 3500–3000 BC, in Egypt and Mesopotamia, there began a political revolution that was second only to agriculture in its power to transform human society. This was the emergence of the state, a new social landscape in which rulers could maintain their control over their subjects, if necessary, by armed force and could also conquer neighbouring societies. But although war and conquest were fundamental duties of kings, they could not rule their people by force alone, and had to be legitimated by descent from the royal line and by religious status. The state was supported by the economic surplus that was extracted from the peasantry by taxes, rents, and compulsory labour. This made possible the monumental architecture of temples and palaces, the magnificent display and expenditure of the king and the ruling classes, and stimulated the development of the arts and crafts and their associated technology.

So a leisured elite of nobles, priests and administrators developed, who in Egypt, Mesopotamia, China, and Central America discovered writing. This, ultimately, allowed the human mind to achieve its full potential in mathematics, philosophy, literature, and all the arts and sciences involved in 'civilisation'. Cities developed for a number of reasons, and provided a new social environment radically different from that of rural life, where elites could co-operate and interact with one another. To be viable, an elite of this kind had to number many hundreds, and since such people were only a minute fraction of the general population (probably less than 1%) only large societies with plenty of peasants could supply the kinds of economic surplus that were needed. Just as civilisations had to be large, they had to be based on social inequality as well. The exploitation of the masses, and the extraction of more taxes in goods or labour from them than they would have been willing to give without compulsion, was the only way in which the high cultures of the ancient world could have been built up. *after consideration*

Those symbolic principles of order that were typical of tribal societies and chiefdoms were inadequate for states: dualism, opposition and alternation, right and left, centre and periphery, the wild and the tame remained only in religious thought and ritual, and there was a powerful tendency for states to reorganise themselves on a more genuinely functional basis. Someone is now in control of the whole society, and this control has to be consciously and deliberately maintained in a number of practical ways. For example, soldiers have to be recruited, fed, and commanded, taxes have to be collected, and records of this need to be kept, land surveyed, perhaps a census taken, local administrators appointed and prevented from rebelling, communications between the royal court and the local districts maintained, large-scale public works organised and paid for, a system of law courts established, and all this requires a rational administrative apparatus rather than the symbolic organization of tribal society based on descent, seniority, age, and gender.

This rationalization of government means that states have also tended to conform to a single organisational pattern rather than displaying the rich variety of tribes and chiefdoms: a hierarchy of social classes with the king and his court of nobles (usually warriors), priests or their equivalents, and administrators forming a governing elite, and below them merchants, craftsmen, a great mass of peasantry, and often slaves at the bottom. The rise of the state also greatly intensified the level of true competition in social evolution – competition in the military, political, and economic areas, and ultimately in that of religion and ideology as well. By this I mean that there were now real winners and losers, ways of life and thought that went down into oblivion while others flourished and spread to other societies.

## 2. How the state first developed.

To put the matter very briefly indeed, the state emerged when a few basic conditions were met: there first of all had to be enough people to produce the surplus on which the whole apparatus of central authority depended. Just having a high population density was not enough, though, and some tribal societies have populations of several hundreds per square mile, which are higher than in many states. [1] (Population pressure only became a factor if, in chiefdoms, it prevented people emigrating, and so strengthened the chief's authority.) There also had to be a foundation of political legitimacy, a ruler who was thought to be entitled to obedience and tribute, and in early states the universal basis of legitimacy was descent in the royal line, and religious status. But the ruler also had to do a number of things in order to keep the people's allegiance and induce them to contribute their surplus. These were, notably, blessing them and being a source of wealth, mediating in disputes and imposing law and justice, acting as spokesman in dealing with other societies, and being, directly or indirectly, a victorious leader in war. This, in turn, gave him the means of repressing internal rebellion by force.

Since there had to be enough people to provide the surplus, early states were therefore always agricultural, and where agriculture was slow to develop so, too, was the state. (While some later states were ruled by pastoralists such as the Scythians, the Mongols, and the Ankole and Tutsi in East Africa, these regimes all depended on farmers and developed relatively late in history.) It quite understandable, then, that no state ever appeared in areas of very low population densities such as the Arctic regions or the deserts of the world. Some environments, on the other hand, such as river valleys are very favourable for the formation of states, because they can support large agricultural populations. They can also provide a convenient system of transport, not only for goods involved in trade and tribute, but also for troops and officials. Not surprisingly, the earliest states often emerged in areas of the world where rivers played a very important part in agriculture (e.g. Egypt, Mesopotamia, the Indus Valley, and China). [2] But these are very broad limitations, and states have appeared independently in a number of regions which have little in common except that they are not in arctic or desert areas. [3] Also, these environmental conditions only *favoured* state development: they did not *cause it*, because other social and cultural conditions had to be present as well.

Although the earliest states – Egypt, Mesopotamia, and China, in particular – emerged in regions of the world where the technology of the day was well advanced, particularly in metallurgy, there is no evidence that improved technology by itself was crucial for the appearance of the state. 'Basic tools, for production in agriculture and craftwork, showed no systematic improvement before and immediately after state formation. The plough and other forms of machinery do not universally either precede or necessarily follow state formation.' [4] Indeed, city-states of the New World achieved literate civilisation without animal power, or wheeled vehicles, and with a poorly developed metal technology. The great buildings and carved stone monuments of the Maya were based on nothing more than men with stone tools, while the pyramids of Egypt, like those of the Maya and Aztecs, were built without the use of pulleys, winches, or cranes. The real importance of the state was as the *stimulator* of technology, and we shall look at this in more detail later.

Warfare was specially important for early states because it was, with sacrifice, one of the most basic means by which a ruler could bestow Life and success on his people, and defend them from their enemies. States usually did not come into existence in isolation, and threats from neighbouring states therefore stimulated military efficiency and discipline. War leadership can certainly enhance the power of the divinely sanctioned ruler from a royal line, and where war leaders arose by force of arms they needed to clothe their rule in these basic forms of legitimacy if it was to become a permanent form of political authority. The conquest warfare of states was a very important source of new land, captives for religious sacrifice or as slaves, and general loot, and so was a very effective contributor to the economic surplus at the disposal of the ruler.

Trade and the control of trade routes could also contribute to the surplus, and provided an alternative source of revenue to taxes from the peasantry. When the king controlled a trade route and could guarantee safe travel, orderly markets, and a regular supply of goods, this would attract foreign commerce from which the king could extract revenue, as well from taxing the markets themselves. In turn, the available prestige goods for distribution allowed the king to attract more followers, and gave his state more wealth and power. [5] The rewards of trade, like the booty from warfare, can greatly increase the wealth available to an increasingly centralized authority, and so intensify the process of state formation by providing support for larger armies, more elaborate temples and other public works, more craftsmen, and more administrators.

State formation was also easier if the population could be prevented from simply dispersing across the landscape to avoid political control, as was typical with chiefdoms. In some cases this dispersal was discouraged by the fact that there was no unoccupied land available, or because it was inhospitable or in other ways less favourable for farming – an extreme case being the deserts on either side of the Nile. In other cases people's investment in their land and the development of facilities such as irrigation, or the greater security offered by the growing power of the state, made it more attractive for people to remain and put up with the disadvantages of increased state control.

There was therefore a great deal of mutual feedback between the different elements of the early state, and also a variety of routes by which it could develop in which the relative importance of warfare, trade, religion, and cities could differ considerably. We can now look at this process in more detail.

### 3. The king, the cosmos, and law.

Early states were normally based on the divine king who was the religious focus of his society and drawn from a royal clan, who was also very often a war leader and the supreme judge and redistributor of wealth. It used to be assumed that the 'divinity of kings' was a cynical imposture by ancient tyrants, smirking behind their hands at the gullibility of the masses. But we now realise that the sacred quality of rulers goes far back into tribal society, where it is intimately linked with kinship and descent from numinous ancestors.

To understand the importance of kings and religious ritual in early states, we must pause and reflect on how people in ancient society thought about the gods and nature. When we read about Egyptian, Aztec or Greek gods we automatically (but wrongly) think of them as scaled down versions of the omnipotent, transcendent God that developed in the Judaeo-Christian tradition – totally spiritual, immortal beings of immense power with no need of man at all. In fact the gods of ancient civilisation were much more like Deng, Garang, and Abuk of the Dinka, hardly distinguishable from the physical form of the sun, the rain, the earth, and the various phenomena associated with these.

A number of gods created the sky, the earth, the sun, plants, animals, and so on, but they did not make them out of nothing: they imposed *order* on chaos, *form* on what had been formless. But order and form were static, and the sun had to rise, the rain to fall, and the plants and animals to grow, and they could only do this because they were filled with vitality. The gods of these elements, however, needed constantly to be strengthened and nourished by man if they were to continue their work, and this vital force came from human worship, and from sacrifice – especially blood-sacrifice – in particular, where the life essence in the form of blood, fat, and marrow were offered to the gods in religious rituals. 'The gods, as the forces of nature, were the source of all life and sustenance, but they depended on the energy that human beings returned to them in the form of sacrifices.' [6]

In India and Iran, the Americas, Egypt and Mesopotamia, China, Greece and Rome, and in African states, it was the special duty of ancient kings to perform the most important sacrifices to sustain the gods of the sun, rain, plants and animals on which their people's prosperity depended. Immense numbers of animals, and often human victims, were needed, while other riches, such as gold and costly wood, might be ritually destroyed as well. The ritual of sacrifice could itself become a kind of magical act that, as long as it was perfectly performed, could compel the gods to confer their blessings on the worshippers. Ritual places were replicas of the cosmos in which 'the sacrificial altar is the utmost ends of the earth'; so ancient temples were not simply like churches in modern religion, places for prayer and the good of the worshipper's soul. They were cosmic models and vital elements in physical survival, and it was thought essential that they should be built of the costliest materials to strengthen the gods. For all these reasons, therefore, religion was one of the major factors, together with warfare, public works, and administration in driving the extraction of an economic surplus from the people.

Temples could be major economic centres because they received the offerings of the people, and might also be endowed with land, employing a large number of workmen and artisans to construct and embellish the buildings. In Mesopotamia especially, they assumed the functions of banks, making loans to the peasants and financing merchants. They were also the base of an increasingly important class, the priests, who conducted the rituals as deputies for the king and developed the doctrines of the gods they served. One of the most important functions of priests was the observation of the heavens and the regulation of the calendar, an essential aspect of that cosmic harmony that had to be maintained. Calendrical calculations, keeping economic records of tribute and expenditure, and the recording of sacred texts were all involved with the invention of writing, as we shall see, so that the development of a literate priestly and bureaucratic elite was closely related to the existence of temples. Even though temples were not associated with the emergence of some early states, as in India, China, and south-east Asia, there was a strong tendency for them to develop in the course of time so that, in most of the ancient agrarian states, a literate priesthood based on temples became part of the standard apparatus of government. [7]

China was somewhat different due to the great importance of ancestor worship. Only the king, not priests, could sacrifice to the royal ancestors; he interceded with the sky god Ti, and was also the principal diviner. [8] While there was a body of diviners who assisted him, and astronomy was always a royal monopoly, the most important sustainers of the royal authority from the earliest period of which we have records were the royal ministers, the forerunners of the Confucian scholar-officials who developed much later, and whose function was ethical rather than religious. While there were important administrative centres,

China was therefore a society without temples and priests. In the city-states of Greece and Italy temples developed, but the priests were not full-time functionaries who occupied a separate status from the rest of society. They might inherit their religious offices or be appointed to them, but while Greece and Rome were just as 'religious' as the rest of the ancient world, and sacrificed vigorously to the gods, they never developed the sort of organised priestly class we find in Egypt, Mesopotamia, Persia, Israel, and India, or among the Inca, Maya, and Aztecs. This had important consequences for the development of thought, as we shall see.

The king also maintained cosmic order by maintaining the social order. The Konso believed that their age-grading system made the crops grow, through the social harmony that it established between the generations, and this association between social and cosmic harmony was just as important in early states, where the king had a central role as the fountain and dispenser of justice, in the sense of settling disputes and restraining the excesses of the powerful, and maintaining the rule of law.

King Hammurabi (about 1800 B.C.) describes himself as called to rule the Babylonian people with justice as a servant of the supreme god Anu: 'Then Anu and Bel delighted the flesh of mankind by calling me, the renowned prince, the god-fearing Hammurabi, to establish justice in the earth, to destroy the base and the wicked, and to hold back the strong from oppressing the feeble: to shine like the sun god upon the black-haired men [the people], and to illuminate the land.' [9] In ancient Egypt the Pharaoh was the god Horus and his rule personified *ma'at*, translated as 'justice' or 'order-truth'. In ancient Israel the association between the king, justice, and national prosperity was equally important, and in India, too, there was a basic association between the righteous king and the earth, and with rain and with plenty. This was in fact general to all the Indo-European peoples, so that among the Homeric Greeks, if the king followed the rules of justice and divine commandments this would bring health and prosperity to his people. Early Chinese kings of the Shang and subsequent dynasties were also responsible for ensuring the prosperity of their people by dispensing justice and maintaining social order.

Settling disputes was an important function of the legal systems of early states: at the local level village councils or clan heads retained an important role, although government officials might also adjudicate. But punishment was the most important aspect of the law, not only against the usual offences of theft, violence, murder, and so on, but to maintain the authority of the state. 'Laws that were administered by the state generally punished crimes against kings, government institutions, deities, and temples more severely than against other individuals'. [10] Legal systems were also one of the basic means of maintaining the social hierarchy, and keeping the lower classes in subjection, and the judges were themselves recruited from the upper classes. Severer penalties for peasants than for nobles for the same offence, or for peasants who assaulted nobles than vice versa were typical.

*4. Warfare.*

Warfare was intimately involved with the emergence of states from their beginning, but the warfare of states was very different from that of tribes and chiefdoms. [11] Primitive warfare lacked discipline, and was unable to conduct campaigns in hostile territory and to conquer and politically subjugate the enemy. One of the essential features of state warfare, on the other hand, was conquest because their much greater resources of manpower, food, discipline, and organization allowed states to put far more effective armies into the field, and keep them there for extended periods of time to fight campaigns of conquest.

Comparatively small professional armies could conquer many times their number of tribal opponents because of their superior discipline and organization, and the development of large organized battles, instead of the endless raids and skirmishes by the warriors of tribal societies and chiefdoms, made it possible for single, decisive battles to change the course of history. The warfare of states therefore led to far more effective competition between societies.

Armies were necessary to defend the state's borders, not only against nomadic tribes but because states usually appeared in clusters, and the rivalry of Upper and Lower Egypt, the city-states of Mesopotamia, the Maya and the Aztecs, and of Greece and Italy, the Hsia, Shang and Chou of China, early states in India, and the early Anglo-Saxon kingdoms in England are all examples of this. [12] The existence of neighbouring states inevitably created conditions of permanent insecurity for all of them, in which attacking a neighbour would often seem the best means of defence. Warfare between states could therefore be as endemic as at the tribal level of organization, although for different social reasons, and with different aims.

From the very beginning of the state, war and conquest, like sacrifice, were royal duties that legitimated the king's rule. War, like building pyramids or digging canals was a life-giving enterprise of the highest value to society, the successful leadership of which was a sign of divine favour.[13] It is not surprising that our earliest image of kingship, in the Narmer palette of about 3000 BC, shows the Pharaoh clubbing prisoners of war to death. The associations between warfare, religion, and kingship were very close, and even where the king was not expected to command his armies himself he was still personally responsible for their success. In early China,

> The Shang kings sanctified war as a hunt for human prey to be sacrificed to their ancestors and the gods. Major Chinese military

campaigns began at the earth altar and ancestral temple of the king or of a major provincial lord, where the war was announced to the ancestors and their help requested. Tablets inscribed with the names of these ancestors were carried into battle. After a successful campaign, prisoners were sacrificed or severed enemy ears and heads were brought from the battlefield and presented at the ancestral temple. [14]

Warfare was not only for conquest, slaves, and sacrificial captives, but to monopolise natural resources such as gold and silver mines or quarries, to control trade routes, and to impose tribute on neighbouring states as an alternative to conquering them. So whereas in primitive society the economic motives for warfare were much less important than vengeance and the need for the warriors to prove themselves, in state warfare they became far more significant. In some cases such as the Roman Republic, warfare could become a self-perpetuating cycle in which endless conquests were necessary to acquire fresh resources, to keep the military machine going, and to supply the slaves that had become an economic necessity. Despite the importance of these economic factors, however, one also needs to remember that the power, splendour, and magnificence sought by military conquerors are not really *material* motives at all, which are only concerned with physical survival: food, fuel, shelter, and security. We should also note the extraordinary ferocity that developed in state warfare and became a more or less permanent feature of it: the slaughter of whole populations, the castrations, the burnings, the impalings, the flayings alive, the pyramids of skulls, and so on, interminably, an immense theatre of death.

Armies were not only used to protect the state against foreign enemies, of course, but to maintain order internally against political rebellion, brigandage, and popular unrest, and to enforce payment of taxes or the performance of corvée labour. They also protected the king, his officials and other members of the upper classes, public buildings, granaries and other state property, so that in all these various ways the army was one of the central pillars of the state. The senior officers were members of the upper class who provided not only leadership in actual operations, but a continuity of military skill, especially if they had specialized weapons such as chariots. It seems likely, however, that initially in early states professional soldiers in the lower ranks were confined to those who provided permanent security for the king and the rest of the state apparatus, while ordinary soldiers for military operations were recruited from the peasantry as an aspect of corvée labour. In some cases, such as the Aztecs, all boys were given military training from puberty onwards. Weapons would initially have been those like spears and bows and arrows already familiar from hunting, but in time more specialized weapons were developed by the state, particularly in relation to siege-craft, and here and elsewhere armies tended to become more professional.

Military conquest was the prime cause of multi-ethnic states, but it had other profound effects as well: it weakened the relations between descent groups and their land when they were conquered, because the land might be given to soldiers of the victorious state as a reward, and even if not dispossessed, the peasants might no longer own their land. Again, they might be deported and resettled well within the victorious state if they were thought likely to rebel, while peasants from the victorious state might be settled on the land of the conquered to increase political control. In situations of endemic warfare between neighbouring states there were also likely to be large numbers of refugees seeking the protection of a friendly king. State warfare therefore became one of the destabilizing factors in the traditional life of village society whose effects we will come back to in Chapter X.

## 5. Wealth and trade.

The state gave rulers and the upper classes unparalleled opportunities not only for splendid buildings, but for luxurious excesses of every kind. Extracting an economic surplus was certainly necessary for running the government – for supporting administrators, priests, soldiers, and so on – and for impressive palaces and temples to bolster the king's legitimacy and support the dignity of the state. But we find in all states and in all periods that the upper classes used their power to finance personal lifestyles of the greatest extravagance in houses, dress, food, and every other aspect of life. This luxury went far beyond the requirements of running the state, and it certainly had nothing to do with material needs: 'the luxury of upper-class lifestyles vastly transcended what was biologically required', [15] and could be positively unhealthy.

In all human societies, wherever there is power there is not only wealth, but the conspicuous display of that wealth. This, quite obviously, is not because of some crude Darwinian competition for resources – such display, as we saw, was irrelevant to physical survival – but was an expression of that human love of beautiful and precious objects that appears in tribal society, and in states it was also an assertion of power and status. Mere wealth without power, as in the case of merchants, was not enough to give high status (we shall look at this in more detail in section 7). In these early states power could not be purchased, and wealth came from power, not power from wealth. Here it is worth recalling the driving force behind Tauade pig feasts and dances. These had nothing to do with eating and everything to do with showing how powerful the tribe was – the herds of pigs they could raise, the size of their gardens, the toughness of the women, the generosity of the men, and the power of the Big Men's oratory. In the same way, the conspicuous displays of personal wealth in states showed the power of upper class families and their command of the state's assets, and therefore their status.

But the conspicuous expenditure by states and the upper classes generally created enormous demands for luxury goods such as gold, silver, feathers, cedar, ebony, spices, ivory, silk, incense, fine vases, and precious stones, and since these, more often than not, were unavailable within the borders of a state, this gave a very powerful impetus to long-distance trade. This, in turn, required full-time, professional merchants, who in some cases were state officials and in others were members of independent merchant associations. Some of these would often take up residence in a foreign country where they would learn the language and customs, to act as brokers between their own merchants and the foreign government and people. Prolonged trading contacts of this sort became extremely important ways of spreading new ideas and of facilitating foreign travel by members of the elite of different states, and trade routes between states were fundamental in the development of international high culture.

The patronage of royal palaces and temples, and the luxurious requirements of the upper classes in general were also essential in stimulating far higher levels of craftsmanship than had been possible hitherto, and the rise of the state led to major technological advances. Some of this was in the fields of warfare, ship-building, architecture, and irrigation, where state investment was crucial, but also in a wide range of non-practical gadgetry that was for the entertainment of royal courts and the wealthy. This growth of craftsmanship, of course, then acted as a further stimulant of trade. Trade and manufacturing in states were also closely associated with urban life, and it is to cities that we must now turn.

6. *The city and the state.*

Cities were intimately associated with early states and the

development of civilisation, but first we must be clear what we mean by 'city'. In the first place, a city is not simply a large settlement of people. The Konso village of Degato had around 3000 inhabitants, with a defensive wall a mile in circumference, so that it was considerably larger than the city of Bath, for example, in Roman Britain. But Degato, like all the other Konso villages, was inhabited entirely by farmers except for a tiny minority of craftsmen, and was simply a large rural settlement. The essential features of a city are not that it must have a population of, say, 10,000, or 50,000 or even 100,000, but that it functions as *an organizational centre in some wider system,* whether this be administrative, commercial, military, religious, educational, or all of these together. We can call these 'focal' [16] activities, and it is through them that cities become large and socially stratified, and increasingly distinct from the rural life around them.

The nature of farming spreads it out across the landscape because that is where crops are grown and animals reared, but focal activities, unlike farming, are most conveniently and cost-effectively carried out when they are located in one place. (Of course, the people actually concerned need not think in terms of cost-effectiveness, but merely take the easy route of going to where other people engaged in this activity are already located.) A very obvious example of a focal activity is a market because buyers and sellers need a fixed and well-known location where they can meet each other. Craftsmen gain no special advantage from living in a dispersed manner like farmers, and are often better off by congregating together in the vicinity of markets. All hierarchical organizations are also inherently focal because, by definition, they must have a head, and he can only operate in one place at a time. The most obvious example is a state with a royal court at its apex of government, together with its administrative apparatus dealing with taxation and the control of subordinate regions. (In some early states, however, royal courts might have no fixed capitals, but moved around the kingdom to maintain order and consume their tribute: Abyssinia is one example, and some kingdoms of early medieval Europe are others.) Religious centres, too, may acquire great wealth from the offerings and pilgrimages of the faithful, and have a large priesthood and administrative structure. As civilisation advances, cities may also become centres of learning which, again, is most convenient and cost-effective when concentrated in key locations. The waging of war is another focal activity because a disciplined military force is most effectively organised by having its headquarters in one place under the direct control of its commanding officer.

It follows from what I have been saying that there were many routes to the development of cities such as mercantile or manufacturing centres, temples or other forms of religious centre, royal courts, or defensive locations, and that it is misguided to think that they could only have originated in one way, such as being storage centres for food surpluses. The Greek city-states, for example, began as the seat of the king in a fortified settlement, the Acropolis, around which grew up the *agora* (market/assembly place) and further settlement. Others, such as those like Mohenjo Daro and Harappa of the Indus Valley, were major trading centres acting as gateways between local centres of population and long distance traders. Mecca, again, was a place of pilgrimage for centuries before Muhammad, to the sacred stone known as the Ka'aba, and was also the major town on the frankincense trade route from Southern Arabia to the Mediterranean. In Mesopotamia the temple and its associated organisation was the essential focus of the city, and a similar pattern of sacred sites can also be found in Central America and Peru. On the other hand, urban centres in ancient China seem to have been basically administrative centres established on the orders of the king. [17]

So, if we look at the list of focal activities we can see that commerce, manufacturing, government and administration, priestly, educational and military functions can not only be carried out most conveniently in one place but are also mutually reinforcing. A religious centre with its priestly schools might well be closely linked with a temple and a royal court, with a market served by long-distance trade, and a complex local organization of craftsmen and artisans not only linked with the market but serving the court and temple. The whole complex could well be defended by walls, with a resident body of troops. The result would almost inevitably be a very large settlement that is not just a place where people have their homes but which performs a new kind of function in society: it is the nerve centre or node in a network of communications by which governmental, commercial, religious, military and other functions are co-ordinated. The result is a population with a variety of occupations, and while these may be largely hereditary we now have a population mainly defined by what they do, and in terms of social class rather than of clan or lineage. Roman Bath was smaller than the Konso village of Degato, but it was a city in the functional sense that it was a religious and commercial centre whose population had a wide variety of occupations and where the bonds between the inhabitants were not primarily those of kinship and local land ownership, but based on membership in a common municipal enterprise.

But here we need to recognise that there were two main routes to early state-formation – the city-state and the territorial state. The city-state began as a settlement of farmers and craftsmen (probably like the large villages of the Konso) that expanded by long-distance trade in particular until it became a city with a religious centre, a king, and many merchants and full-time craftsmen, and controlling the surrounding territory. Many of the farmers would have lived within the city, for defensive reasons, and the rest in the neighbouring settlements.

City states of this sort appeared in many different parts of the world, such as in Mesopotamia, in the New World among the Aztecs and Maya, in the Mediterranean (Greece, Italy, and the Carthaginians), in south east Asia, and West Africa (Yoruba and Benin). [18]

Many city-states typically appeared together and trade between them was of particular importance, especially for exotic raw materials brought by long-distance traders for the full-time craftsmen who worked for the king, the temple, and the upper classes, but who also sold their products in the local markets. Warfare always developed, over water rights and the control of trade routes, for example, and in these military struggles for power the most powerful city, or alliance of cities, would force weaker cities to pay tribute rather than subjugate them absolutely, but trade tended to continue regardless of warfare. These regions of city-states would also share a common religious and artistic culture and their royal families and upper classes would intermarry. The powers of the king varied widely, but tended to be shared with councils of leading families, temples, and other institutions, and they were ascribed rather less divine powers than the rulers of territorial states. City-states, however, being inherently small, were always vulnerable to conquest by territorial states.

Territorial states, such as Egypt, Persia, the Shang kingdom of China, the Inca, many kingdoms in Africa (the Ankole, the Buganda, the Zulu), and in India (Kosala, Maghada, and the Maurya empire) developed by absorbing militarily their neighbouring chiefdoms and petty states. In Egypt, for example, warfare was an important factor: King Menes or Narmer, the first Pharaoh, is credited with uniting the kingdoms of Upper and Lower Egypt in around 3100 BC and making the city of Memphis his capital. It seems that Memphis was essentially a religious and administrative centre with no trading functions, and urban settlements seem generally to have been considerably smaller than they were in Sumeria.

In territorial states, then, royal authority was the unifying factor and the capital was essentially the royal court, comprising the king, his administrators and councillors, priests, soldiers, and craftsmen, not the peasantry. The royal palace might also be a temple, as at Memphis in Egypt. The area of territorial states was much larger than that of city-states, and so the major problem was how to maintain political control at a distance. This involved dividing the kingdom into a hierarchy of provinces, districts, and so on and delegating subordinates to run them. One of the simplest and cheapest methods, followed for example by the Shang kings, was to give these positions to clan relatives, and another was to grant land to local magnates in return for military and other services. But this made rebellion rather easy, and a more efficient method, followed by the Egyptians from the beginning, and later in China, was to appoint officials who owed their rank only to the king, and to move them around periodically to prevent them developing a local power base. Even better was the practice of separating the functions of government so that, for example, tax collection, military command, and the administration of justice came under different local officials.

It was also necessary for territorial rulers to have much more control over the economy than in city states, and Inca and Egyptian rulers in particular controlled foreign trade and all mining and quarrying. Territorial states were also able to extract a much larger surplus from the peasantry, whether in goods or labour, than could the city-states. While the conquests of territorial states might involve the displacement of large groups of people, village life as such was not significantly different from what it had been in tribal society, except for the need to contribute foodstuff and labour to the government. For this reason there was a much greater difference between the lifestyles of the peasants and the upper classes in territorial states than there was in city-states.

But in the course of time the administrative centres of territorial states also became centres of trade and manufacturing to which local people had daily access, just as the victorious city-states extended their power over much larger regions – Rome being the classic example – so that from these different beginnings kingdoms and empires developed that covered great expanses of territory in which were large numbers of urban centres that performed all those focal activities we mentioned earlier. City-states and territorial states therefore both tend to evolve into what we may call regional states, with extensive territories and important urban centres.

The city and its way of life thus came, for its inhabitants, to define the highest form of human society. The city was the dwelling-place of the gods and the king, the source of social order, and of knowledge and all the arts and crafts, and the Mesopotamians, for example, contrasted their way of life with that of the nomads and the dwellers in forests and mountains, primitive beings living in disorder and speaking barbarous tongues. These attitudes became applied to farmers as well in all the ancient civilisations, and a basic distinction developed between sophisticated urban-dwellers (poor as well as rich) and the rural peasantry – slow-witted, ignorant, and uncouth.

### 7. *The status of merchants and craftsmen.*

A social hierarchy was therefore a fundamental part of state societies which, if anything, became more and more clear-cut over time. We have already seen that the difference between urban sophistication and rustic ignorance became a profound social division that remained a permanent feature of states until modern times, but I would now like to look at two other aspects of class status, the first being that of the merchants. (Here I have in mind those merchants who specialise in long-distance trade, as distinct from the craftsmen who also sell their own wares in a shop or local market.) The pillars of the state were the king, whose status derived from his descent and his divine attributes, the priesthood or some equivalent body, the landowning nobility that often had military functions as well, the army, and provincial governors and other senior officials. These upper classes were wealthy but they controlled this wealth because of their central importance for the state, and they disposed of it lavishly both publicly and in their private capacities, and so their wealth was seen as legitimate, just as the force exercised by the king was legitimate, because it was part of that essential social and cosmic order which they sustained. (In the same way, the wealth of the lineage head and the chief had been seen as legitimate in tribal societies.)

In this context, the wealth of merchants was bound to be problematical for a variety of reasons, so that 'commerce was often a low status occupation in societies as distant as ancient Greece and Tokugawa Japan'. [19] While wealth is inherently more prestigious than poverty, and merchants acquired it by bringing useful or prized commodities into the country, trade was not central to running the state in the same way that being a ruler, general, landowner, priest, or administrator was. Even though merchants in ancient states were usually members of associations, their wealth was individual, private profit which had no social function at all, and was spent for purely personal gratification. They might be envied, but were not particularly respected just for being rich. In this fundamental sense traders were outsiders in relation to the Establishment, especially if, as was normally the case, they spent much time on their travels abroad, or were actual foreigners living in merchant enclaves in a host country.

Making profits by trade might also be regarded as contemptible, especially by comparison with glorious military exploits. This was the view of the Aztecs, for example: 'Whereas individual warriors proclaimed their high status publicly and proudly, however rich a merchant became . . . he was required to appear in public as a simple commoner, barefoot, and dressed in [coarse] maguey-fibre clothes'. [20] For Homer, too, the honourable source of wealth was warfare and plunder, whereas the riches of the merchant were ignobly gained by cunning, often by cheating – indeed Hermes, the god of merchants, was also the god of thieves as, later, St Nicholas was in the Christian Middle Ages. [21]

Peasants and craftsmen, earning their daily bread by the sweat of their brows, resented those who could amass great fortunes with what seemed so little effort, while officials might regard large amounts of private wealth with suspicion, as potentially subversive of authority. In centralized territorial states such as Egypt and the Inca the government itself controlled long-distance trade, while in China it seems that official control of trade steadily increased, and by Confucian times there was distinct government hostility to mercantile enterprise. Even if rulers were not actively hostile to merchants, their wealth was always a tempting target to be plundered when more taxes were needed, especially as they had no power base with which to resist, unlike landowners with their peasantry, and this was especially true in the case of foreign merchants.

The second aspect of the social hierarchy on which we should reflect is the opposition that developed between brain-work and manual labour, between theoretical and practical knowledge. In tribal societies there had been no literate elite, no higher forms of theoretical knowledge except, perhaps, knowledge of myths, while craftsmen might be admired for their skill and the powers it gave them. But in literate states, manual labour was only for the poor, an unpleasant chore done by peasants, servants and slaves, and was of low status in all ancient civilisations, especially in contrast to the life of the rich, and the many official occupations in states that involved telling other people what to do. Its low status was shared by most of the crafts such as pottery, weaving and metal-work, although luxury craftsmen, especially those working for the king, did enjoy a higher status. At the opposite end of the scale was the sort of knowledge involved in administering the cosmos, such as calendrical studies and astronomy, and administering the state – general, more theoretical knowledge about 'the big picture', of the kind we can call 'architectonic', about designing and planning – and this was given an enormous boost by literacy.

Once writing could express the whole range of a language, literacy then became the basis of a kind of knowledge that was inaccessible to anyone without years of special education (and leisure), so that religious texts, government, philosophy, and literature became embedded in a high culture that was fundamentally different from craft knowledge. The knowledge of literate high culture, especially religion and philosophy, acquired from books, was speculative, self-consciously intellectual and concerned with the general nature of things rather than with practical detail. Craft knowledge, on the other hand, could only be acquired by years of direct, hands-on experience in the workshop that was passed on by direct teaching, and in this sort of apprenticeship book learning would have been irrelevant. Craft knowledge was not normally transmitted by writing, and practitioners were largely illiterate and had no access to higher learning about mathematics, for example, while those from the educated literate classes, whether officials, landowners, or merchants, had no occasion to collaborate with craftsmen. This was a perfectly viable state of affairs that, like the status of the merchants, was neither adaptive nor maladaptive, and was certainly not selected for, but was merely the result of a common set of circumstances and continued for several thousand years. It had profound effects on the development of scientific thought, however, that will be examined in later chapters.

### 8. Core principles in the development of the state. [22]

Just as the evolution of tribal societies was influenced by their different core principles, so too was the evolution of states. A good example of this can be found in the city-states of Greece and Rome which, besides having kings early in their histories, and councils of nobles or men of similar status (the Roman Senate and the *boule* of Greek cities), also had popular assemblies that were of equal antiquity (the *ekklesia* of Greece and the *comitia curiata* of Rome). But while councils of eminent men were functionally necessary for running city-states, popular assemblies were certainly not functionally necessary at all, and were highly unusual in early states of any kind. To understand why they should have existed we have to understand that they were in fact specific features of ancient Indo-European culture that can also be found in early India and among the Germanic peoples, Hittites, [23] Celts, and Scandinavians.

The Indo-Europeans emerged from the steppes of what is now the Kazakh-Kirghiz region of southern Russia. The dominant features of their way of life were migration and warfare, but although they had horses which were the basis of warfare, plunder, and conquest, they also practised farming, with wagons and cattle. In the second millennium BC their migrations took them into India, and also across Persia and into the Mediterranean region and Western Europe as far as Ireland, taking with them their languages and basic social institutions and culture. While these had developed in the context of their migratory life on the steppes, they could still be adapted, within certain limits, to the city-states of the ancient world and to a sedentary agricultural economy, rather as in East Africa age-systems, that had been developed by pastoralists, could also survive among farmers.

The tribe was divided into clans, but unlike, for example, the Arabs, for whom clans always remained the basic social units, the Indo-European clans were part of a wider social order because their clans were divided into castes of priests, warriors (who provided chiefs and kings), and the farmer/herdsmen who were the food-producers. (Outside this order were slaves and captives.) All three divisions were integral parts of the social order, and in India they became the basis of the three 'twice-born' castes, the Brahmins, Kshatriyas, and the Vaishyas, and the idea of the Three Estates of priests, warriors, and farmers, although in much modified form, survived into the European Middle Ages. While patrilineal clans and lineages were clearly important in ancient Indo-European society, cognatic ties were important too, as were those of neighbourhood, while the war-bands seem not to have been recruited on the basis of kinship at all, but of sufficient wealth to support the lifestyle of a mounted warrior. By comparison with Africa, Polynesia, and especially China, patrilineal descent groups were therefore relatively small and weak.

It is easier for nomadic and semi-nomadic peoples, especially when constantly engaged in warfare, to travel greater distances than settled farmers, and they typically develop some form of general, tribal, assembly or council in which matters of peace and war can be debated, and important legal disputes resolved. While ordinary men may participate at the village level, we typically find that higher level councils are only composed of clan or lineage heads without the participation of ordinary men, such as the tribal *kuriltai* of the Mongols, the *jirga* of the Afghans, or the Bedouin *majlis*, for example. This use of the hierarchical principle is obviously an efficient way of organizing such gatherings, and the Indo-Europeans were unusual, therefore, by involving free male householders at the highest level of tribal council, perhaps because there was no well-developed clan structure like that of the Bedouin and the Mongols.

In northern India the tribe not only had the *sabha*, the council of eminent men, but also the popular assembly, *samiti*, which comprised the adult men who were heads of families; it is important to note that these were very definitely freemen, proud of their status as fighting men as well as farmers and herders, and very different from the large contingent of slaves, many of whom would have been captured in war. There was, however, a distinction of status between the mounted warriors of the war-band and the ordinary farmer-herdsmen who fought on foot. The *samiti* was a sacred institution that decided matters of peace and war, and acclaimed as chief or king the man chosen from the royal line by the senior men. [24] They did not *elect* the king from a slate of candidates – that was unknown in ancient society – but simply acclaimed him, and he ruled only for life, with no right of succession for his sons. The popular assembly was also the principal court of law, and cases involved debates in which displays of eloquence and powers of oratory were much admired. (Competition in general, in debate, and litigation and a variety of sports, as well as warfare, was greatly admired and a major source of high status.) The Senate and People of Rome, at least under the Republic, and the *boule* and *ekklesia* of the Greeks, were in fact cultural cousins of the *sabha* and *samiti* of northern India, and the Roman historian Tacitus (c.56–c.115 AD) describes very similar institutions among the Germanic peoples of his day.

Warfare was glorious, and the basic source of plunder and honour, and the king was its leader, but as a shedder of blood he could not function as a religious leader. The priests alone, for whom warfare and bloodshed were forbidden, had to perform the sacrifices and the other religious ceremonies, and they were also guardians of the laws, which were proclaimed at the assemblies. What we would call civil law, particularly the contract, the covenant, the oath, and the ordeal, was sacred and fundamental in Indo-European society from ancient times. The king had important religious status, however, and by ruling justly and according to law he gave his people plentiful harvests and prosperity, but the assembly and council were nevertheless more fundamental than the kingship, (in some cases there was no king, and the tribe was ruled by council and assembly.) Greece (except for Sparta) and Rome were therefore able to get rid of their kings without people feeling that the world as they knew it had been overturned.

In the case of China, however, we are dealing with a very different, long-settled farming society with no pastoral tradition, which was based on very large and powerful patrilineal clans, *tsu*, within which were senior and junior lines, while the *tsu* themselves were ranked in a hierarchy of status. They seem to have been the basis of military organization and of the allocation of farmland, and were also the primary residential units of about a hundred households. In particular, there is absolutely no trace of anything like a popular assembly in Chinese history, and the primary restraint on the king was provided by the royal ministers who had a very high status from the earliest times, and may originally have been from his *tsu*. The patrilineal clan was not only the foundation of Chinese society in a much more fundamental way than among the Indo-Europeans, but was also the basis of an elaborate system of ancestor worship.

This transferred the kinship hierarchy to the supernatural realm, and meant that the king's ancestors were much more important than those of any other *tsu*, so that he was 'the one man' sacrificing on behalf of his people. The royal ancestors were believed to intercede with Ti, the Sky God, in particular, who became of increasing importance. We saw that it is difficult in a system of ancestor worship for priests to take over these religious functions, so that a priestly class did not develop in China, and the status of the Chinese king had none of its Indo-European ambivalence. But ancestor worship, in combination with the kinship system, had a profound effect on Chinese culture as a whole, suggesting that:

> the powerful model of social order which we find in ancestor worship may have profoundly coloured the entire 'elite cultural' religious view of both the socio-political and cosmic orders. Within the family, the kin members both here and in the world beyond are held together in a network of role relationships ideally governed by a spirit of peace, harmony, and ritual decorum. Here the value of order is central. As a metaphor for the cosmos, it suggests a world of entities and energies held together in familial harmony under the authority of the high god. As a model for the sociopolitical order, it projects the picture of an immanent order based on networks of clearly defined roles and statuses held together by a system of sacred ritual. [25]

The image of the family, of peace, harmony, and ritual decorum, and lack of competition and litigiousness, sums up much, at least of the ideal, of Chinese society, in which the king was seen as a father to his people, setting them a moral example reinforced by punishment. Law was always essentially penal in China; civil law did not develop, and civil litigation became increasingly disapproved of, and it was preferred that such disputes should be settled by clan elders, or later by Confucian administrators, the 'father-mother officials' of the people. While warfare had originally had a sacred character, over the centuries it became increasingly deprecated as barbarous and wasteful, and only to be resorted to when there was no alternative. As we shall see, the whole tenor of Chinese culture was also to prove hostile to the merchant class and to commercial initiative, and it is clear that the core principles that I have sketched exerted a major influence on the general development of Chinese society, as did those of the Indo-Europeans on the history of Western Europe.

*9. Conclusions.*

So while there were different pathways by which the state could be reached, such as city-states or territorial states, once states had developed they had a number of those basic similarities we have been looking at in this chapter, and which marked a decisive break with tribal societies. To be sure, in almost all early states, descent was still of great importance, but it was so within the broader structure of the social hierarchy, and class was now more important than descent or, for that matter, gender or even ethnicity. Although traditional rural life could allow kin-based patterns of behaviour to survive, even here this was liable to disruption by warfare and the movement of people that this could produce, while cities were likely to involve frequent contacts with strangers between whom kinship obligations were irrelevant. The abolition of hereditary rights to office, and making all appointments dependent on the king's pleasure strengthened his power, while the growing number of administrative functions, and the increasing demands for a high level of individual competence, created an irresistible pressure for the choice of administrators on the basis of their personal abilities, rather than on their descent alone. [26] As military organization increased in scale and complexity it became increasingly specialized, and, as in other areas of administration, there was a corresponding pressure to select war leaders on the basis of personal ability rather than on birth.

The emergence of the state is even more destructive to age-systems, because councils of elders necessarily conflict with centralized authority, and if such councils survive at all they are reduced to the village level. Some states, notably in southern Africa, retained the warrior grades as age-regiments for purely military reasons, but generally speaking age-systems are inherently incompatible with centralized political control except at the local level.

Territorial boundaries become more significant at the expense of kin groups and their relation to land. In tribal society, group boundaries will be relevant in, say, the allocation of hunting or grazing rights, but in many other aspects of social relations such boundaries may have little importance, because ties of kinship and marriage typically extend across many local groups. The rise of the state, however, involves a marked change in the significance of boundaries, and hence of territoriality in general. A central government has to be able to impose its authority on a specific *range* of individuals – the government's subjects – by coercive force, with corresponding obligations on them to support the government by tribute, corvée labour, obedience to its officials, and so on. Centralization of those relations around some focal point of authority therefore requires that they must be sharply bounded geographically. For these reasons, external boundaries become far more important than they are in pre-state societies. Similarly, internal boundaries also change their significance when societies develop centralized government. From being defined simply by kinship or residence, they now become the expressions of an administrative hierarchy, and take on much more precise functions in relation to the purposes of the centralized bureaucracy.

The state has also been one of the grand simplifiers of human society. A centralized system of political authority with a hierarchy of officials and specialized departments is much easier to understand than the sort of diffused authority we find in tribal systems of authority. There, lineage heads, elected councils, leaders of age-sets or various grades of big men may all exercise different aspects of political authority in some degree. Again, a hierarchy of social classes drastically reduces the complexity of a social order based on kinship; coinage as the medium of exchange, and a unit of accounting for the bureaucracy, drastically simplifies the economy, as do standardised systems of weights and measures. Writing simplifies communications, by freeing them from the social nuances and associations of the spoken word; and the world religions, when they appear, simplify all the local variations of ritual and belief by standardized models of worship and doctrine. Indeed, the idea of a standardized model sums up a great deal that is involved in this business of simplification, because it combines the notion of universality with rationality of organization, and ease of communication between strangers of different cultural backgrounds. Although, as the importance of core principles reminds us, we must not minimize the major cultural differences between early states, nevertheless, because all states had to deal with a common set of organizational problems, and because they were all based on such general precepts as the role of the king as the source of justice and Life for his people, they provided a kind of standard social model over large areas of the world. The operational demands of the state also had a fundamental impact on the whole process of creating a more articulate, generalized, and simplified social order, because it was much easier to reflect on the workings of the state than on the institutions of tribal society.

Although states were much more powerful than tribes, they were also much less stable, not only because of the prevalence of conquest warfare, but because the hierarchical structure of the state depended on a surplus of production to sustain the rulers and the elite, and also on a supply of manpower for military purposes and public works. So crop failure due to drought, insect pests, or soil degradation had far more serious consequences for states than for tribes, because it threatened that surplus of production on which the state depended. Tribal societies, on the other hand, were much better placed to survive famines, because they only operated at a subsistence level of production, so that their clans, age-organization, and other institutions could carry on even with a substantial death rate. Moreover, the intensive agriculture often found with early states inevitably led to population pressure and to environmental degradation simply as a result of the law of diminishing returns. The most fertile land, the easiest routes, the best sources of water and so on are naturally the first to be used by any population, and it follows that as pressure on resources increases it will have to spend proportionally more time and effort in maintaining their supply. The burden of taxation, corvée labour, and military service, the perceived injustice of oppression and corruption, administrative failure and financial incompetence, soil-degradation or drought [27] could lead to rebellion and the collapse of the state. So too, of course, could barbarian invasion.

But while early states often fell, successors usually rose again on the rubble of their predecessors. The combination of divine kingship, imposing temples and palaces, great wealth and luxury, and military power has always had an overwhelming influence on the human imagination. Even when a particular state has collapsed, it has been easy for later rulers to set themselves up as its successors within the same culture area, often appealing to the ancient glories of an archetypal ruler or Empire – Rome, or Teotihuacan, or the Western Chou, or the Old Kingdom of Egypt. Despite the fall of particular regimes or dynasties the state could therefore be re-established over and over again in its cultural area, so that in the long run the state, despite its inherent instability, has become increasingly dominant in the course of world history, and was the indispensable foundation of the modern world.

While the state was unplanned, it is not a convincing example of a random mutation from, say, the chiefdom that was then selected for. States are not simple 'traits', but complex systems that developed from that *combination* of necessary conditions we have been examining, in a process that was strongly constrained and far from random.

Applying the Darwinian notion of adaptation also has problems: for whom, in the first place, was the state adaptive? Obviously, for rulers and the upper classes, but much less obviously for the great mass of the peasants, who had to give up far more of their crops in taxes than they would have done in tribal societies, and also had to perform arduous labour for the state as well. Generally speaking, ordinary people were better off, both materially and in terms of personal dignity, in tribal societies than in the agrarian states. In the sense of individual well-being the state was fairly maladaptive, then.

The Darwinist might reply, however, that the state was clearly better adapted for military competition than tribal society, which is quite true, but we have now lost touch with real people, and are talking about an *institution* instead. Slavery, too, was a very successful institution, in the sense that it was widespread and persisted for millennia, but it was obviously not adaptive for the slaves. We are, in fact, discussing two different kinds of adaptation here. One is essentially 'usefulness to real people', and the other is the ability of an institution to survive because some of its features fit in well in a particular social system. Especially in systems of radical inequality like the state, it is therefore possible for some institutions to survive very well, despite being of little benefit to the majority of the people. In view of all this ambiguity and confusion, we therefore do better, once again, to avoid adaptive explanations, and simply ask instead, 'What factors led to the appearance of X, and what were the consequences of this?'

But before we trace the further development of the state, we need first to reflect in more depth on the significance of technology in the evolutionary process.

# CHAPTER VIII

## TECHNOLOGY AND INVENTION

*1. Necessity and invention.*

In the Victorian mill-towns of the Age of Steam, it was natural for Mr Gradgrind and his ilk to assume that it was technological innovation that drove society forward, and that human history was primarily the history of inventions. 'Necessity is the mother of invention', they would say, man's response to the need to improve his material conditions of life. A classic example of this was when Brunel's engineer, Francis Humphreys, needed a special machine to forge the huge axle-shaft for the paddle-wheels of the SS *Great Britain*, because the tilt-hammers of the time were too small. He put the problem to manufacturer James Nasmyth, who took only half an hour to come up with the design for his famous steam-hammer on the back of an envelope. [1] In this sort of industrial milieu, with highly focused inquiries into very specific problems, and where inventors could reap immense financial rewards for success, technology was naturally seen as driven by necessity. But this world of Mr Gradgrind was a revolutionary environment that tells us nothing about the history of technology in pre-modern society. Prehistoric man did not go around saying that it would be a good idea if we could find an alternative material to stone for making tools and weapons, in the way that Humphreys looked for an alternative method of forging large metal axles.

Some new ways of doing things, such as using leaves as shelter from the rain, or cutting down a tree for a bridge across a stream, are so straightforward that we can reasonably describe them as responses to a material need. Clearly, there is a basic human ability to solve new problems up to a certain level of difficulty. Hunter-gatherers have an intimate knowledge of the animals and plants in their habitat, and as we saw in Chapter III, the first agriculture would simply have required an application of this knowledge to changing circumstances or to satisfy new aims. But once we get beyond this level of obviousness, to, for example, weaving and pottery, the unpredictable factor of creativity starts to show itself, and the point about creativity is that one can't know what a discovery will be before it has been made, or predict what can be invented before anyone has actually done it. What people would like if only they knew about it can be of no help in actually discovering it, and in this sense it was obviously not the 'need' to make clay vessels that led to the discovery of pottery, or the 'need' for an alternative to pottery that led to the invention of glass, or the 'need' for metal tools that led to the smelting of ores. [2] Here I would like to stress again what I said in Chapter I, that the environment does not present man with problems that have to be solved: it is simply there, and is, to a considerable extent, as we define it, by our creative imagination and theories on the one hand, and by ignorance and apathy on the other.

Again and again in the history of technology we find that what we would consider an obvious need or problem has been ignored, either because the people concerned did not think there was a problem, or because their mediocre solutions seemed adequate. For example, the Egyptians, Mesopotamians, Greeks and Romans between them had thousands of years' experience of moving materials around on construction sites, and they were all familiar with the wheel, but none of them thought of the wheel-barrow. This was a Chinese invention [3] of about the first century AD, apparently as a conveyance for people, which only appeared in Europe around the beginning of the thirteenth century AD on castle and cathedral building sites. The Egyptians never discovered the winch, the crane, or the pulley, and there are many other examples of inventions that would have been extremely useful throughout history but were never made. [4]

In fact, our 'needs' are often revealed to us only *after* we have found a use for something. Even in modern society, there was no 'need' for the aeroplane, or for the laser, but, having been invented people have certainly found uses for them. A society's technology is not therefore a set of responses to material needs, *but a set of discoveries about the uses of things.* We rearrange our lives to take advantage of these uses and then cannot give them up, so that they become needs. In other words, an invention itself changes the world irreversibly and becomes the mother of necessity. But how do such discoveries occur? It is no use saying, in Darwinian fashion, that novelty is bound to occur for many different reasons, so that we can treat it as random and concentrate on selection, which is what really matters. [5] As I said in Chapter I, that is just mental laziness, and a camouflage for ignorance; inventions are not random mutations, and we can actually say a good deal about the origins of technological discovery and innovation.

Rather than trying to explain new technology as a response to needs, or as blind variations, we should look instead at *the sort of social environment that fosters creativity and invention.* (This, of course, is something entirely outside the Darwinian world-view.) While one can't predict discoveries before they are made, or invent something to order, it is certainly much easier to discover and invent in some social situations than others, where long-distance trade brings in new ideas, where there are opportunities to play about with things, where attention is focused on something as a problem, and where innovation is rewarded and admired. The problem with traditional crafts as a source of innovation, certainly at the village level of production, is how unexperimental and conservative they tend to be. They simply have to produce standard articles, and in this situation there is no incentive for the craftsman to spend much time exploring different possibilities that may turn out to be a waste of effort and materials.

There was obviously much more technological innovation in the urban civilisations of Greece, India, and China, especially in warfare and ship-building, but even the Roman water-wheel, that appeared in the first century BC, took another three or four hundred years to become quite common. Once people had devised a way of performing the practical task that they wanted to do, they could not usually imagine how it might be improved, so what further incentive would they have to change? Contrary to the belief of Adam Smith, Marx and other economists that technology is naturally dynamic, in the pre-modern age it had a tendency to stagnate. Far from always wanting to improve their living conditions, in the absence of competing alternatives it was easy for people to be satisfied and to rest on their laurels, or rather on those of their ancestors: 'Technical progress, economic growth, productivity, even efficiency have not been significant goals since the beginning of time. So long as an acceptable life-style could be maintained, other values held the stage'. [6]

Technology, then, has often needed the stimulus of *non-practical motives*, such as aesthetics, intellectual curiosity, magic and religion, pride and status, or entertainment, that will encourage people to focus their attention on physical objects, and play around with them in ways that they would never think of in the ordinary work of daily life, or in relation to material needs. In the ancient literate civilisations, too, royal patronage supported the production of curiosities for the court and a wealthy clientele, gadgetry for temples and theatres, and astronomical observatories. These were all very important parts of those environments that foster creativity, and lay outside the ordinary technology of their society.

Secondly, things have multiple properties, and some of these will often be more obvious or attractive than others, and so will provide a relatively easy way for people to get to know more about a particular thing. But once it is in their cultural repertoire, these other properties can also become available over time in the right circumstances. Fire, for example, presumably attracted early man as a means of keeping warm and of cooking food, but it has far more evolutionary potential than that. It converts clay into pottery, it allows metals to be softened for working, ores to be smelted and alloys melted, and the making and blowing of glass; it is the basis of a vast range of chemical processes and reactions including distillation and gunpowder, and it causes water to expand to about two thousand times its volume as steam, and so is the means of converting heat into mechanical work. We must think of inventions and discoveries less as isolated things, and recognise that they are simply the ends of particular strings, as it were, that can lead to all sorts of other natural phenomena.

Thirdly, because of these multiple properties of things, it follows that the richer a society's technology becomes, the easier it will be for further discoveries to be made, so that technological advance can become a cumulative process, feeding off itself. But, fourthly, there will come a limit to this, because the more complex the technology the more it costs, and the greater the need for substantial economic investment, which will depend on vested interests and cultural priorities. Finally, and very significantly, the more remote and hard to understand some aspect of nature may be, such as electricity, the more dependent men must become on pure science to be able to reach it, and pure science has been remarkably difficult to achieve. To develop these points we can look at a few crucial inventions, beginning with the origins of metals and glass. These are good examples of the multiple properties of things, and of discoveries arising from non-material motives.

## 2. The discovery of metal and glass. [7]

There is a popular belief that as soon as our ancestors became aware of metals, they must have seen at once how much better they would be for tools and weapons than stone. In fact, our ancestors had taken stone technology to a very high level of excellence, and stone and flint can even compete in efficiency with steel. In the words of a modern expert on stone tools, 'I've been out elk hunting with Native American friends where they have the latest carbon-steel knife for butchering and skinning, but when they see what my stone knife can do they like it better and want one'. [8] The gold, copper, and iron our ancestors would have found in their pure or 'native' state would have been useless for practical purposes, being softer and blunter than flint, and also required the development of a whole new technology. So why did our Neolithic ancestors bother with metals at all? Mr Gradgrind would certainly have tossed them aside with contempt, and it is only because gold and copper are beautiful and rare that people initially treasured them, and were sufficiently motivated to explore their properties further. In the words of R.J.Forbes, 'Metal made its first impression as a fascinating luxury, from which evolved a need.' [9] It was man's aesthetic sense, his love of self-decoration, and his desire to own rare and precious objects that was responsible for the early development of metallurgy and, as we shall see, of glass as well, long before their practical possibilities became obvious.

Gold was the first metal to be worked by man, and this was certainly not for practical, but for aesthetic reasons. It can be found in its native form as grains and nuggets in streams, and also in veins of quartz embedded in rocks. But because it is very unreactive it does not occur in the form of an ore, and so never required smelting, although the Egyptians roasted alluvial sand to extract particles of gold. Nuggets of copper, too, occur in streams as greenish nodules that, when rubbed, look not unlike gold, and in the Middle East, as early as 9000 BC, men had begun making small copper objects that, like gold, were originally used as beads, pins, and other small ornaments but, unlike gold, native copper is difficult to work because hammering makes it harder. It was only when men discovered, by 5000 – 4500 BC, that heating it ('annealing') makes it softer and workable again, that it was possible to beat larger amounts into daggers and spear-heads. These were still relatively rare prestige items, however, because there was not much native copper available, and flint and obsidian (volcanic glass) were much harder and sharper than copper. In rare cases, such as in North America, native copper can occur in large outcrops, but normally to produce useful quantities copper has to be smelted from its ores, and this requires much higher temperatures than the 600° C or so of domestic fires which were used for annealing. Many of these ores, such as malachite, are beautiful green, blue, or sometimes red stones with which men had long been familiar in their search for precious objects, and in fact most metallic ores were originally used by man either as precious stones or as pigments for body decoration. By around 3500 BC the potter's kiln, which could reach over 1000° C, had been adapted in the Middle East for smelting malachite and some of the other ores. This not only made much more copper available, but, just as important, allowed it to be melted and cast into the shapes people wanted. [10]

Once copper could actually be melted, and not just softened, it was possible to add other metals to produce alloys that were superior to pure copper, and the addition of tin, in particular, was the basis of the most famous of these alloys, bronze, first made around 2500 BC. This was much harder than copper, and was also better for casting, but the cassiterite ore from which tin is obtained is rare: it does not usually occur in veins like copper, but often as heavy pebbles in streams, where it has to be panned like gold. (It was this association with gold in many streams that would have drawn early metal-workers' attention to it, although it was confused with lead.) Bronze was therefore extremely expensive and remained a prestige item for more than 1500 years. When it reached China, for example, in about 2000 BC, it remained a luxury commodity confined to the upper classes, while the peasants, there and elsewhere, continued to use tools of wood, bone, and stone until iron became available. 'The failure to manufacture copper and bronze farming tools where these metals were known reflects the relative scarcity of copper and tin throughout the world. By contrast, the abundance of iron ores made iron an appropriate material to manufacture farming implements once it was known how to smelt them.' [11]

But iron was first encountered as meteoric iron, (and in the Americas only meteoric iron was known in pre-Columbian times). Like copper in its native condition, meteoric iron is not only scarce but hard to work so that it, too, was initially a precious rarity that could only be used as jewellery, amulets, or small ceremonial daggers for the upper classes. It is possible that the Egyptians first discovered how to smelt iron ore in the process of extracting gold from the gold-bearing sands of the Nile and Nubian gravels. These also contain magnetite, a very rich source of iron, and when the Egyptians were heating these sands for gold, iron would also have collected on top of the gold in a form ready for forging (some Egyptian iron and gold jewellery has been found). While only small decorative objects could have been made from such amounts of iron, at least this process would have identified magnetite as a source of iron. The large-scale smelting of magnetite and other ores seems to have occurred elsewhere, in the mountains of Cappadocia and Armenia, around 2000 BC, and initially only produced wrought iron which, without carbon, is softer than bronze and was initially used only for the usual amulets and ceremonial objects. 'The new metal represented no improvement over copper and bronze tools, it was much less easy to work, the forging demanded much expensive fuel, and the cutting edge made by hammering blunted quickly'. [12]

It took many centuries to discover that it could only be hardened by repeatedly heating it over charcoal and hammering it to raise the carbon content, which at between 0.3% and 1.8% is steel, and much harder than bronze. This, too, seems to have been discovered in the same Armenian region around 1400 BC, giving the Hittite Empire that controlled it the monopoly of steel production for a couple of hundred years. From here it spread eastwards into India and China, the Indians in particular becoming masters of fine steel production by using crucibles in which the carbon content could be carefully controlled ('wootz steel').

So the *initial* driving force behind this long accumulation of experience with metals clearly had nothing to do with their practical usefulness. They were very difficult to make, often softer and blunter than alternatives such as flint, obsidian, or bamboo (in Asia), and were only available in very limited quantities to make correspondingly expensive prestige items, so that it was not until around 500 BC that metal, basically iron, started to have a really significant technological impact.

It should also be remembered that, even though metal-workers were eventually able to produce useful objects such as tools and weapons, the craft was never purely materialistic. Men believed that metallic ores grew like embryos, they were male and female, caves and mines were like wombs in the maternal earth, and the smelting of ores could be regarded as a sexual act. Among the Zulu, for example, the smelting furnace is a female, which remains unfruitful until it is fertilized by the male. The air of the bellows corresponds with semen; the nozzle of the bellows is phallus-shaped, and the aperture at the base of the furnace into which it is inserted is shaped like the female vulva distended as in childbirth. [13] This magical view of metals would later be of great importance in the development of alchemy in which, of course, gold was of primary importance.

We have to apply similar reasoning, based on the human love of beauty and display, and the ownership of precious and prestigious objects, to the origin and development of glass, which for thousands of years was of no practical significance whatever. The basic material from which glass is made is silica, found primarily as quartz, which can occur in numerous forms, from transparent rock crystal, through varieties of sands, to a range of gems such as amethyst, onyx, jasper, bloodstone, and agate. Like the bright ores of copper, these stones would have fascinated our ancestors, and the first appearance of glass, in the fourth millennium BC in Egypt and Mesopotamia, is in the form of glazed beads in imitation of gems, and was then applied to pots, 'faience'. 'The earliest coloured glazes and the earliest glass were used to produce imitations of precious stones for use as beads, inlay in jewellery, coffins, furniture and other objects. They embrace imitations of such natural materials as turquoise, lapis lazuli, red jasper, etc.' [14]

It took a long time for actual glass objects, instead of glazes on stones and pottery, to be made, and even so, only small bottles and amulets could be produced. Significant glass production using kilns does not occur in Egypt, for example, before about 1500 BC, and the glass was typically greenish and often opaque because it had not been fired at sufficient temperature. It was only after the invention of glass-blowing in Syria, in the first century BC, that transparent glass vessels of significant size and in large quantities could be produced, and only then did it become more than a luxury product. But the real evolutionary potential of glass was not on people's dinner tables, or even in windows, but as an essential ingredient of the scientific revolution. This only began to emerge much later in the Middle Ages when magnifying lenses [15] were first used in spectacles, and later as the telescope and microscope, whose importance for science needs no further emphasis. The transparency of glass, its ability to be blown into a variety of shapes, and its lack of reaction to most chemicals also made it the ideal material for conducting an immense range of chemical experiments, which would have been impossible without it, and it also has innumerable electrical applications as well, not least as the light bulb.

Metals, too, were not only necessary for any machinery that went beyond water- and windmills, but also for the development of steam power, electricity, and the whole industrial and scientific revolutions at the basis of modern society. These early developments in the knowledge of metals and glass all occurred, of course, in the hands of practical craftsmen, but I should now like to look at how gunpowder developed, because it is an example of a theme that will become increasingly important later in the book: how men of learning – in this case the alchemists – made an essential contribution to a fundamental item of technology that would not have been discovered by craftsmen alone.

*3. The discovery of gunpowder. [16]*

All across the Old World from early in the first millennium BC, fire was used in warfare by Assyrians, Greeks, Romans, Indians, and Chinese. Inflammable materials such as sulphur, pine resin, tar, powdered charcoal, and in some areas petroleum products, were all hurled in various ways at the enemy on land and sea. It might therefore seem that it could only have been a matter of time before one or other of the military engineers responsible for concocting these incendiary mixtures would inevitably have stumbled on the formula for gunpowder by trial and error. But this could not have happened, because the crucial ingredient in gunpowder, saltpetre, was not among them, and had to come from another source.

The three ingredients of gunpowder are saltpetre (potassium nitrate, $KNO_3$), sulphur, and carbon, ideally in the form of charcoal. Charcoal burns easily, and sulphur has a low ignition point of 250° C, but the essential function of saltpetre is to supply large amounts of oxygen to give full effect to the combustion of the sulphur and carbon, making these ingredients highly inflammable when mixed together and ignited. What, though, was so special about this mixture? Petroleum, after all, which was well known in warfare, is highly inflammable too. The point about saltpetre, however, is that when it reaches about 75% of the mixture, combustion becomes so fast that, when confined, it produces an actual explosion capable, among other things, of driving a projectile from a tube.

To understand the inclusion of saltpetre in the gunpowder recipe we have to leave the military engineers, and go to the Taoist alchemists of China in the first millennium AD, who were attempting to find elixirs that would prolong life and even achieve immortality. Gold, and a number of other metals and minerals were central to this project, and their use in Chinese medicine was already ancient. One of the properties of saltpetre that greatly interested the alchemists is that, when combined with many metals and minerals, it converts them into salts that are soluble in water, and so can be drunk as elixirs. Saltpetre occurs naturally, as white crystals produced by the decay of organic matter such as excreta, especially in warm, humid climates. But it has to be distinguished from the sodium, calcium, and magnesium nitrates, which resemble it, and the basic test was to heat it on charcoal where it burns with a distinctive purple flame.

The alchemists, in their experiments, were always heating different mixtures, and by 300 AD are known to have heated saltpetre, sulphur, and charcoal together. In around 850 AD Tao Tsang warned alchemists not to heat saltpetre, sulphur, realgar (an arsenic sulphide) and honey together (honey would have produced carbon), because the result could be a devastating fire that might burn down the laboratory. The special properties of the saltpetre, sulphur, and carbon mixture, when ignited, had by now become familiar as the 'fire-drug', *huo yao,* and the Taoist alchemists had links with military engineers. These took up the fire-drug because there had been a long tradition of incendiary warfare in China, and its first military uses were in bamboo tubes, in the tenth century, to produce the fire-lance: 'The gunpowder which it contained was emphatically not a high-nitrate brisant explosive mixture, but more like a rocket composition, as in a "Roman candle", deflagrating violently and shooting forth powerful flames, not going off suddenly with a mighty bang'. [17]

In 1044 AD Tseng Kung-Liang, in a military encyclopaedia, for the first time in any society, published some formulas for gunpowder to be used in fire-bombs launched by catapults. These formulas still had a fairly low proportion of saltpetre, around 50%, but now that the significance of saltpetre was known it did not take the military engineers very long to discover, doubtless by trial and error, that increasing its proportion in the mixture would produce an explosive, and the first Chinese guns appeared in about 1280 AD.

Trial and error clearly played a part, then, in the discovery of gunpowder, but only in the context of some highly focused activities, and it would have been quite impossible for the Chinese to have discovered it by randomly mixing all the substances known to them. It was the multiple properties of saltpetre that were crucial here: the alchemists were initially interested in it because of its importance in preparing elixirs, and only later discovered its importance for combustion, but this property was highly relevant to the interests of military engineers.

In the same way, metals first attracted man's interest because of their aesthetic qualities, and only later did their other properties become apparent. This theme of multiple properties – entirely different from biological mutation – is of fundamental importance in the evolution of technology, and a further example is that of the wheel.

*3. Evolutionary potential: the wheel. [18]*

The first evidence for wheeled vehicles is from Mesopotamia, where a Sumerian pictograph of about 3500 BC shows a sledge with wheels, and the wheel had spread to China in the form of the chariot by about 2000 BC. Well before 3500 BC, people in many parts of the world had in fact been familiar with the wheel in the form of the spindle-whorl. This is a disc of wood or clay on one end of a small shaft of wood, rotated in the hand, to act as a fly-wheel, to which is attached wool, cotton, or some other fibre for spinning into thread. The spindle-whorl is a miniature wheel and axle, and in Mesopotamia, where pottery was also highly developed, the potter's wheel appeared at the same time as the vehicle wheel, so it is possible that the potter's wheel and the vehicle wheel were both inspired by the spindle-whorl. [19]

In the popular imagination, the wheel stands only second in importance to fire among the crucial inventions of the human race, so that a society without wheeled vehicles is regarded as almost at the caveman level of development. This is wild exaggeration: there is no point in mounting a vehicle on wheels unless there is also a decent road, or at least fairly level ground such as steppe or veldt, for it to run over, and also animals to pull the vehicle itself. For these reasons wheeled vehicles were never used at all in the Americas or in sub-Saharan Africa, or in mountainous areas of the world until modern times, because wheeled transport in general is much less efficient than boats or pack animals until road-building has reached a fairly high standard. Even in the

Roman Empire it cost less, for example, to send a load of grain from one end of the Mediterranean to the other by ship than to send it a few miles by wagon. Roman roads were primarily used for moving troops and for postal services by horse, and only for relatively short distances to transport goods by wagons. Generally speaking, therefore, the ancient agrarian empires depended far more on boats, pack animals, and porters than on vehicles for transporting goods, and the Maya, Aztecs, and Inca were able to develop complex civilisations without using the wheel at all.

But although our basic image of the wheel is the vehicle wheel, it had even greater evolutionary potential in machinery and the harnessing of power. Human beings and all animals basically use reciprocating motion because their skeletons are systems of levers, but machines (with the exception of the piston and cylinder) must operate on the basis of rotary motion and therefore rely on various forms of the wheel. The first and in some ways the most important development of the wheel, was the water-wheel. This appeared more or less simultaneously in China and the Graeco-Roman world in the first century BC, and obviously had profound implications for the future of technology because it allowed man to harness natural power for the first time, and to use it to drive machines. (While wind power had first been applied to boats by the Egyptians around 3000 BC, the wind mill was not invented until the twelfth century in Europe.) In China and the Mediterranean the water-wheel was used for driving grain mills, (and for metallurgical purposes in China), and this was possible because in both these areas animals had for some time already been used to turn grind-stones which it was easy to adapt to water power. [20]

The gear and the pulley were fundamental applications of the wheel for transmitting power that were also developed in the ancient world, together with winches and cranes. The first use of gears in power transmission occurred in relation to water-wheels, but gears had already been developed in the Hellenistic world by the fourth century BC and at around the same time in China. Initially, it seems that they were used for small devices such as astronomical calculators, elaborate toys, cross-bows, and in China south-pointing carriages [21]. (The Chinese did not discover the worm-gear or the screw-thread, about the only gap in their technological equipment.) The drive belt seems to have originated in China where there was a special need for spinning-wheels to rotate spindles that could wind up the very long silk threads from the cocoons, and, together with the axle, drive-chain, and gears it formed one of the essential forms of power-transmission.

So while the New World was able to develop complex literate civilisations without the wheel, they could never have harnessed water or wind power, or have developed any form of machinery. And in the absence of a well-developed metallurgy, and of iron and steel in particular, and without glass, no real scientific advance was possible at all. It is therefore this evolutionary potential of discoveries such as the wheel, and substances such as metal and glass, that has been of such fundamental importance in the history of technology and science, and in that accumulation of necessary conditions for further development.

*5. Diffusion.*

Some inventions, such as gunpowder, the smelting of metals, and writing are undoubtedly much harder to make than others, so that the easy ones will obviously be made in many more places than the difficult ones. The domestication of plants and animals occurred independently in numerous places and similarly, if people are building stone walls it needs no great insight to think of bridging the gap over a water channel, or a window or a door, with a lintel of stone or wood. But the solution of the corbelled arch is less obvious, while there only seems to have been one source of the true arch – Mesopotamia in about 3000 BC. [22] Twisting animal or vegetable fibres to form a thread seems to have been a very common invention, but hand weaving was much less so, while the varieties of loom had a very patchy distribution both in the Old World and the New. Pottery was certainly discovered in many different places, but the potter's wheel was only invented once, in Mesopotamia. Gunpowder, the stirrup, efficient horse-harness, the stern-post rudder, printing, and the compass also had single places of origin, mostly in China.

So the more complex the invention, in the sense of the number of stages or components that it requires, the less likely therefore it was to happen. If it did, it would probably be in those creative environments with large populations and where other conditions were most favourable, such as a high level of technology, good communications, a culture that favoured innovation, and so on. In other words, the more difficult the invention, the longer it would take for it to occur independently elsewhere, and therefore the greater the probability that it would arrive from one of these primary centres of civilisation before any local inventor would have been able to think of it. [23] Trade and good communications have therefore been of growing importance in the history of technology.

Technological innovation, then, quite unlike the evolution of new biological forms, gradually became less and less widely distributed, and more and more concentrated in an ever-decreasing number of centres, so that by the beginning of our era the dominant regions, the overwhelmingly creative environments, were China, India, and the Graeco-Roman world. It was diffusion between these creative centres that ultimately made the industrial revolution possible, but the whole of the Americas, as I have said, were now effectively out of the technological 'race' because of their almost complete lack of animal power, the wheel, and developed metallurgy, as were Australia and Oceania, and much the same was also true of sub-Saharan Africa.

## 6. Conclusions.

The Greeks and Romans thought that man had raised himself out of savagery to comfort and affluence because of his disgust with his miserable conditions of existence. In the eighteenth century, economists like Adam Smith also believed that technology is naturally progressive, because of this alleged need for man constantly to improve his material conditions of life: 'The typical argument was that man is subject to certain "natural" and "insatiable wants"; wants which . . . serve to "rouse and keep in continual motion the industry of mankind".'[24] We have seen, in fact, that this was a crucially false assumption because no such 'need' exists: accepting one's traditional way of life, especially in a primitive society, is much easier than trying to change it, so that technological stagnation was more normal than progress in much of pre-modern society. But on the basis of this assumption, technological advance was believed to produce a series of stages of social and political organization, and systems of belief. One can see why the experience of the industrial revolution, the great manufacturing cities centred on factories powered by steam, the urban proletariat, the organization of trade unions, political revolution, and socialism should have made this scheme of development even more believable. Modern materialists claim that technology has this determining power over social organization and belief systems because we are animals and must therefore eat before we can do anything else, or because we cannot change the laws of nature, or because technology is the interface between culture and nature, but these are vague philosophical platitudes that don't really get us very far.

The real reason for technology's importance is that fundamental discoveries such as agriculture, the smelting of metals, and the steam engine can *potentially* affect some basic features of society: examples are population size, trade, transport and communications, manufacturing, urbanization, warfare, political power, and wealth and social status. But it does not do all these things simultaneously, everywhere, and all the time. Those who are involved in the Search for Extra-Terrestrial Intelligence (SETI), for example, believe that once beings of sufficient intelligence have evolved, natural selection and the predominance of material interests must ensure that something like modern technology will inevitably develop. But there was no seamless progression from the stone axe to the nuclear reactor, because as technology becomes more complex, innovation will increasingly depend on how far the social milieu encourages creativity, and on two factors in particular. One is cost, because any advanced form of technology is expensive, and technology cannot build itself. It requires investment, and decisions about what to invest in will be taken by a very small minority, notably rulers and rich men. Their decisions will reflect their own interests and values, and these will not necessarily be favourable to technological innovation.

The main obstacle to the highest levels of technology, however, was the lack of a scientific understanding of nature, and the Industrial Revolution depended on the discovery of atomic theory and the basic principles of chemistry, electricity, the laws of motion, and steam power. The modern experimental science that made these discoveries possible was the culmination of two thousand years of 'natural philosophy', and we shall see how this led to modern science in the later chapters. Our immediate task, however, is to understand how writing and mathematics developed, without which natural philosophy would have been impossible.

# CHAPTER IX
## Writing, Mathematics and High Culture

*1. The significance of writing.*

From the remotest times, people all over the world have carved designs such as crosses, lozenges and zigzags on rocks, ('petroglyphs') and this had led to claims that writing was actually invented far back in the Stone Age to express some sort of primeval philosophy. 'The cross is an ideogram for wholeness, cyclical time, and the renewal of life…Another sign in the system, the spiral, symbolises the universal snake, which in turn is the embodiment of the dynamism, life force and regenerative powers of nature.' [1]

These speculations are not supported by serious evidence, but even if we were to accept, for the sake of argument, that the cross and the spiral were ideograms in a primeval philosophy, they would not be examples of writing. Ideograms are signs that represent *ideas*, like the hammer and sickle of Communism, but the essence of writing is the use of signs to represent *the actual words* of a language, so that someone with no previous knowledge of the message can read back from a string of these signs what the writer originally said.

We get a good deal closer to the origins of writing by studying living peoples who use pictures to tell stories. For example,

> In the seventeenth century, the Huron Indians of Canada left pictographic messages along trails reporting how many people were travelling, by what means, from what villages (identified by their symbols), and some of the problems they might be encountering. . . Yet they regarded the ability of the French to transmit precise verbal messages over long distances by means of writing as being wholly different from anything they themselves could do . . . [2]

They were quite right to think so, but the use of pictures to record sequences of events (which is entirely different from the carving of petroglyphs) was a crucial step in the development of actual writing, because it seems that nowhere have people been able to make the conceptual leap directly to signs for the *sounds* of words. They first have to represent words by the *things* that the words stand for, so that pictures of things were the crucial gateway into writing. Writing was a highly counter-intuitive achievement, because for people in non-literate societies the spoken word is an integral part of the person, and of the relations between real people, not something that can be magically dissected away and given a physical form.

Not surprisingly, for people who have never encountered writing before, such as the Huron or the Maori, the idea of 'paper that talks' is absolutely astonishing – perhaps rather the same sort of bewilderment that is felt by some older members of our society trying to understand how pictures can be sent down telephone wires by a fax machine. The ability to convert speech into permanent, visible form was one of the most fundamental inventions in all of history, and it was only during the fourth millennium BC that true writing began to make its appearance. I say 'began' because at first it could only be used to make abbreviated records, using names of people and places, nouns, verbs, and adjectives, and these were not enough to reproduce sentences of actual speech. Writing seems to have begun independently in at least four different places, and in each case in societies at the state level of development: in Sumeria and Egypt around 3500 BC, [3] in China after 2000 BC, and by the Maya of Central America early in the first millennium AD, or even before. (The Aztecs, too, had a script that was at this 'recording' stage of writing, but it developed no further.)

All these societies maintained specialist and leisured classes of administrators and priests, which provided a very appropriate environment for the development of a complex skill such as writing. But it had to have some purpose, and it is tempting to think that early states must have needed writing for tax records, planning monumental public works, and administration in general. Certainly, in Mesopotamia, temple accounts do seem to have been the most important early use of writing, but to judge from Egypt, China, and the Maya we should also include inscriptions on monumental architecture about kings and their deeds, religious texts, calendrical science, and divination as additional areas where writing found important uses from the very beginning in early states. [4] But complex states could also function perfectly well without writing. The most famous example is that of the Inca, who had a system of knotted cords, *quipu*, to keep account of taxes and tribute, and a special group of professional rememberers who were employed to interpret them. Although this was an evolutionary dead-end – knotted cords were going nowhere in the development of human communication – it obviously worked adequately for the Inca state. In sub-Saharan Africa, too, states were able to organize tribute, military service, and the other aspects of centralized government without writing, and Sir Richard Burton found that the kingdom of Dahomey kept tax records by means of pebbles deposited in gourds.

The first Sumerian script of Mesopotamia, like that of Egypt, China, and the Maya, was pictorial, but developed away from this in a few centuries into 'cuneiform', wedge-shaped impressions in clay made with the end of a stylus, and this script was used not only for administrative purposes but for religious inscriptions. Egyptian hieroglyphs, however, were not only works of art, 'an exalted mode of communication within a formal ritual setting', but 'the words of the gods' [5], and their use was always primarily religious: for temples, tombs, and cemeteries, and anything related to the Pharaoh, who was a living god. The names of people and gods were magical, and contained their essence, so that putting names in hieroglyphics on images gave them an identity and extra magical power; it brought the statue of the god to life and embodied his essence in it, and ensured that rituals benefited the right person in the next life. So Egyptian kings, for example, had to be able to read the spells and the description of the route to the afterlife carved inside their tombs. [6] But as well as pictorial hieroglyphs, the Egyptians also developed, at roughly the same time, a separate non-pictorial script, hieratic, for administrative purposes, that was easier to write.

In China writing seems first to have developed as a coherent system around 1500 BC. We do not know if it was also used for economic and administrative purposes, but in the Shang dynasty the examples of writing that have survived from around 1200 BC were used for divination: questions were written on the shoulder blades of deer or on tortoise-shell and the cracks produced by applying red hot irons were supposed to give the answers from the ancestors. By this time Chinese writing was well developed, which indicates that its origins lay many centuries earlier. The Maya of Central America also developed a true writing system independently of the Old World, but here the trigger does not seem to have been public accounts but the opportunity of carving inscriptions about calendars and rulers on monumental architecture, although they were also painted on tree bark. Mayan scribes had noble status, as guardians of knowledge about astronomy, the calendar, and royal history. We should note that it took a long time before early scripts were able to reproduce speech efficiently – about 1000 years for Sumerian, around 600 years for Egyptian, and probably similar periods for the Chinese and the Maya.

Writing therefore began in the easiest available way, by using pictures of things to stand for words, and it is actually much easier to learn to read by pictures than by letters. So it is far simpler, for example, to recognise 🐈 as 'cat' than to read the separate letters c-a-t, and then identify each as standing for the sounds 'ker' – 'a' – 'ter'. (Tests have also shown that children recognise words more readily by the distinctive *clusters* of letters than by forming the sounds of each letter individually to construct the word, and we ourselves, once we are used to reading, actually recognise the *words* on the page rather than the letters.) [7]

In early scripts, then, the signs primarily indicate specific words, 'logograms', but a script cannot be based entirely on logograms. While it is easy to find a sign for 'cat', there are many words, especially the names of people and places, but also abstract words such as 'truth', and grammatical words like 'in' and 'but', that can't easily be indicated by a picture of anything. So some method has to be found to express the *sounds* of these words, by 'phonograms', and an easy way was in fact available through the pun. Puns, which presumably occur in every language, are words with the same sound but different meanings ('homophones'), such as 'sun' and 'son'. So a name like 'Nielson', for example, might be written with two knees followed by the sun, as 'kneel-sun'. [8] Here, the signs for knees and sun are acting as phonograms because they are read for their *sound* values, not for what they represent. Again, the Sumerian word for 'arrow' was '*a*', which also meant 'in', so here the sign for 'arrow' could also be used for the preposition 'in' because they both shared the same sound. (Scripts developed the phonetic principle in different ways that were greatly influenced by the various languages of the writers.) [9]

In some contexts it would have been obvious if the sign meant 'arrow' or 'in', but in others it was necessary to have some way of indicating which particular meaning of the sign was intended. Let us take the homophones 'son' and 'sun', for example. Here 'son' could be written by the sign for 'sun', plus another sign telling us that the word refers to a kinsman; this is not pronounced but indicates the type of thing being referred to, and is called a 'determinative'. As a result we have three basic types of sign: logograms, standing for particular words; phonograms, representing sounds; and determinatives, telling the reader which category of thing the logogram or phonogram belongs to. All early scripts used various combinations of these types of sign, though in different and changing proportions over the centuries. We can now see that writing did not develop a wild variety of forms, from which natural selection then winnowed out a few successful ones, but was constrained by a very limited range of possibilities from the outset.

While there are many thousands of words in every language, these are made up of only a relatively few sounds, 'phonemes', so that English with something like a million words only uses 46 phonemes to do so, (and an actual alphabet of only 26 letters). The phonetic principle used by phonograms was also, then, potentially an extremely powerful means for simplifying scripts, cutting drastically the number of signs that were needed, with the alphabet as the simplest script of all. But in none of the scripts we have been considering was there any tendency to do this.

This is very puzzling for Mr Gradgrind; surely, he would say, since all these people understood the basic principle of using signs to represent sounds, then there must have been an overwhelming selective pressure to develop simple, phonetic scripts, to make them easier to learn and more accessible to everyone, and generally quicker to write and more convenient to use. It certainly took much longer to learn these scripts than it takes us to learn the alphabet, but our modern notion that an efficient script should, above all, be easy to learn for ordinary people, is completely out of touch with the conditions of ancient society. Writing began as a highly specialized craft skill confined to a minute section of the population and was quite irrelevant to the masses. For example, in Mesopotamia literacy 'was so complex and cumbersome a skill that it could be mastered only by an elite, and the expense involved in this was so considerable that it could be borne only by the ruling groups.' [10] Scribes and the literate elite therefore had a monopoly of a difficult, beautiful, and prestigious script that set them above the illiterate masses. In Egypt, for example, a famous text, the *Satire of the Trades*, has survived in which the scribes ridicule manual workers, such as potters, fishermen, laundrymen, and soldiers. As a result they had no incentive to make their scripts easy to read or to teach (and the peasants had no need to learn it), whereas we unconsciously assume that simplicity should be the basic requirement of any script, and so cannot understand why the alphabet was slow to develop and in some cases, as in China, was never accepted. 'In China, Korea, and Japan, more simplified, phonographic versions of scripts have traditionally been regarded as appropriate only for the use of women, children, and the lower classes.' [11]

The major reason in the history of writing for the increasing use of the phonetic principle was not to make writing easier to teach to the masses, or to improve the efficiency of communication. It occurred when a script was adopted by the members of a different culture who could not be bothered with all the cumbersome details of hieroglyphic scripts, and therefore only retained the phonetic element:

> ...while the Mesopotamians, Babylonians, and Assyrians accepted almost without change the Sumerian system of writing, the foreign Elamites, Hurrians, and Urartians [who conquered Mesopotamia] felt that the task of mastering the complicated Mesopotamian system was too heavy a burden; they merely took over a simplified syllabary and eliminated almost entirely the ponderous logographic apparatus. [12]

The Semites of Palestine and Syria developed a script inspired by the consonant signs of the Egyptians, (in which vowels were not indicated), and this West Semitic consonantal system was then borrowed by the Phoenicians and other Semites in the latter half of the second millennium BC. Greek traders in around 800 BC then seem to have adopted the Phoenician script, but modified it by using some of the consonant signs that they did not need, in order to represent vowels. This was important because in Greek, unlike Phoenician, vowels often occur at the beginning of words, and so the first alphabet was developed, meaning a script that indicates vowels as well as consonants. In parallel fashion the Japanese and the Koreans later used the traditional Chinese logographic script as the basis of their radically simplified syllabaries. 'In all cases it was the foreigners who were not afraid to break away from sacred traditions and were thus able to introduce reforms which led to new and revolutionary developments.' [13].

While writing never developed spontaneously in pre-state societies, this is not because there were no possible uses for it at this level of social development, and the borrowers of scripts, such as the early Greeks, unlike the inventors of scripts, did not need to be members of well-developed states at all. There are a number of recorded cases where members of tribal societies who have heard of the existence of writing then developed their own scripts which were used for such purposes as writing letters or recording sacred texts. The basic social situation in which the Vai syllabic script [14], for example, developed and spread in nineteenth century Liberia, from contact with Europeans and Muslims, was an extensive social network of traders. They needed to keep in touch with one another by letters far beyond the confines of village life, and it was possible for them to learn the new script by informal instruction within the space of a year. There also grew up a scholarly class of men with sufficient leisure to begin writing clan and family histories, and record traditional oral mythology, proverbs, and so on. It seems likely that the alphabet would have spread in a rather similar way from the Phoenicians to the Greeks, and then to the Etruscans and Romans, in the course of trading relations. The requirements of state administration and law would have had nothing to do with literacy at this stage, but would have been much later developments. Similar considerations apply to the development of Runes during the Roman period among the Germanic peoples, and Ogham by the Celts.

Again, in Polynesia there were sometimes special schools where boys were taught to memorize myths by priests, but this was not enough to stimulate anyone to develop a form of writing, until the Spaniards made a very brief visit to Easter Island on 20th November 1770. They obliged the Rapanui to sign a treaty ceding their island to Spain, and so the chiefs who had to put their 'signatures' to the document were introduced to the wonders of writing. This very short exposure was enough to convince them of the powerful *mana* that lay in the written word; once they had grasped the basic idea of writing as word-signs, they rapidly developed a script, *rongorongo*, that took over some traditional petroglyphs, but used them to express words and sounds by essentially the same methods as first appeared in the Egyptian and Mesopotamian scripts. [15]

Once a script had been invented or borrowed, with the passage of time more and more uses were normally found for writing in every society. The Minoans were unusual in apparently restricting their use of writing to economic records, and in most ancient civilisations it came to be used for law codes, and a variety of legal documents such as wills and contracts, civil and military administration, private correspondence, religious texts, mythology, poetry, and literature generally. In particular, writing greatly enhanced the power of the state, not only because it enabled accurate and detailed records to be kept that did not depend on human memory, but also because it allowed rulers to obtain written information about local conditions and to issue precise administrative instructions to their subordinates. In particular, laws could now be written down, and the codification of laws is a characteristic activity of literate states.

Not only was literacy closely bound up with the emergence of administrators, legal scholars, judges, and priests, but writing intensified intellectual co-operation generally because members of the intelligentsia did not have to meet one another in person but could share ideas at a distance. Writing also allowed a vastly greater store of knowledge to accumulate than was possible in purely oral cultures, particularly in the keeping of astronomical records and historical chronicles. But more fundamentally still, until the appearance of writing, people are only conscious of language through their experience of speech in all the concrete situations of daily life. The revolutionary importance of writing for thought is that words, for the first time, can be detached from speech and given a permanent form, unlike the spoken word which vanishes as soon as it is uttered. Writing therefore allows people to study their language as a thing in itself, and to start thinking of statements as made up of words that are put together according to certain rules. It is no accident that the Greeks began the systematic analysis of grammar in the fifth century BC, which was taken up by the Romans in the second century, or that the study of Sanskrit grammar by Panini and others appeared in India during this period as well.

Thinking about language in this way also makes it easier to think about thinking; grammar was closely related to formal logic, and the development of ancient philosophy and the world religions would have been impossible without written texts whose meaning could be pondered and analysed over and over again. Once such texts exist there is then the further opportunity for commentary and analysis by men of learning, and there is no doubt that philosophical reflection generally in Greece, India, and China was enormously stimulated by the availability of writing. (As we shall see, however, philosophy also needed intellectual debate as well, and without this, literacy itself was not enough for the development of philosophy.) Ancient writing also had a very close connection with mathematics, especially in the development of geometry by the Greeks, for example, where precise and elaborate proofs would have been quite impossible using the spoken word alone. Mathematics generally could only have developed beyond its elementary stages when it became possible to express its operations in written form.

Although the alphabet made literacy much less time-consuming and difficult to learn than the hieroglyphic scripts, one has to resist exaggerated claims that the alphabet was the main factor responsible for the rise of Greek culture, abstract thinking and philosophy, and, because it was easier for ordinary people to learn, for democracy. Predominantly logographic scripts such as Chinese have been able to survive quite well in the cultural milieu of modern science and industrialism, and while Japanese and Koreans have been familiar with syllabaries for many centuries they have still retained complex logographic scripts as well.

Even with the alphabet, it is most unlikely that more than 5–10% of the adult population of the Roman Empire in, say, the first century AD, would have been able to read and write *with ease,* although there were degrees of literacy, [16] and in pre-modern societies people might be able to read to some extent, but not to write. Schooling was time consuming and expensive, and the vast mass of the illiterate rural peasantry, and the urban poor, would not have been able to spare either the money or the labour of their children to send them to school, nor would they have had any particular reason for making this sacrifice. (Historically, important factors encouraging popular literacy were paper, printing, commerce, and sacred texts like the Protestant Bible and the Qur'an, or the Confucian classics, which people were expected to know, and which might be taught in charitable or religious schools.)

While some craftsmen in the ancient civilisations, particularly the more prosperous, would have been literate, knowledge of craft skills was basically handed down by the practical apprenticeship of hands-on experience in the workshop, not by book-learning, and in none of the ancient civilisations do we find books written by or for craftsmen. Writing was therefore not only one of the main factors in creating the increasingly significant class division in early civilisation, between the 'dark-faced masses', sunburnt from their toil in the fields, and the literate elite. It also powerfully reinforced the division between 'theory', the work of professional scholars and philosophers, and 'craft-knowledge', the practical skills of metal-workers, glass-blowers, potters, masons, ship-wrights, and so on which was seldom if ever put into writing. It was this division, as we shall see, that had profound consequences for the subsequent development of civilisation.

## 2. The development of mathematics.

The first writing always included some form of number notation. Just as logographic systems of writing developed the phonetic principle into the simpler and more economical forms of the syllabary and the alphabet, so number notations also evolved into simpler and more economical forms in ways that we shall see. Here too, just as systems of written language use a very small number of basic principles – logograms, phonograms, and determinatives – so also do systems of written numerals. [17]

Writing, however, ultimately contributed much more to mathematics than efficient number notations. The real evolutionary potential of being able to write numbers down was not just to record them, because eventually written notations made it possible to liberate mathematics almost entirely from words, and use symbols as a pure technique for manipulating *ideas* on paper. We must also remember here the fundamental importance of geometry, as well as purely numerical problems, because problems of area and volume, square and cube roots, fractions, proportion, trigonometry, and equations, for example, were all closely involved in the development of written mathematical calculations.

Practical problems involving measurement and calculation were clearly the starting point of early written number systems. But in tribal societies people do not need to do much measuring to construct simple and traditional huts, and it is only with the practice of erecting large buildings that some more formal system of measurement than sticks and lengths of cord became necessary. Public accounting and estimates of food supplies for the labourers on public projects, land measurement and surveying for taxation purposes, and the standardisation of weights and measures for markets, all involved measurement in the complex economies and material cultures of early states. The Rhind Papyrus of Egypt, for example, which probably goes back to at least 2000 BC, sets students such problems as: the excavation of a lake; building a brick ramp; the number of men needed to transport an obelisk; and the provisioning of a military mission.

The human body provided an easy and obvious basis of measurement, and measurement was fundamental to the development of operational thinking. As in the case of number systems based on ten, there is no reason to imagine that people only decided on the body after some prolonged process of trial and error with sticks, cords, animal bones, and so on. The human body was the obvious choice from the beginning because it was readiest to hand and was reasonably uniform. Standard units of measurement were finger joints, the distance from the elbow to the middle-finger tip, the width of the palm, the height of a man, and the pace, for example. [18] Seeds of various kinds also provided another very common basis of measurement for weight and length, because they are also uniform and readily available. [19]

In discussing the origins of written number systems, we must remember that initially they were only used for writing down the totals, not for working them out. The actual process of calculation, particularly when multiplication and division are involved, is something else again, and with numbers larger than ten typically involves using things like small stones or other counters, especially when dividing heaps of things into shares. [20] When it became necessary to add, subtract, multiply and divide large numbers of things in the developing economies of early states, the piles of small stones used in tribal societies developed into the abacus. Here, counters were placed in different columns, drawn in the sand or on counting boards of wood or stone, (the abacus with beads and wires on a frame was a much later development) and the columns had the same values as the verbal number system, usually units, tens, hundreds, thousands, and so on. This meant that the numerical values of the counters varied, depending on which column they were in; in other words, the counters had what we call 'place-value'. So seven counters, for example, could stand for 7, 70, 700, or 7000, depending on their column. But when writing down the results of these calculations on clay or papyrus or bamboo, it took people a very long time to transfer this use of place value into their written system of notation.

Most verbal number systems of course are based on 10, so naturally written number systems, too, use the same base 10, or a multiple of it (20 in Mesoamerica and 60 in Mesopotamia). Some systems also have a sub-base, often 5, which is multiplied by the base to give important numbers such as 50 and 500, e.g. Roman L = 50, D = 500. It was easiest to represent small numbers by repeating simple signs: |, ||, |||, or •, ••, •••, etc., but by about 5, and certainly by 10, this became extremely tiresome, so further signs were used, sometimes for 5, and always for 10 and some of its higher multiples. The familiar Roman system, for example, used V = 5, X = 10, L = 50, C = 100, D = 500, and M = 1000. So writing down a number involved creating a string of signs, and these were *added up* to reach the total: MDCCCLXVIII = 1868. We call this type of system 'cumulative', because it simply repeats the signs for the same number values, as here with three Cs for three hundred. The first Egyptian hieroglyphic system worked in just the same way, with even fewer signs for the higher numbers. [21]

But the Egyptians also developed a more economical system for use in their hieratic script, in which single letters were used for the numbers 1–9, 10–19, 20–99, 100–999, requiring 36 letters in all. [22] The great advantage of this alphabetic or 'cipher' system is that only *one* sign is needed to express tens, hundreds, thousands, and so on, although one has to remember 27 or 36 or more signs. It was therefore an extremely important device for simplifying number notation. This cipher system was in all probability adopted by the Greeks, [23] who were in close contact with the Egyptians in the middle of the first millennium BC, for their alphabetic number system:

|  | 1 | 2 | 3 | 4 | 5 | 6 | 7 | 8 | 9 |
|---|---|---|---|---|---|---|---|---|---|
| 1s | A | B | Γ | Δ | E | Ϝ | Z | H | Θ |
| 10s | I | K | Λ | M | N | Ξ | O | Π | Ϙ |
| 100s | P | Σ | T | Y | X | Φ | Ψ | Ω | Ϟ |
| 1000s | ͵A | ͵B | ͵Γ | ͵Δ | ͵E | ͵Ϝ | ͵Z | ͵H | ͵Θ |
| 10,000s | $\overset{A}{M}$ | $\overset{B}{M}$ | $\overset{Γ}{M}$ | $\overset{Δ}{M}$ | $\overset{E}{M}$ | $\overset{Ϝ}{M}$ | $\overset{Z}{M}$ | $\overset{H}{M}$ | $\overset{Θ}{M}$ |

[from Chrisomalis 2003b]

Here, for example,

$$843 = ΩMΓ$$

This sort of system is much more convenient for doing actual written calculations when each position, from left to right or right to left, is associated with only one range of values – units, tens, hundreds, and so on.

```
Ψ M Γ        7 4 3
P   Z        1 0 7
-------      -------
Ω   N        8 5 0
```

In this example of addition, it will be seen that the system allows the blank spaces between letters to act as zeros. But while these cipher systems have a positional quality, position is a by-product and they are still operating on an additive basis.

In the number systems we have considered so far, all the meaning is concentrated in the *signs* themselves. While they are also written in a conventional order from highest to lowest, and this order can be from right to left, left to right, or top to bottom (but never bottom to top), a sign's place in the order *itself* has nothing to do with the value of the sign. These types of system are still 'additive', because the values of all the signs have to be added up to reach the total. In positional systems of notation, however, now their *order* also becomes of fundamental importance to their value because their position in that order indicates whether the sign represents units, tens, hundreds and so on. (These number signs may not be single ciphers, but quite complex combinations.) It will be objected that a positional notation actually needs ten signs, not nine, because the zero must also be used, but historically this was not the case. One cannot have the zero without a positional notation, but positional systems have existed without the zero by the use of blank spaces instead.

By about 1800 BC the Babylonian successors to the Sumerians had developed a notation in which position was fundamental, although ciphers were not used. (It is important to note that this notation was used exclusively by astronomers and not even by ordinary officials.) The Babylonians were unique in using a base of 60; the earliest Sumerian texts show that 60 was the weight of 1 *mana,* divided into 60 shekels. [24] So it seems likely 60 was chosen because it has more divisors than any other number below 100 (1, 2, 3, 4, 5, 6, 10, 12, 15, 20, and 30), and would therefore have been exceptionally convenient for dividing into shares or fractions. They retained the Sumerian base of 60 so that the place values were not $10^1$, $10^2$, $10^3$..., but $60^1$, $60^2$, $60^3$...; (from 1 – 59, however, numbers were still written *in the old cumulative manner* on a decimal base). A space was used to indicate the absence of numbers, but by 300 BC Babylonian astronomers were using a zero sign. Although they did not use ciphers, the zero with positional notation made multiplication and division, and also the handling of fractions, much easier.

It is very significant that the Maya also developed their sign for the zero because of their astronomical interests and their complex calendar, which required them to produce long strings of number signs to represent dates. In the same way, while the Indian use of a zero sign (usually O or •) is not known before the eighth century AD, there is plenty of evidence that Indian astronomers, who calculated with very large numbers, used *words* such as *sunya,* 'void', to refer to zero long before this. For actual calculations they used the abacus with its columns, but these developed a crucial difference. The counters began to have number symbols, ciphers, written on them, as opposed to being blank, and this seems to have been the decisive factor in the use of only nine ciphers (instead of 27 or 36, for example) and whose value now depended entirely on place. On the abacus itself a blank space would initially have been left for zero, but this, too, seems to have acquired its own cipher by the eighth century. [25] The combination of ciphers that included the zero, together with position, which allowed the complete replication on paper of the abacus, was therefore an Indian achievement. It produced a notation that was exactly the same in principle as our modern number system, although their actual ciphers were different.

The Chinese had also developed a positional number notation, based on counting rods, and the idea of the zero naturally fitted into this very easily, [26] and the whole Indian notation was brought to the Muslim world in the eighth century by Indian astronomers. The concept of the zero, of course, was to prove of central importance in the development of mathematics, not least in equations, but we will return to this in Chapter XV. To conclude so far, then, systems of number notation, like systems of writing, can only be constructed in a very limited number of ways (five), and these are summarised in note 27.

Manual systems of reckoning were also significant in the discovery of negative numbers. The idea of negative numbers was as counter-intuitive as the idea that zero could be a number, and the Greeks certainly regarded them as absurd, but by the first century BC the Chinese, in their commercial calculations, were using black rods to indicate a debt and red rods to indicate a positive balance (red was a sign of wealth and good fortune). The same association between negative numbers and debt also developed among the Indians, and the idea was transmitted, with the zero, to the Islamic world and then to the West, where their importance was finally understood in the Renaissance.

While number notations originated in practical problems of land surveying, designing monumental buildings, organizing public works, and keeping commercial accounts, it was astronomy and the calendar that led to positional notation and the zero. Astronomy and calendars were to have a profound effect in the development of more advanced mathematics during the course of history, and we need to make a short digression to understand why calendars and astronomy were so important in ancient civilisation.

The disciples of Mr Gradgrind have always believed that the first farmers needed calendars to plan their yearly cycle of activities in the fields, and so it seemed obvious to them that the great interest of the ancient literate civilisations in astronomy, and their highly elaborate calendars, were to allow the state bureaucracy to regulate agricultural production even more efficiently. Indeed, some claim that astronomers were so successful in this that they became overconfident in their abilities to predict the future, and this led them into the absurdities of astrology. [28] But we saw in Chapter IV that tribal farmers can survive perfectly well with very simple calendars or with no calendars at all, because agriculture actually depends on the seasons, and these can vary by weeks each year. While the rising of the Pleiades, for example, is very widely taken as a harbinger of the rainy season, this is only a rough guide, and no farmer who wanted to survive the year would rely on that alone in judging when to sow his crops. So the suggestion that Ptolemy, for example, calculated the year as $365^d\ 6^h\ 9^m\ 48.59^s$ to allow Egyptian agriculture to be regulated more accurately is totally implausible. His calculations would have been of no practical use whatever, and astronomy was studied for entirely different reasons.

In order to understand the ancient world's obsession with calendars, and the exact astronomical observations on which they depended, we therefore have to ignore the simple requirements of agriculture, and the needs of administrative efficiency, and remember what was said in Chapter IV, that the seasons were not only the basis of agriculture, but of religious festivals and rituals. In the Christian world, the calculation of the date of Easter was so difficult that a special science, the *computus*, was devoted to it, and in India 'The Vedas are revealed for the purpose of performing sacrificial rites; these rites are laid down in order of time. Therefore, he who is versed in astronomy, the science of the reckoning of time, knows the sacrifices'. [29] More generally, the movements of all the heavenly bodies, the sun, the moon, the planets and their conjunctions with the constellations of the zodiac, and especially lunar and solar eclipses, were seen as bound up with Life and prosperity, and so it was in fact astrology that was the motive for astronomy in the ancient civilisations. The calendar was a kind of map of destiny and cosmic order, and only when it is thought of in this way can we understand its importance to the ancient civilisations.

Astronomy has three parts: the sun and moon ('luni-solar' astronomy); the fixed stars; and the planets, and we can begin with the problems of the sun and moon. In a rough and ready way, people in simple farming societies can base their calendars on named lunar months, and keep these in a more or less stable relationship with the seasons and the solar year. This only involves adding or repeating a month from time to time, combined with simple observations of weather patterns, and little or nothing in the way of mathematics. But to construct a really accurate calendar is much more difficult because, of course, the lengths of the day, the month, and the year are completely unrelated to one another, and so can't be expressed in convenient whole numbers such as exactly 30 days in a month and 12 months in a year. A month is actually 29.530589 days, and 12 of these months produce a lunar year of only 354.36706 days, whereas the solar year is actually 365.242199 days. If we imagine days, months, and years as gear wheels of different sizes meshed together, the marks painted on each one at the beginning will not be aligned again until many rotations or cycles of the wheels. If, for example, the New Moon falls on 1$^{st}$ January, it will not do so again for another 19 years, or 235 lunations (the period from one new moon to the next), or 6940 days. This is the Metonic cycle, discovered by Meton of Athens in 430 BC, while the Callippic cycle (330 BC) was four times longer – 76 years, 940 lunations, 27,759 days.

Even very small discrepancies in measurement ultimately produce very long cycles: the Egyptian civil calendar was of 365 days due to the deliberate omission of the ¼ day, but this meant that their year took 1460 (4 x 365) years to return to its original relation to the solar year (the Sothic cycle). Eclipses of the sun and moon – of not the slightest practical significance, but very important as omens – were a further complication, but also depend on their relative position to the earth, and the Babylonian discovery of the *saros* cycle of 18 years (6585½ days) enabled them to predict lunar eclipses from 747 BC, and this was made more accurate by use of a 54–year (19,756½ day) cycle.

The Maya had a civil year of 365 days intermeshed with a ritual year of 260 individually named days in months of 20 days, and the resulting cycle took 52 years (18,980 days), the Calendar Round, to complete. The Maya were especially concerned with dating royal and religious events, and in order to link historical events over a period greater than 52 years they used what is known as the Long Count, based on days, *k'uns*:

<p style="text-align:center">20 k'uns = 1 winal (20 days)<br>
18 <i>winals</i> = 1 <i>tun</i> (360 days)<br>
20 <i>tuns</i> = 1 <i>k'atun</i> (7200 days, approx. 20 years)<br>
20 <i>k'atuns</i> = 1 <i>bak'tun</i> (144,000 days, approx. 400 years)<br>
1 Great Cycle = 13 <i>bak'tun</i> (1,872,000 days, approx. 5125 years)</p>

The Great Cycle in which Maya culture reached its high point in the first millennium of our era began on 13th August 3114 BC; this was a purely notional date since Maya society did not exist then, but it is a graphic example of the time span in which their scribes calculated. [30] It is also obvious that none of these elaborate calendars was inspired by the need to farm more efficiently.

A great deal of early astronomical mathematics was therefore a matter of discovering that '*s* intervals of one kind equal *t* intervals of another kind' [31], and we have seen how this generated enormous numbers. [32] The astronomical observations needed to calculate all the various periods and cycles could only be made by professional, literate observers, because it required accurate records to be kept for hundreds of years. Ptolemy, for example, in the second century AD, used astronomical records from King Nabonassar of Babylonia 746 BC to the Emperor Antoninus in 137 AD, or 907 years to calculate the length of the year.

Men had observed the yearly cycle of the stars from time immemorial, for the practical purposes of agriculture, or navigation, or telling the time at night. But correlating the movements of sun and moon with the stars raises a whole range of new problems, particularly because the sun rises and sets throughout the year in relation to different constellations, which we know as the twelve signs of the Zodiac. Because the earth's axis inclines at 23½° from the vertical, the sun appears to trace a curved path through the Zodiac that goes from 23½° north of the celestial equator to 23½° south, crossing the equator twice in the year, and charting all this requires very complex calculations that contribute nothing to agriculture.

The irregular motions of the planets, however, made them even more useless for any practical purposes, and the original interest of astronomers in them was purely astrological, because from early times the planets, especially Venus and Mars, were regarded as omens. In Babylonia, for example, c. 1800 BC, in the period of Hammurapi, the appearances and disappearances of Venus as Morning and Evening Star were recorded: 'important events in the life of the state were correlated with important celestial phenomena, exactly as specific appearances on the livers of sacrificial sheep were carefully recorded in the omen literature' which became enormous over the centuries. [33] As we shall see, while the much smaller irregularities in the motions of the sun and moon could have been fudged, it was the attempt by the Greeks to fit the erratic planetary motions into circular orbits around the earth that drove their astronomers to extraordinary feats of geometrical ingenuity. But if early astronomers had not been interested in the planets, it is much less likely that they would ever have begun debating whether the earth or the sun is at the centre of the cosmos.

Another enormously important factor in the development of early mathematics was geometry. While rectangles, circles and triangles had been used from the earliest times for decoration, there is no doubt that the actual study and manipulation of geometrical shapes was particularly stimulated by the planning of buildings. (The accurate measurement of area was also a basic problem in the surveying of land for tax and administrative purposes, as among the Egyptians and Mesopotamians.) In tribal societies the floor plan of round huts, for example, is marked out by a cord fastened to a centre peg, and the much larger and more elaborate buildings of early states would have required techniques for setting up accurately squared corners, as well as calculations of area and volume. We know, for example, that the Maya and Aztecs, constructors of major monumental architecture, had compasses, levels, plumb lines, and squares, and these and surveys of their buildings imply that they must have been familiar with the basic properties of the different types of triangles, circles, squares, right-angles and so on [34]. In India, as in Egypt and probably the other ancient civilisations, ropes (*sulba, sulva*) tied to poles were used to set up geometrical constructions on the ground and the term *sulba* was given to Indian geometry and the calculations of areas and volumes. In particular, it was known to them, and to the Egyptians and Babylonians, that a triangle whose lengths were 3, 4, and 5 units was a right-angled triangle, and so very useful for laying out the square corners of plots of land and large buildings.

But the practical problems of designing houses and other structures, and surveying, do not take geometrical knowledge very far, and in India geometry was greatly stimulated by the need to construct elaborate brick altars with great precision. Just as the words used in rituals had to follow an exact formula without the slightest deviation so, too, it was considered necessary for the altars to be built in accordance with precise specifications. The shapes of these altars were numerous and complex, depending on their ritual purpose, such as hawks, herons, chariot-wheels, wooden troughs, chariot-poles, funeral pyres, and villages, all of which were symbolised by particular geometrical shapes. [35]

The *Sulva-Sutra* of Baudhayana (5th– 3rd centuries BC), for example, contains detailed geometrical instructions for generating squares, oblongs, triangles, trapezoids, rhomboids, and circles, and ritual rules required them to be transformed into one another while keeping their areas constant, or doubling or tripling their areas. Techniques were also therefore needed for calculating the areas of these figures, and all these requirements were extremely important for a deeper understanding of geometry. [36] For example, one problem (that also appeared among the Greeks) was how to construct a square altar that was twice the area of a given square. This is achieved by drawing a diagonal across the given square and then using this as the first side of the new square (see diagram). We can see that the first square contains 25 tiles, whereas the second square contains 50. But by drawing a diagonal across the first square we have also produced two right-angled triangles, and we can also see that the area of the square on the hypotenuse, the diagonal, is equal to the area of the squares on the other two sides when these are added together, for example 3, 4, 5 ; $3^2 + 4^2 = 5^2$ ; 9 + 16 = 25, which is the famous so-called 'theorem of Pythagoras'. Other problems were considerably more difficult than this.

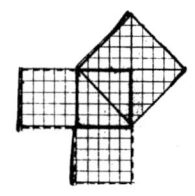

The fact that the altars were built out of baked bricks, which can be standardized, undoubtedly made this kind of precise construction much easier than if they had been of natural materials, but the materialist argument [37] that therefore Indian geometry grew out of brick technology, not the studies of the Brahmins, is too simplistic. The specifications for the altars, and the need for exactness, came entirely from the scriptures, whose study was a Brahminical monopoly, and only Brahminical scholars could therefore have developed the pure geometrical techniques for generating the different shapes the altars required, which were quite unrelated to any normal architectural problems. It is certainly true, however, that the *Sulva-Sutras* also discuss the many different shapes of bricks required, specify their dimensions in detail, and work out the very complex mathematical problems of designing different patterns and sizes of bricks which would produce the shapes and areas needed. At the same time, however, most of the brick shapes were given symbolic meanings, just as the layers of bricks symbolised the social structure, all of which indicates Brahminical thinking, and there is no archaeological evidence that such bricks were used for anything besides constructing altars. The Brahmins, of course, knew quite well what the different altars *looked* like, and it also seems very likely that there was a collaboration between Brahmins and literate architects, who would have needed a thorough grasp of what was being required of them in order to instruct the foremen-bricklayers.

But there is a long way from this to the conclusion that the mathematics in the *Sulva-Sutras* 'is inconceivable without being connected with the contributions of the manual workers – the craftsmen and technicians, specially the brick-makers and the brick-layers'. [38] No practical knowledge of how to make or lay bricks would have been needed by the authors of the *Sulva-Sutras,* and actual construction could have been left to the architect and the foremen on site. It was, however, a very specialised tradition of geometry that died out with the demise of brick altars, and Indian geometry only revived much later in the 5th century AD in relation to astronomy.

In these calculations, the actual lengths involved are less important than the procedure for manipulating them, which constructs *relationships* between the lines of squares and triangles regardless of their *actual* measurements. This is the basis of a whole range of what we know as algebraic concepts, such as squares, cubes, square roots, cube roots, proportions, and especially equations. The great value for mathematical calculations of replacing actual numbers by symbols is that they allow us to write down very complex mathematical relationships without the need for words, making it far easier to solve equations. In modern mathematics we call the sides of a right-angled triangle *a* and *b*, and the hypotenuse, *c*, and so the relationship between their squares is $a^2 + b^2 = c^2$, which is an elementary example of an algebraic equation. In the ancient civilisations they did not have any convenient notation for doing this, and while the Babylonians, Greeks, Indians, and Chinese knew perfectly well, for example, that the area of a circle was $\pi r^2$ (in our notation), they could only express problems in cumbersome verbal form, instead of a neat formula. For example,

'Divide 100 loaves among 5 persons in such a way that the shares are in arithmetical progression, and 1/7$^{th}$ of the sum of the first 3 shares is equal to the sum of the other 2'. This was a huge encumbrance on mathematical reasoning, rather like having no musical notation and having to rely on the 'doh, ray, me,' method instead, and the Greeks in particular solved equations for hundreds of years by geometry. Only with Diophantus (second century AD) did some algebraic notation make a brief appearance among the Greeks, [39] but his work was only seriously taken up again at the Renaissance, when it combined with the Indian positional notation and the zero acquired through the Islamic world, to be the basis of the immensely powerful algebraic notations of modern mathematics.

So while arithmetic and geometry began with the practical problems involved in building, administration, and commerce, these could only take it so far, and its higher development needed the stimulus of astronomy, and ritual activities such as the construction of sacrificial altars.

*3. Conclusions.*

The development of writing, calendrical science, and mathematics in the ancient world were very impressive and important achievements, but they were quite compatible with an essentially mythical mentality that was not significantly different from that of tribal society – the Mesopotamian *Epic of Gilgamesh* [40] is a good example. A literate class of priests and administrators by itself, then, was not sufficient to generate any critical, philosophical thinking about man and how society should be organized, or about the natural world. That needed further social changes that we shall discuss in the next chapter.

# CHAPTER X
## Social Crisis and the Need to Think

During the first millennium BC people in China, India, Iran, Greece, and Israel, [1] experienced a great deal of social turmoil – political instability, warfare, increased commerce and the appearance of coinage, and urbanization – that in various ways eroded traditional values and social bonds. All this produced confusion and even despair, when people questioned the whole meaning of life and tried to find the secret of happiness and tranquillity of mind. This period has been called the Axial or Pivotal Age [2] because it led a new breed of thinkers and philosophers to search for a more transcendent and universal authority on how one should live, that went beyond the limits of their own society and traditions, and beyond purely material prosperity. 'Everywhere one notices attempts to introduce greater purity, greater justice, greater perfection *and a more universal explanation of things.*' [3] The result was that for the first time there appeared rational, unified, general theories about society and the state, human nature, and our inner mental life, and in this chapter we shall examine the social conditions that produced these new thinkers, and how they applied formal operational thought to man and society by developing the moral and political sciences. This new type of philosophical thought was also applied to religion, where it led to a complete revolution that we will come to in the next chapter.

*1. The Axial Age.*

In China rebellions had brought the Western Chou empire to an end in 722 BC. While the religious supremacy of the emperor remained, collapse of political control allowed a large number of states to establish their effective independence, but this proved highly unstable. During this 'Spring and Autumn Period', members of royal clans rebelled and established new states; leaders of noble clans overthrew rulers; and ministers succeeded in making their offices hereditary. Younger sons of the nobility, the *shih*, well-educated but unable to inherit office, travelled around offering their services to any state that would employ them. By the sixth and fifth centuries BC violent factional warfare within states had led to the mutual extermination of many noble families, while growing warfare between the states had eliminated about a hundred of them. In this process rulers and ministers of conquered states could find themselves reduced to slavery; the aristocracy had ceased to command armies, which were now composed of mercenaries whose professional generals were often men of low birth, and who had been able to use these new opportunities to rise to great status and wealth. Commerce, too, greatly increased during this time and merchants could acquire large fortunes.

By 463 BC, the beginning of the period of the so-called Warring States period, the many small states had been amalgamated by conquest into only twenty-two; the class of feudal nobility and hereditary ministers had been decimated, and kings were much freer to appoint whoever they wanted as ministers. These large states, with professional and mercenary armies, needed a much higher standard of public finance, administration, and generalship, so that ability became more important than birth as the basis of office. At this period the *shih* formed professional schools of government, with masters such as Confucius, and their disciples, offering not only different philosophies of government to rulers, but theories of man and society.

India and Greece, on the other hand, represented a very different situation, of the rapid growth of new and complex civilisations out of tribal chiefdoms, that had acquired literacy much more recently. In India, the ancient literate civilisation of the Indus Valley had collapsed a thousand years before, followed by the Aryan incursions from the north, about 1400 BC, that had brought a wholly new culture, centred on the Vedic scriptures and the three-fold order of priests, warriors, and food-producers. The 'second urbanisation' of India began in the Ganges Plain in around the seventh century BC when trading kingdoms developed. Increased trade, together with the stimulus of iron technology that led to a much more sophisticated level of crafts, were associated with the growth of substantial cities of tens of thousands of people, and an enormous increase in wealth and luxury and the social status of the merchants.

The Brahmins strongly disapproved of the new urban life, and of the rising merchant class, which they saw as a corrupting influence on traditional village life, but new religious movements grew up in the cities, notably Buddhists and Jains. They rejected the authority of the Brahmins and the Vedas, were strongly opposed to the whole principle of caste, and were generally supportive of merchants and of commercial life in general. They also developed new theories of the state as part of their general critique of Vedic society and the Brahmins. The Brahmins were also involved in a struggle for social supremacy with the chiefs and kings of the Kshatriya caste, and both the Buddha and Mahavira, founder of the Jains, came from leading Kshatriya families. Conquest warfare dominated this period, and a succession of increasingly powerful states, notably Kosala, Maghada, and then the Mauryan empire of Chandragupta established control over an ever larger area. In this process these states came into strong conflict with the political values of the old 'republics' and especially the popular assembly, the *samiti*.

At roughly the same time in Greece, monarchy had been overthrown in almost all the city-states, to be replaced by various types of government including tyranny, oligarchy, and democracy, that involved prolonged conflict and debate about how the state should be organized. Class struggles between the nobility and the people focused in particular on debt and landlessness, and representation in government. New legal codes and very elaborate constitutions and voting systems were promulgated, and in all this the assembly and the courts were of central importance as centres of debate about fundamental social issues. The great increase in commerce had led to a breakdown of the network of feudal duties and the challenge to inherited status by wealth, because men of humble birth could become richer than the nobility. By the fifth century B.C. the foundation of colonies as far afield as the Black Sea, and long distance trade had given the Greeks a much wider knowledge of different places and peoples, so that people began to question the assumption that their own customs were right and natural. War against the Persians, and then between the city-states also raised profound issues of the morality of conquest and slavery, although Greek political thought took place, of course, in a very different context from that of Chinese and Indian thinkers who were living in powerful centralized monarchies.

One important new factor in these societies was coinage, which developed in the course of trade during the middle of the first millennium BC in the Mediterranean, India, and China. [4] Coinage essentially involved states putting their official seal of authentication on standard weights of precious metals that had long been used for trade, and its consequences were felt throughout society. As it became increasingly common, it made the collection of taxes and the payment of armies very much easier – the conquests of Alexander, in particular, would have probably been impossible without the vast quantities of coins used to pay his soldiers. Banking and credit and long-distance transactions between merchants, [5] rental agreements, and contracts in general are all more easily paid in coinage than in kind, and it is more convenient to borrow cash than stock, land, or produce. So when corvée labour, feudal duties, and a whole range of personal services could be commuted into cash payments, and legal rights and privileges could be bought and sold, traditional obligations based on social status became increasingly transformed into commercial relations between individuals on the labour market.

In the ancient world generally, the urban society based on trade was marked by increased opportunities for choice in career and mode of life, and also became the main centre for the rootless and the dispossessed. These people were cut off from their rural origins and kin structures, and cities were therefore the potential focus of class antagonisms between rich and poor, who were all the more dangerous for being concentrated in one place. The politically revolutionary nature of towns has been one of the most important factors in the history of civilisation, and this revolutionary potential has not been confined to politics: intellectual ferment and debate in general has made towns the major source of the most significant religious and philosophical movements in history. It was no coincidence that the major world religions developed in an urban rather than a rural context. "The notion of the peasant as truly religious is a fairly modern idea. On the contrary it was the townsman who was much more likely to be numbered among the devout, and Max Weber has pointed out the great fecundity of the urban middle strata in religious innovations throughout the several great historical traditions.' [6]

The search for happiness and tranquillity of mind led thinkers into a deep study of the human personality and its essential components, of the need for wisdom and the control of the emotions and desires, and schemes of the basic virtues. The nature of society itself was questioned in a fundamental way, and new models of the state were proposed, because the state was obviously basic to human welfare. In this crisis of civilisation, some thinkers believed that it was possible to reform society and that it was the duty of wise men to try to do so. Their theories will take up most of our attention, but we should also notice that others concluded that, since society was thoroughly corrupt, happiness could only be found by withdrawing from it.

Those who advocated withdrawal fell into two basic categories, the first of which were the ascetics and world-renouncers. These were common in India, where Buddhists in particular believed that all suffering was caused by attachment to worldly desires. In China the Taoists held rather similar views, and retreated into isolation because they regarded the state as inherently corrupt, and looked forward to an age when people would revert to the simple village life of their primitive ancestors, when all had lived on terms of equality. In Greece there were many who thought that the good man should not enter the corrupt world of politics, but the earliest world-renouncers were the members of the Orphic cult about whom I shall say more in the next chapter. Other world-renouncers were the Cynics, meaning 'dog-men' because they rejected all social conventions both in theory and in practice. They regarded themselves as citizens of the world, and taught that we should be content with the barest necessities of life because only in this way could we achieve true detachment and tranquillity of mind. The Stoics were influenced by their teachings, and their ideal of the wise man as one who is able to remain unmoved by the passions was later very influential among the Romans. The other class of those who withdrew from society were extreme materialists who also considered society and its rules and obligations a ridiculous sham, and maintained that the only thing worth doing was therefore to enjoy oneself as much as possible, especially with the pleasures of the flesh. These were the Yangists in China, and the Carvakas in India, to whom the Greek Epicureans were somewhat similar.

## 2. Debate and the discovery of the mind.

Severe social problems were not enough by themselves, however, to produce the sort of articulate, philosophical response we are talking about, nor was the existence of a literate elite. There also had to be the kind of society where speculation and intellectual debate could take place. The Mesopotamians, for example, in their long history of conquests and re-conquests by different empires, had had as much opportunity to reflect on the problems of government, and the miseries of the human condition, as had the Chinese, and the opportunity to write down their reflections. But the scribes, bureaucrats, astronomers, and other *literati* were first and foremost officials in the employment of the state, whose responsibility was to keep its different departments functioning, not discuss the meaning of life. 'They were all, ultimately, in the service of the prince, who demanded absolute obedience. . .But such habits of subordination did not favour free discussion and, in Mesopotamia, we do not find an awakening of rational philosophy', and 'There is no arguing against opposing views; we find here none of the revealing dialogue, which in Greek life and thought finds expression in the court, in the theatre, and in the lecture room . . .'. [7] Highly militaristic and conservative cultures, such as Sparta, would obviously not encourage philosophical questioning either, and the Romans, too, who had plenty of experience of the human condition to think about, only developed a late and uninspired interest in philosophy under Greek influence. [8]

In Greece, China and India, however, there was vigorous debate within a wider educated class which had no specific links with government or priesthood, but tended to be 'persons in a precariously independent, interstitial – or at least exposed and somewhat solitary – position in society'. [9] In China there was no effective priesthood to impose religious orthodoxy on thought; in Greece, while the official gods of the state had their own cults, they were served by particular families that did not form a distinct caste in society, with a monopoly of religious learning. In India the Brahmins were engaged in an intense struggle with Buddhists and Jains, and with the Kshatriya caste, to impose their views on society.

Thinkers could meet and communicate in a new way, and form groups of masters and disciples so that new ideas could be handed on as schools of thought. In Greece, philosophy developed in a culture where public debate was normal in the law courts and the assembly, and public speaking in general was becoming increasingly professional. Here the Sophists were of great importance: [10] they were travelling teachers and lecturers from all over the Greek world who appeared in response to a demand for education that went beyond the poetry that was the traditional curriculum. In particular, they taught rhetoric and the techniques of argument, which were vital in law courts and the assembly, but this practical political concern led many to an interest in political and moral theory, and Athens, especially, became the meeting place of thinkers from all over the Greek-speaking world.

In the society of Northern India, too, during this 'second urbanization', and into which the Buddha was born, there were different schools of hermits and also the 'Wanderers' (*paribbajaka*), celibate but not ascetics, laymen (and sometimes women) rather than priests:

> They were teachers, or sophists, who spent eight or nine months of every year wandering about precisely with the object of engaging in conversational discussion on matters of ethics and philosophy, nature lore and mysticism. .And they were in the habit, on their journeys, of calling on other wanderers, or on the learned Brahmins, or on the hermits, resident in the neighbourhood of the places where they stopped. [11]

In China, by the end of the Spring and Autumn Period, the

wandering scholars of the *shih* class, and others from humbler backgrounds such as the artisan Mohists, had developed into groups of masters and disciples with different theories of government and ethics. Whereas the Greeks and Indians positively relished debate, traditional Chinese culture, and especially the Confucians, were strongly opposed to argument and disputation, but the existence of differing schools of thought left them no alternative: 'The Warring States, Ch'in, and Han periods therefore witnessed a remarkable growth in the influence of the arts of disputation and rhetoric on the lives of the ancient Chinese... Debate and argumentation came to play a central role in how the society and government of the time resolved difficulties and determined proper policy.' [12] During the Warring States period, debates were held in the courts of vassal kings, and some princes founded schools of learning, the most famous being the Chi-hsia academy. In the fourth and early third century, debates at royal courts seem to have been central in the intellectual life of the times, and this was known as the period when a Hundred Flowers bloomed. But the unification of China swept away the independent courts and their debates, and after this time scholars address themselves to the Emperor; [13] these great periods of debate were all relatively short-lived in Greece and India as well.

Debate in Greece, India, and China on fundamental issues of religion, statecraft, ethics, and human nature had widespread effects on thought as a whole. First of all, it inevitably led to the development of standard techniques of rational argument and even of formal logic. The technical Greek term for 'logic' was in fact 'dialectic', from *dialegesthai*, 'to discuss'; the dialogue was a basic form of philosophical text, and Aristotle was the first philosopher in the Western tradition to set out the formal principles of logic in a systematic way.

Debate had somewhat similar effects in China:

[the Confucian] school does not enter into rational debate until it begins to be challenged by other schools, first of all by the Mohists. The early Mohists are ignorant men, excluded from the best culture of their time, but compelled to give reasons for their tenets, because they are new. Each of the ten triads of chapters defending their ten doctrines is a laboriously assembled collection of arguments to convince doubters. Some of the argumentation is very crude . . . nevertheless this is the start of rational discourse in China. Within a century or so the Mohists will have developed into the most sophisticated of all the ancient Chinese thinkers. [14]

In India, 'As a system passed on it had to meet unexpected opponents and troublesome criticisms for which it was not in the least prepared. Its adherents had therefore to use all their ingenuity and subtlety in support of their own positions, and to discover the defects of the rival schools that attacked them.' [15]

The result was that formal techniques of logic and rational argument in general became well developed. Whether they were arguing about political theory, virtue, or religious doctrine is not the point: 'Beyond the particular sphere of ethics, rationalistic methods are normal in inter-religious controversy, because if the adversaries are to be able to discuss religion at all they must find common ground and not presume the truth of their own faith. This is an idea that recurs constantly in the history of religions'. [16] Rationality has therefore been essential for transcending the limitations of one's own culture: 'What we call rational discourse is not a cultural speciality of the West but a necessity for any complex and mobile society'. [17]

We saw in the previous chapter that thinkers also became increasingly concerned with grammar, and the nature of language and meaning, topics that are closely related to logic. Aristotle, for example, begins his study of logic, in the *Prior Analytics*, by discussing the grammar of propositions, terms, predicates, and negation. The development of grammar in India, as in Greece, was also of great importance as part of a general science of language and meaning, especially by the Nyaya school who were notable for their contributions to logic. [18] This reflection on language and meaning also removed primitive confusion about names. Plato, for example, 'wrote a whole dialogue (the *Cratylus*) to dispel the belief that everything has a name which naturally belongs to it and embodies its real nature.' [19]

The precise definition of names was also of special importance because this was regarded as essential for rational discussion and analysis. The BBC long ago had a programme called *The Brains Trust*, where listeners sent in questions to a panel of celebrated thinkers, one of whom was Professor Joad. He became famous for beginning his replies with 'It all depends what you mean by . . . ' and the thinkers of the Axial Age would have entirely agreed with this. The definition of such concepts as justice and law, for example, were basic for Greek political thought, and Socrates in particular established a reputation for proving to people that, literally, they did not know what they were talking about, and so could not hope to think effectively about fundamental questions. The same emphasis on rigorous classification and the definition of terms applied in India and China, and we can now see how the logical, taxonomic classification discussed in Chapter VI became essential for all forms of rational thought. [20]

The study of argument and the development of logic, the clear definition of terms, and the analysis of grammar not only involved a far greater awareness of language, but of thought processes in general, of the nature of meaning and of how we can attain truth and detect error and illusion. In other words, for the first time, thinkers became truly aware of the mind itself and how it obtains knowledge of the external world. A far clearer distinction could now be drawn between the rational, thinking element in man, and the body and the senses, and this in turn had an extremely important impact on traditional ideas of the soul, which for the first time becomes associated with reason. I said in Chapter VI that our notion of the mind has not been a constant of human experience throughout history, but was constructed in specific social circumstances, and it is now clear what these were. The discovery of the mind, however, was only part of a general process of introspection that led to new understanding of the inner life of the self.

### 3. *The reconstruction of the self.*

The search for peace of mind was a basic pre-occupation of the age, and the solution was to ensure that one's equanimity was not disturbed either by suffering and disappointment, or by the temptations of luxury and power. Peace of mind was obtained by realising the ultimate unimportance of worldly possessions and social status in the scheme of things, and by controlling one's feelings and desires by reason and self-discipline. The importance of self-knowledge and control of the desires and the passions led thinkers to analyse those basic features of human character that were essential for performing well as a moral being – the cardinal virtues, which were very different from traditional thinking about virtue.

A good example of this is the traditional Roman notion of *virtus*, which in its original sense meant 'manliness' from *vir*, a male, and included two other qualities besides courage; these were *pietas*, meaning not 'piety' but proper behaviour towards parents, the state, and the gods, and *gravitas*, meaning 'dignity' or 'presence', especially when performing the ancient offices of state. Traditional Roman *virtus* was therefore thoroughly embedded in the social roles of a man, especially an upper class man, in Roman society; it was not concerned with the essential qualities of the good human being as such, regardless of age, sex, or nation. But philosophers of the Axial Age based their ideas of virtue on a universal model of the human psyche, and it was generally agreed that there are four or five virtues of special importance for the complete human being. Wisdom or reason controls the emotions and desires and keeps them in balance, and the basic virtues are often compared to the organs of the body, (very much in line with their organic analysis of the state). The cultivation of these virtues puts man in the proper relation to the cosmos and society, and will produce a state of harmony or health within him, so leading to tranquillity of mind and invulnerability to changes of fortune. Virtue can only be attained by a long process of training and self-scrutiny, involving a lifetime of struggle to bring one's heart into conformity with the right.

For the Greeks and Romans there were four cardinal (or 'pivotal': *cardo* = hinge) virtues. They first appear in Plato, and are related to the basic components of the human psyche: the rational part of the soul rules, and the principle of courage is its ally, and both co-operate in restraining the appetites, whose sobriety is the result of being in proper subjection to the superior principles of courage and wisdom. Justice, however, is the proper coordination of all three principles of wisdom, courage, and temperance. 'Virtue, then, as it seems, would be a kind of health and beauty and good condition of the soul and vice would be disease, ugliness and weakness'. [21] Aristotle added enormously to the discussion of virtue, but the four cardinal virtues of justice, wisdom, temperance, and courage passed into the orthodoxy of Classical culture, notably among the Stoics, while Cicero popularized them for the Romans.

In China, Mencius related the basic dispositions of man to four basic virtues, which were benevolence, *jen*; righteousness or sense of duty, *i*; propriety, the spirit of *li*; and true moral knowledge or wisdom, *chih*. [22] Just as the four cardinal virtues of Mencius are compared to the four limbs, the five cardinal virtues of Buddhism are called organs, *indriya*, of moral practice, analogous to the five sense organs. [23] The generally organic concept of virtue in the ancient world also included the idea of moral 'health' as a condition of equilibrium, in which extremes are to be avoided, and no quality is present to an excessive degree. The inner life of the morally developed individual required constant self-examination as well, because all would have agreed with Buddhism that 'True morality is not confined to the external act of the doer but, rather, relates to his mental purity.' [24] 'Know thyself' was therefore a maxim in all the literate civilisations of antiquity, [25] and it is during this period that the idea of conscience finally becomes fully developed [26], and the importance of intention in legal responsibility and the whole mental element in law becomes increasingly recognised. The understanding of why other people behave as they do, what effects our actions are likely to have on them, and how it would feel to be in their place also leads for the first time to various statements of the Golden Rule, 'Do to others as you would have them do to you'.

*4. A science of the state.*

The state was subjected to the same sort of penetrating analysis as the individual, and for the first time came to be thought about in a naturalistic way, as a social arrangement to enable human beings to live an orderly and virtuous life. While kingship might still be regarded as divinely approved, it was approved for certain reasons that could be understood, and was not simply an unquestionable part of the nature of things derived from the cosmic order. (As we shall see, this naturalistic and rational analysis of the state also had profound implications for religion, because it required a complete revision of the role of the gods in the prosperity of the state.) The political problems of the Greeks, of course, with their city-states, were very different from the powerful centralized monarchies of the Indians and Chinese: the assembly, for example, was basic in Greece but was suppressed by Indian monarchies, and unknown in China. Understandably, there is no discussion there of democracy, oligarchy, and tyranny which so concerned the Greeks, but some basic issues were still common to them all. Generally speaking, most thinkers agreed that the state was analogous to the human body, with each social class making its distinctive contribution to the good of the whole, and just as reason should rule the emotions and physical needs of the individual so, too, the state should be governed by the wise. This might seem obvious to us, but it marked a drastic revision of old ideas of sacred kingship.

In fact, instead of talking about a hierarchy of social 'classes', which suggests the informal economic divisions of modern society, it would be more accurate to refer to 'orders' or 'estates'. 'An order or estate is a juridically defined group within a population, possessing formalized privileges and disabilities in one or more fields of activity, governmental, military, legal, economic, religious, marital, and *standing in a hierarchical relation to other orders*'. [27] There was a general tendency for these hierarchies to become more rigid as a form of social control, although *inheritance* of status was increasingly challenged by the arguments for individual merit, especially in appointments to office. Since the state was seen as a kind of organism, like the body, the hierarchy also took on a moralistic quality, because high status became associated with education, wisdom and therefore with greater virtue, while low status became associated with ignorance and brutishness, and the merchants with dishonesty and greed.

But how should the wise king rule? To answer this question, one first had to decide if man is basically selfish and aggressive, or sociable and co-operative. If he is selfish then the state should primarily be an instrument of punishment and terror, but if he is sociable then it may be possible to rule more by moral example. A major issue was therefore the relation between morality and political expediency, between the function of the state to make its members into better human beings, and its function as an efficient administrative organization that could develop a prosperous economy and a powerful army.

In India the Brahmins argued that people are naturally aggressive, so that it was necessary for the gods to give them kings who would use punishment to stop 'the stronger roasting the weaker like fish on a spit', and to use punishment to maintain the social hierarchy of caste and the ritual order. Warfare, in particular, was the caste duty of the king, and there was no conception of a moral order that should govern the relation between states. Buddhists, however, maintained that selfishness, greed, and violence were not the result of innately depraved human nature, but of their attachment to private property that had been created by agriculture. The people therefore met in the assembly and chose a virtuous and able man as king to maintain social order, for which he was recompensed by tribute from the people. Although Buddhists accepted that monarchy was the political reality of the day, they tried to influence it in a morally responsible direction. So while the king had to rule justly and punish wrongdoers, he should also be a moral example and father to his people: 'For [the Buddhists] the state is not merely a punitive instrument but primarily an agency for the moral transformation of man as a political animal'. [28]

According to the Brahmins, justice and punishments should be graded in severity according to the caste of the offender, but in the Buddhist view, while social and economic inequality were accepted as facts of life, 'The goal is to prevent hierarchical relations from restricting equal opportunities for moral and spiritual development and in the adminstration of justice.' [29] The Indian ruler who best exemplified the Buddhist ideal of kingship was Ashoka (c. 273-232 BC), successor to Chandragupta, who stated 'that it was the remorse and pity aroused in his mind by the horrors of [his conquest of Kalinga] – the killing, death by disease, and forcible carrying away of individuals, to which non-combatants and even peaceable Brahmans and recluses were exposed – that resulted in his conversion [to Buddhism]'. [30]

But these Buddhist ideals did not last, and something like the model of government that had been laid down by the Brahmin Kautilya ultimately prevailed. Kautilya, who was chief adviser or chancellor to the Emperor Chandragupta Maurya in about 300 BC, wrote his *Arthashastra*, basically a treatise on politics and economics, because he wanted to show that government could be made into a genuine science with clear principles and rigorous definitions. [31] He argued that the aim of the state was prosperity, and this depended on the centralization of absolute power in the hands of the king. The essential components of the state were the king, the chief minister, the peasantry and the agricultural base, the fortified city, the treasury, the army, and the ally. Most important, apart from the king, were the agricultural economy and the treasury; this was not because he had the modern idea of increasing the GDP, but because they were essential for supporting a large and well-equipped army to defend the state against invaders and to extend its power by conquest. It was also essential for public works such as irrigation, roads, bridges, and dams, to provide food reserves in time of famine, and to assist the poor and helpless. The state also engaged in production and trade, and controlled the artisans by guilds.

The king also needed an army of spies and informers to acquire as much information as possible to protect the state from plotters (including his own ministers) and from popular rebellion. Kautilya's government therefore required an immense bureaucracy to regulate all aspects of people's lives; this included a clear but drastic penal code that was to be applied mercilessly to wrongdoers. People should be arrested on suspicion, and torture and assassination were legitimate tools of government. There should be no public meetings, and the old tradition of the assembly, the *samiti*, was firmly prohibited, although it had already disappeared from the powerful states by 500 BC. He disliked the merchants, regarding them as thieves and probably considered any large accumulations of private wealth as threats to the power of the state, since he also disapproved of large landowners. But while he was determined that the Brahmins should not exercise political power, he generally supported the caste hierarchy as a bulwark of society.

Kautilya was certainly not a cynical amoralist, but he combined morality with efficient statecraft in terms of enlightened self-interest. He realised that the king could not rule by terror alone and that benevolence, generosity, and mercy were in his own interest because they would earn him the affection and support of his people, especially if it were attacked by another state. The king himself should be a righteous man and educated in the moral qualities of leadership, especially in the control of lust and anger. Kautilya accepted that, in private life, the king like everyone else should abstain from injury to living creatures, be truthful, upright, free from malice, compassionate and forbearing. But to rule a state effectively the king sometimes had to ignore private morality, and he would have entirely agreed with Machiavelli: 'A prince cannot observe all those things which are considered good in men, being often obliged, in order to maintain the state, to act against faith, against charity, against humanity, and against religion'.

This was especially true in foreign relations where every state would necessarily pursue its own interests; he had no idea of a balance of power, so for him the only choice was between conquering or being conquered. A state's immediate neighbours would automatically be the most hostile, so that *their* immediate neighbours were potential allies, on the principle that the enemy of my enemy is my friend. This is simple enough, but Kautilya took it to mathematical extremes: so, if one's own state is 1, in order of distance potential enemies would be 2, 4, 6, 8 etc., while potential allies were 3, 5, 7, 9 etc. While allies were essential, one only kept them as long as they were strong, not out of good will or gratitude, and all alliances and treaties were to be ignored when there was an advantage in doing so. Kautilya wrote extensively on the science of warfare, not only on the strategy and tactics of open warfare, but on guerrilla warfare, and on 'silent' warfare conducted by secret agents, spies, and assassins. But while he advocated killing the rulers of conquered states, he again applied the principle of enlightened self-interest in his policy of being merciful to conquered armies and peoples, because this would make them more compliant in defeat. Not surprisingly, he believed that if Chandragupta could extend his empire over the whole of India this would achieve peace and prosperity for all.

In India the caste hierarchy was defined and enforced with increasing rigidity as a response to the social turmoil of the age: 'The complexity of the new Indian society is clearly reflected in the need for codifying the laws of the various social groups, which is what is aimed at in the Brahminical *dharma–sutras*.' [32] But the caste hierarchy also had an association with wisdom and virtue. The *Laws of Manu* [2.136] say, for example, that learning is the most important basis of respect, then actions [e.g. government], then age, then kinship, and last of all wealth. [33]

In China the conflict between ethical ideals and political efficiency produced disputes about government that were similar in many respects to those between Buddhists and Brahmins in India. Confucius wanted to maintain the ancient Chinese model of society as regulated by the paternalistic family and the norms of kinship, in which people did the right thing spontaneously because that was their role in the system of things, and they were motivated by benevolence. In this system, although criminal law (*fa*) and punishment were necessary, government was inherently paternalistic and the ruler as a good father-figure established the norms of conduct (*li*) for the guidance of all, which promoted virtue and harmony throughout society. Mencius argued that man is naturally compassionate, and gave the example of someone who sees a child about to fall down a well. He will spontaneously reach out to save it through compassion, not because he wants the thanks of the parents or the good opinion of other villagers. The Confucians therefore argued that government was most effective when rulers led by their moral example and by *li* to which the people would naturally respond, and used their enlightened judgement when applying punishment. Publishing clear legal codes would only encourage the people to find ways of evading them and weaken the moral authority of the ruler and his officials. They also accepted the traditional doctrine that rulers held the throne by the Mandate of Heaven, which favoured good rulers but punished the bad by allowing them to be overthrown by violent rebellion.

The Legalists, however, replied that this was ineffectual idealistic waffle: like Thomas Hobbes, they believed that without the state there could only be the war of all against all, with the strong slaughtering the weak. The people were naturally rebellious and disorderly, with no more idea of their real needs than children; they were only motivated by fear of pain and hope of reward, not by moral example, and the only way of governing them was by clear legal codes mercilessly enforced even on the smallest disobedience.

Their theory of government was therefore in many ways very like Kautilya's, but unlike Kautilya it had no moral dimension and was only interested in enhancing state power by fear. The Legalists came to realise, however, that enforcing a penal code by itself was not enough, and in the fourth and third centuries BC in the semi-barbarian state of Ch'in they developed a managerial programme, *shu*. This involved the complete rationalization of the bureaucracy and the army, and concentrating all power in the hands of the king, in very much the same spirit as Kautilya. Agriculture was regarded as the basic source of wealth, on which the army and the security of the state depended; they were as hostile as Kautilya to private initiative, and banned private commerce, useless crafts, and dissident intellectuals like Confucians. The harsh penal code was applied to all, regardless of rank, and even nobles had to earn their position in the state by their performance on the battlefield.

They also realised that a ruler can only carry out his plans if he has complete control over his officials and can be assured of their competence. So the programme of *shu* included an impersonal, objective mechanism for ensuring the ruler's control: officials were appointed and promoted only by merit through tests; the 'job description' of each position was tightly defined and no departure from it was permitted; and the ruler had a range of powerful sanctions ranging from demotion or dismissal to execution.

Warfare was a key aspect of the Legalist programme, as it was for Kautilya. The famous treatise by the Legalist Sun Tzu on the art of warfare begins in a thoroughly rational and deductive spirit by establishing some basic principles. The most important is that war always costs more than peace, and so a ruler should only go to war when he has no alternative methods of weakening the enemy. 'The fundamental question the *Sun Tzu* seeks to respond to is how does the enlightened ruler achieve victory at the minimum cost? The answer is that the ruler must give free reign to the consummate military commander'. [34] To achieve victory at minimum cost, the good commander first has to ensure the integrity of his own forces by a clear chain of command, and rigorous discipline. He always takes the active role, so that the enemy is forced to act according to his plan, which requires speed of manoeuvre, and the avoidance of long campaigns. On these theoretical foundations a whole set of rules of strategy and tactics are then developed.

Han Fei-tzu, the most important teacher of the school, recognised that *fa* (the rigorous penal code), warfare, and *shu*, while essential were still not enough because, in addition to coercion (and reward), people still had to be persuaded to accept state authority as legitimate, and they would not do this if they thought that bad kings could be deposed with the approval of Heaven. He therefore abolished the Mandate of Heaven and substituted the doctrine that the people must be taught to regard the kingship itself as sacred, not the king, so that it could never be justified to overthrow a king, however bad he might be.

Han Fei-tzu became adviser to the young ruler of Ch'in, Ch'in Shih Huang-ti, destined to be the first Emperor of a united China when the state of Ch'in achieved military supremacy in 221 BC, but Han Fei-tzu did not live to see this because by then Ch'in Shih had executed him. Han Fei-tzu would have been even more disappointed to learn that his theory of government soon failed (in 202 BC when the Ch'in dynasty was overthrown), because the rest of China was not prepared to be governed by Legalist principles, and there was a reversion to a modified Confucianism under the Han dynasty.

But the Chinese during this period had had to rethink the nature of a society in which the old distinctions of nobles and commoners were no longer workable: 'Opposition to hereditary privilege is the single theoretical position common to all the philosophical schools of the Warring States era'. [35] As in India, the state was constantly compared to the human body, in which all the different parts had to work together for the common good under the ruler. [36] By the beginning of the second century B.C. the Legalists had developed a new hierarchy of social orders with the scholar bureaucrats (*shih*) at the top, followed by farmers (*nung*), artisans (*kung*), and merchants (*shang*) at the bottom, and this also reflected their relative evaluations of the moral worth of the different orders. The Legalists' aim of building up the economic and military power of the state by an impersonal bureaucratic machine made it 'natural that they, more than other thinkers, should be interested in classifying the population along socio-economic lines.' [37] Although the Legalists fell from power, their hierarchy became firmly established, as did their system of bureaucratic examination.

For the Greeks, the city-state, not powerful centralized monarchies, was the political reality of the day; while Plato, like Confucius, wanted to produce philosopher-statesmen, in reality theories of the state were largely confined to philosophical speculation, with little scope for figures like Kautilya or the Legalists. (Although Aristotle was tutor to Alexander the Great, and is supposed to have written a treatise on kingship for him, it has not survived.) Plato's theory of the state is essentially cooperative, and he presents it as the solution to man's diverse needs which can only be met by co-operation: 'The origin of the city-state is to be found in the fact that we have many needs that we cannot satisfy by ourselves, but need the help of other people, and the state therefore exists to organize this'. [38] His ideal society, as described in the *Republic*, is the best example of the state as a moral organism analogous to the individual, and is very similar to the Indian model, except that the Brahmins are the rulers. [39] Others, particularly the Sophists, argued like Thomas Hobbes that men are only interested in their own security, and ideally would like to inflict all manner of injury on others. Therefore they set up rulers by a social contract who can at least protect us from other people, even though we are no longer able to harm them.

Aristotle, however, said that human society had evolved naturally from simple village life to the city-state of his day, in which the social needs of man were best provided by the law and justice of the state, without which he was actually worse than the beasts. The function of the state was therefore to ensure the welfare of its citizens, physical and moral, and so it should engage in all aspects of their lives – law and justice, education, commerce, defence, and the public worship of the gods. He distinguished between the virtuous minority who could be persuaded to behave well by argument and reason, and the majority who were only influenced by fear of punishment. But his theory of the state was supported by observation, and he collected details of the constitutions of 158 cities, analysing how states develop and decay, and how different forms of government are overthrown.

In his research he paid particular attention to the different social classes in the state, and concluded that democracy is unstable and has an innate tendency towards mob rule, whereas oligarchy is also unstable but has a natural tendency towards tyranny. He argues this particularly on the ground of virtue: both the arrogant rich and the ignorant poor are the least amenable to reason. So he concluded that the best form of government is where a middle class, those with a moderate and adequate property (by which he did *not* mean shopkeepers), neither envying nor being envied, and the most rational section of the population, are the majority of the citizens. [40] (This is his doctrine of the mean transferred from ethics to politics, and it brings us, once again, to the close association between the model of the individual and the state in ancient thought about the social hierarchy.) He also discusses the tyrant – the ruler who has seized power – very much in the spirit of Machiavelli or Kautilya: 'The tyrant must retain the power to govern, or else the tyranny is at an end. But in everything else he should act, or at least appear to act, the part of king. He should be a wise administrator, use taxes for the public benefit, exercise restraint in his personal behaviour, avoid giving personal offence, adorn his city, and show zeal in the public cult of the gods'. [41]

In the ancient world generally, the development of a broad education among the elite added a new dimension to class attitudes, because it raised the question of teaching those qualities of mind, as opposed to wealth and status, that distinguished them from their social inferiors. In Greece the liberal arts – literature, philosophy, grammar, logic and so on – which were appropriate for landowners and educated professions such as law, medicine, and architecture, were considered vastly superior to the mechanical arts of the artisan. By the time of Plato and Aristotle some degree of wealth was regarded as necessary not just for social status but to lead an independent life without subservience to anyone. Nevertheless, the temptations of luxury were morally dangerous, and how that wealth was obtained was also important: its ideal source was landed property, and merchants were still regarded as dishonourable.

Orders of noble and commoner had once existed in Greek society, but in 594 Solon divided the citizens of Athens into four orders, defined by property-holding. The rise of democracy in particular eroded this sort of hierarchy into something more like the class distinctions of our own society. But Philip of Macedon's final conquest of the Greek city-states at Chaeronea in 338 BC, and then by Alexander, brought their independent development to an end.

In Rome the ancient orders of patricians and plebeians had disintegrated under pressure of the new economic circumstances and political expansion, and was replaced during the Republic by the senatorial order, the equestrian order, freemen, freed slaves, and slaves. The Romans, like the Greeks, associated social status with moral worth, distinguishing the Optimates, the 'best people', from the Populares, the masses, and still more from the barbarians outside Roman civilisation. Here, too, education was a crucial factor, because the essence of a civilised person was the ability to control one's animal passions by reason, and this was developed by the full mastery of Latin grammar and rhetoric, and familiarity with literary classics and the moralistic works of writers such as Cicero and Seneca. Only the wealthy could afford this, and when the Republic was replaced by Imperial rule,

> Augustus never dreamt of altering these conditions [of the social orders]; he took them for granted. What he did was to sharpen the edges, to deepen the gulf between the classes and to assign to each its part in the life of the state. If Roman citizens were to be the masters and rulers, each group of them must have its special task in the difficult business of ruling the world-empire. [42]

To sum up, then, the articulate, organic model of the state that developed across the Old World was the birth of political science. People thought very much in terms of a functional hierarchy of social orders, in which the contribution of each order to the good of society was also related to what were thought to be its moral qualities. In practice, there was inevitably a strong hereditary element in such a hierarchy, but there were means of promotion for those with notable ability.

*5. A universal view of man.*

When thinkers were debating the origins of the state they had to discuss the essential nature of man, and what distinguishes him from the animals; this inevitably laid the foundations of a universal view of man that transcended social and cultural differences. [43] The idea of a basic human nature also helped to undermine confidence in the 'naturalness' of inherited rank as the basis of moral worth. Socrates, for example, 'regarded high birth and wealth as irrelevant, and the body in general. The psyche [soul] is our vital part, and its primary feature is the logos [reason]'. [44] Wisdom and virtue were the most important qualities of the soul and provided a new universalistic standard for assessing human beings: 'The vertical classification of men into sages and fools... had as its necessary complement a horizontal grouping of them as kinsmen, it being a matter of no consequence whether they were Greek or barbarian, rich or poor, free or slaves'. [45] This was reinforced by the increasingly relativistic view 'that custom, regarded in each locality as fixed and absolute, is in fact variable from place to place. The thought is expressed several times by Pindar: "Custom is king of all", he says, and "Different peoples have different customs, and each praise what is right as they see it" '. [46]

The Cynics and Stoics, in particular, rejected any obligation to the conventions of their own states, and saw themselves instead as citizens of the world, because there is a fundamental harmony between all wise men that transcends the limitations of political and cultural boundaries, and of social class. Euripides wrote 'Every quarter of the sky is open to the eagle's flight; every country is fatherland for a man of noble mind'. [47] For intellectual Greeks, Alexander's conquests of so many different societies reinforced the idea of the whole known world as our natural home rather than the particular city we come from, and the human race as being one, rather than divided into Greek and barbarian. The idea of a universal moral order that had a higher authority than one's own society was also strengthened by growing awareness of conflicts between the laws of particular states, and between the letter of the law and basic notions of fairness and mercy.

This led the Greeks and Romans in particular to formulate the difference between law and morality, and to the notion of natural law that transcended the laws of particular states. The Greek word for the laws enacted by a state in its code was *nomos*, but *nomos* became increasingly thought of as merely conventional and arbitrary, producing the famous antithesis between *nomos*, codified law, and the law of nature, *physis*. Aristotle distinguished between these as the particular laws of each city-state, and the universal law of natural justice that is common to all, regardless of the society they are living in, and this natural law is eternal and superior to human law. He gives the example of Antigone in Sophocles' play who defies King Kreon's decree that her brother Polyneices' body shall lie unburied because he is a traitor. She buries him because she is his sister, and by natural law a sister's duty to her brother outweighs her duty to obey the order of a king.

Ordinary laws must always be unjust in some individual cases because they have to be framed in general terms and need to be corrected by fairness and mercy, or equity. [48] Equity became of great importance in Roman law, too, which also developed a very similar idea to natural law in their *ius gentium*, the law of nations:

> The traditional origin of this 'natural law' was the increasing residence at Rome of merchants and other foreigners, who were not citizens and therefore not subject to Roman law, and who wished to be judged by their own laws. The best that the Roman jurisconsults could do was to take a kind of lowest common denominator of the usages of all known peoples, and thus attempt to codify what would seem nearest to justice to the greatest number of people. [49]

## 6. Conclusions.

Throughout this chapter we have, of course, been mainly concerned with the ideas of the intellectual elite, not of ordinary people, and we can see that these philosophical ideas about society and the individual have a close resemblance to those features of formal operational thought discussed in Chapter VI. Politics and ethics have become rational sciences. A society can be explicitly thought of as a total system to which its institutions stand in a part-whole relationship; conventions are understood as arbitrary rules that might have been different, adopted for the general good of society, while moral principles are distinguished from custom and law. One's own society can become the subject of criticism, and hypothetical social orders discussed. The individual is distinguished from society, and it becomes possible to think of moral obligations to all human beings, regardless of the society to which they belong. The idea of justice has advanced to the principle of the Golden Rule: doing to others as one would like them to do to us. The idea of the self becomes predominantly defined by psychological and spiritual attributes, and is much more differentiated and integrated; the cognitive functions of the mind are realised, and the self can be the judge of the self. The idea of conscience is fully developed and the mental element in legal responsibility becomes much more important.

This active and creative response by ancient thinkers to new social conditions can, of course, be described as an adaptation to their social environment, like a great deal of thought, but it was conscious adaptation to which Darwinism therefore has no relevance. The strong resemblances between the moral and political ideas of the different civilisations were not the result of random variation and selection, that produced similar outcomes in the competitive process. They appeared because the human mind develops in the same basic way regardless of cultural differences, so that formal operational thinking will have certain common features wherever it occurs. These include the ability to think about systems in an abstract way, to engage in self-consciously logical argument, and to think about thinking and the mind. As we have seen, when this formal operational thought is applied to society, the state, and the individual – to the political and moral sciences – it also discovers that there are a few fundamental problems, and that the ways of resolving these are strictly limited. In these circumstances, then, it is more or less inevitable that similar theories will appear.

We must now see the effects of this new philosophical type of thinking on traditional religion.

# CHAPTER XI
## The New Religions

*1. The structure of the new religions.*

The traditional religions of early states still had the ancient pagan emphasis on success in this life, on victory in battle, and wealth, and fertility and general prosperity. The gods of the state, on whom this prosperity depended, were an integral part of the political regime, and were sustained by blood-sacrifices performed by an hereditary priesthood or its equivalent. Participation in these rituals, and in private domestic rituals, rather than any spiritual states of mind or belief, was sufficient for the religious life of the ordinary person. While there were sacred texts that were the preserve of the priests, there were no creeds, of the kind that we are familiar with, in which the basic doctrines are set out for the believer, and no idea of converting the members of other societies to the worship of one's own gods. (That, if it happened, was the result of conquest, not of the inner convictions of individuals, or of some belief in absolute truth.) The gods punished evil-doers in this world, and there was generally little idea that the soul of each individual would be judged in the next life for his or her moral failings in this one. Indeed, while people believed that some part of the person survived and went to the abode of the dead, the notion of the soul as we know it had not really developed.

But we saw that the philosophers of the Axial Age had come to think of the state as devised by human beings for the common good, not as part of the cosmos, and this meant that the gods could no longer be an integral part of the political regime, as they had been in all the archaic states. These crudely human pagan gods also seemed increasingly absurd to the intellectual elite, who searched in various ways for some universal source of cosmic order instead of the multiple gods of archaic religion. (We shall come back to this in section 2.)

The thinkers of the Axial Age also rejected the traditional values of archaic religion:

> For them the divine no longer dwelt in the manifestation of power, wealth, and external glory the [Chinese] description of those evil tendencies which impede the achievement of the good is strikingly similar to the diagnoses made by prophets, wise men, and philosophers in all the high civilizations of this period. The unbridled pursuit of wealth, power, fame, sensual passion, arrogance and pride – these themes figure centrally as the source of 'the difficulty'. [1]

A new distinction between the spiritual and the material had therefore grown up, that stressed the radical inferiority of the body to the mind and especially the soul: one found tranquillity of mind and salvation through moral purity associated with world-renunciation and the rejection of luxury and self-indulgence. In Buddhism, Hinduism, Taoism, and later in Christianity the highest path of all was withdrawal from the world entirely, into an ascetic life of self-denial.

The soul now came to be seen not as some crude vital essence, but as the highest, sometimes the most intellectual, part of the person, whose moral perfection was the route to salvation. If the most essential quality of a God is his directing mind, there is an obvious similarity here to man, who alone of all creatures has reason and can impose order on society and the natural world, though obviously to a vastly smaller degree. All over the ancient world, it was believed that there was indeed a unique association between God and man because of this shared mental and rational element in man, and this had profound implications for the development of religion. The soul was not only immortal, but actually shared in the divine, so its aim was union with the divine after death, but it could also draw closer in this life by ascetic practices and by prayer and meditation. On the other hand, punishment after death for a sinful life in this world also became a standard belief.

Here I need to draw attention to a very significant change in religious experience. From the most ancient times, it seems that a variety of techniques, such as hallucinogenic plants, drumming, and dancing, had been used by shamans to produce trance states in curing rituals, while dreams and vision quests relating to supernatural beings are also reported in many tribal cultures. While all these techniques for producing altered states of consciousness persisted, and have continued into modern society, the growing awareness of the inner self was also accompanied, in all the major religions, by new sorts of spiritual exercises in which the divine is approached by meditation and prayer, leading to mystical experience of a kind that is not reported in archaic or tribal religion.

The new type of religion was not only profoundly individualistic, offering salvation for the soul of the believer, but also created the possibility of new forms of social conflict, in which for the first time 'religion' emerges as something distinct from society in the modern manner. The great teachers all had to get their message of truth accepted, since they did not speak from any prior position of authority, but emerged in an atmosphere of controversy, criticising accepted ideas, and this meant that they had to convince enough of their contemporaries for their ideas to survive. Religion was now a system of explicit beliefs that were held to be true, rather than implicit assumptions, and because it developed in reaction to traditional religion we find the development of creeds, sets of doctrines that could bind believers together, regardless of their social identity, and this had very important consequences. Apart from the new possibilities of converting foreign unbelievers, which we shall come to later, it also created new loyalties and new divisions within societies. Because religion was now distinct from society and the state, and its doctrines were claimed to be true in an absolute sense, it provided a formidable basis for a new type of religious elite to criticise rulers, and

> ...brought a new level of tension and a new possibility of conflict and change onto the social scene. Whether the confrontation was between Israelite prophet and king, Islamic *ulama* and sultan, Christian Pope and Emperor, or even between Confucian scholar and his ruler, it implied that political acts could be judged in terms of standards that the political authorities could not finally control. [2]

But while philosophy and debate [3] were essential factors in the new religions, ordinary people were not interested in theological discussions, and the pure theory generated by reason was quite inadequate as the basis of any religion that was to have general appeal. We can see this very clearly in the case of Greece, where philosophical and ethical schools such as the Stoics and Epicureans were confined to the intellectual elite:

> What Jacob Burckhardt said of nineteenth century religion, that it was 'rationalism for the few and magic for the many' might on the whole be said of Greek religion from the late fifth century onwards . . . As the intellectuals withdrew further into a world of their own, the popular mind was left increasingly defenceless . . . The loosening of the ties of civic religion began to set men free to choose their own gods, instead of simply worshipping as their fathers had done; and, left without guidance, a growing number relapsed with a sigh of relief into the pleasures and comforts of the primitive. [Such as magical healing, and 'foreign cults, mostly of a highly emotional, "orgiastic" kind, which developed with surprising suddenness during the Peloponnesian War'.] [4]

Orphism, however, appealed to a wider range of people in a more universal manner. It was supposed to have been founded by the legendary musician Orpheus, and seems to have begun in Thrace, reaching Greece in the fifth century BC. Besides a theology and written scriptures, and a set of doctrines of salvation, it also had the rich appeal of myths, rituals, and common rules of life for the community of believers:

> The soul, it seems, to the Orphics (and this is a new idea in the Greek world) was an immortal god imprisoned in the body and doomed, unless released by following the Orphic way of life, to go round the wheel of reincarnation in an endless succession of lives, animal and human (so that all living things are akin, and to kill an animal is to murder one of one's own family). By ritual purifications, by an ascetic life of which the most important feature was abstinence from animal flesh, and by knowledge of the correct magic formulae to use on the journey after death, the Orphics hoped to win release from the body and return to the company of the gods. The next world was to them more real and important than this, a place of joy for the blessed initiates and of torment for those who were not of the company of the elect. This other-worldliness and the ascetic life which went with it were very different from normal Greek beliefs and religious practice and had an effect of the greatest importance on later Greek philosophy and religion. [5]

Many of these features of Orphism are familiar to us because they were also found in those movements that became the universal or world religions, initially Buddhism, and then Christianity and eventually Islam. They, too, had a strong missionary aspect, and sought their converts primarily in the towns, but among the people as a whole, not just the intellectual elite. To do this they had to promise individual salvation and integrate this with a communal life and a new social identity provided by ritual, special codes of conduct, and especially by myths and stories, which are extremely useful in this respect, as stories can be interpreted at many different levels of intellectual sophistication.

Another striking element in the universal appeal of the new religions was their frequent stress, in one way or another, on the benevolence of God or Heaven to ordinary people, who are called to love God in return.

World religions typically had an individual founder, with a non-hereditary group of experts at the core, interpreting the basic written texts in which the belief system was expressed and transmitted. By setting out to convert people as individuals, regardless of nation, and so not connected with hereditary social groups such as clans or castes, they were quite different in this respect from the priesthoods of the ancient literate civilisations because they were communities of *believers*. Belief systems need creeds, in the sense of clear summaries of doctrines, and this was again quite different from the state cults of archaic religion, where it was the cult and its ritual that was central, and one was a 'member' of the religion because of one's social identity and not because of one's personal beliefs, whatever they might be. Since the creeds proclaimed by the new religions were *true* for everybody, this made the attempt to convert foreigners at least reasonable, if not inevitable.

Not all the potentially universal religions were in fact interested in converting other peoples, because they did not effectively break this link between creed and society. Zoroastrianism was absorbed into the institutions of the Persian Empire; and in Israel the universal vision of Jeremiah, Second Isaiah and the other major prophets, who saw Israel as 'a light to lighten the Gentiles' was more or less extinguished by the subsequent struggles of the Jewish people to maintain their independence against the Hellenistic and Roman Empires. Only with Christianity were its most important ideas included in a universal religion. Hinduism spread very widely by migration throughout southeast Asia, but the intimate association between its doctrines and the caste structure clearly limited its ability to convert members of alien societies.

Early religions, including Buddhism and Christianity, were first spread by missionary activity, but once kings could be converted a new relationship was set up between religion and the state, whose rulers still needed religious legitimation and took advantage of the new religions to provide this. This was one obvious reason why Constantine declared Christianity the official religion of the Roman Empire, and why we later find the Russians and the Mongols deliberately choosing between world religions in their search for an official state religion too. But as we saw, the new religions, even though used to legitimate the rule of kings, were now in a different and potentially critical relation with them from the priesthood of archaic states. 'Even though notions of divine kingship linger on for a very long time in various compromise forms, it is no longer possible for a divine king to monopolise religious worship.' [6]

War was another means of converting unbelievers, but its use varied greatly from one religion to another. Buddhism relied on it least of all, Christianity to some extent, while Islam stands out from all the other religions, from the very beginning, by its reliance on armed conquest, and holy war against unbelievers, to spread the message of the Prophet, the only major religious teacher to lead his followers in battle. Indeed, this war against non-Arabs was a basic strategy to unite the disparate tribes of Arabia under the banner of Islam. [7] But whether they spread their message by force or by missionary activity, the world religions are the first examples of ideological competition in history. The religious wars that resulted between Muslims and Hindus, or between Muslims and Christians, and later between Catholics and Protestants have created the impression, however, that religion has a unique ability to stir up human belligerence. But this was merely because they were the first examples of universal ideologies to appear, and the French Revolution, Communism and Nazism have subsequently shown us that political ideologies based on equality, atheism, nationalism, or Social Darwinism can be responsible for even more violence in 'holy' wars of a different kind. It is not religion in itself, then, but a fanatical adherence to ideology of *any* kind, regardless of other human considerations, that has such potential for violence and warfare.

The world religions, like the state, have clearly been very successful institutions, but is this because they were better adapted than the archaic religions they replaced? Here again, we have to distinguish between institutions and people. I said earlier that religion is how people find meaning and order in the world around them, and the new forms of religion were adaptations by which they could make sense of the new social conditions of the Axial Age, and attain peace of mind. This could be done in very different ways, from the consolations of philosophy to personal experience of the divine in prayer and meditation, though we must also recognise that, then as now, there were many people who were largely indifferent to all this, and simply followed the dominant fashions, or hoped that religious observance would bring them good luck. *Institutionally*, these religions were successful for the quite different reasons that, unlike the archaic religions, they had the intellectual depth to gain the support of the educated classes, and secondly, they could provide legitimacy and support for the king or emperor, who in turn could support them. Thirdly, they were not limited to any one society, and were therefore able to spread very widely, which made it impossible for any ruler to close them down.

*2. God, Man, and Natural Law.*

All this shows, yet again, that 'religion' is not a single stage in the development of thought, but itself evolves as part of the general increase in human cognitive abilities from intuitive to formal operational thinking. Notions of cosmic order were always as important as beliefs in supernatural beings, and in the new type of religion order was of special importance. My next aim in this chapter is therefore to show how the central importance of cosmic order helped to lay the foundations for early scientific speculation. In the archaic religions, the gods had behaved as they did because it was their nature, or because it pleased them to do so, and the Greek myths are a good example of archaic thinking about the gods and the origins of the world. According to these myths, Heaven and Earth had given birth to the Titans, the youngest of whom, Kronos, with the encouragement of his mother Earth, had castrated his father who then went far away from Earth. Kronos married his sister Rhea and had a number of children by her, but swallowed all his sons except Zeus, whom his mother hid and gave Kronos a stone instead. Later, Earth managed to get Kronos to vomit up his other sons who defeated him in battle, leaving Zeus to reign on Mount Olympus over a large number of subordinate gods and goddesses. These engaged in the kinds of activities that nowadays we associate with the 'celebrities' of football and the entertainment world.

The new breed of Greek intellectuals regarded all this as contemptible rubbish, and as the Ionian Xenophanes (c.545 BC) put it, 'Homer and Hesiod have attributed to the gods everything that brings shame and reproach among men: theft, adultery, and fraud…Mortal men imagine that gods are begotten, and that they have human dress and speech and shape…If oxen or horses or lions had hands to draw with and to make works of art with as men do, then horses would draw the form of gods like horses, oxen like oxen…' Instead of this naively human image of the gods, said Xenophanes, 'One god there is…in no way like mortal creatures either in bodily form or in the thought of his mind… effectively, he wields all things by the thought of his mind.'

As the ideas of Xenophanes illustrate, there was now a new awareness of God as a rational being who sustained the cosmic order by divine law. I am using 'God' here as a convenient and very general way of referring to a single supreme 'entity' of some kind for explaining the cosmic order and why everything is as it is, and the Creator God of the Old Testament is simply one example of this. Once we start thinking in terms of a single universal entity of this sort we are no longer in the world of ancient polytheism where, in a sense, 'anything goes', but in a far more constrained intellectual landscape, rather like the relatively few possible ways of organizing the state. We should first remember, in trying to understand to what extent these Gods should be regarded as persons, that it is extremely difficult to separate, even in thought, the idea of an orderly system from conscious purpose and design, since the two are linked so closely in our own experience. Even today, orthodox scientific biologists, when explaining some ingenious adaptation, frequently talk about what 'nature' has designed, as though it were some God-like being with conscious purposes. In ancient cosmology, therefore, it is not surprising that thinkers found it almost impossible to separate the idea of order from consciousness, and therefore from some Being who is the conscious source of that order.

Just as there were a few fundamental issues in the design of the state, so too we find that the idea of a unified entity regulating the cosmos also involves a few central problems. In the first place, matter, at least, may always have existed and so it is not absolutely necessary to have a Creator such as the God of Genesis, conjuring up the heavens and the earth out of nothing. So the essential function of a God might have been to impose order on matter in place of chaos, rather than producing the matter itself. Secondly, for ancient thinkers, the obvious order of the cosmos implied some co-ordinating power akin to mind or reason, and this raises the question of whether such a mind or co-ordinating power is in matter itself, or is something entirely different from matter. A Creator God, more or less by definition, has to be entirely different from the cosmos he creates, outside it and 'transcendent'. At the other extreme, if God is the rational, co-ordinating power within matter then he is an 'immanent' or 'pantheistic' God. But matter is then no longer the inanimate, senseless 'stuff' that we think of, but has an innate rationality, and in its highest and most refined forms such as man is capable of thought and reason. God and Nature are then one, as the Stoics believed.

Order itself is not enough, however, and there also has to be some dynamic principle to make things actually happen and keep the whole system moving. Whatever this is, it must be self-sufficient, some sort of First Cause, to avoid the infinite regress of 'What caused the First Cause, and what was the cause of *that*, and so on?' But it is possible to argue that the order of the cosmos has always existed, like matter, so that the First Cause is then simply what keeps the system working in the present, and not some historical event at the beginning of time.

Finally, can God, whether transcendent or immanent, violate the basic principles of order – for example, can he ordain that two plus two equals five, or that good is bad and bad is good? In other words, are there absolute standards of reason and goodness to which a perfect God must conform, and does this limit his omnipotence, or does omnipotence allow him to do anything he likes? And if God is perfect and also omnipotent, how then are we to account for death, disease, destruction, and evil in general? Some, such as Aristotle and especially the Stoics, argued that these are illusions, and that nature orders everything for the best, but others argued for various forms of dualism, a tension between opposite forces of good and evil that are held in precarious balance.

Universal order of the sort we have been discussing is obviously very similar to the natural law that in the previous chapter we saw developing in Greece and Rome. The Stoics, for example, defined the universe as a well-ordered state, 'City of Men and Gods', and treated the order of nature as the result of universal law, *koinos nomos*. A very important aspect of universal law is what we may call its 'intelligibility': since the cosmos is permeated by a rational order, it can be comprehensible to man as a thinking being. The rational nature of God therefore involved the very important idea that the essential and distinctive mental element in man is akin to the creative and ordering element in the cosmos, of man as microcosm in relation to the macrocosm.

So the world religions had the same tendency to the rationalization, simplification, and unification that occurred in thinking about the state. This is not to be explained by some crude model in which thought merely reflects the social structure, and monotheism is the mirror-image of empire. The Chinese, the best example of empire, did not have the idea of a Creator God in the Judaeo-Christian sense at all; the Romans were polytheists for most of their history; the Zoroastrian God of the Persian Empire was in a dualistic relation with an evil deity, while in ancient Israel, which by the time of the Axial Age had produced the best example of a transcendent Creator, the whole idea of monarchy was problematical, and kings were subject to the claims both of the Mosaic law of justice, and of the prophets. It was debate on a range of social, ethical, and religious issues that developed those abstract intellectual skills of formal operational, philosophical thinking, which led to the broadly similar results in social and religious thought we have seen. Nor was monotheism a simple, random mutation from polytheism: it was not random, but the product of a new style of thought that arose from a combination of factors in some unrelated societies; and it was not simple, but involved a complex world view.

Having sketched out the basic intellectual landscape, we can now look in more detail at three of these cosmic, unifying principles and how they were related to man: For the Greeks this was the Logos, for the Indians Brahman, and for the Chinese Tao.

The notion of Logos was central to the new Greek idea of the universe. It meant not only 'word', but 'language' and by extension 'reason', 'proportion' and 'order'. Heraclitus (c.500 BC) seems to have been the first thinker to make extensive use of Logos as a philosophical idea. He regarded the cosmos as governed by a permanent tension between opposites, like the Yin and Yang of the Chinese, and the Logos is the basic law of proportion and order which produces harmony between these opposites. The Logos doctrine was later to have a profound impact on Stoicism. [8] For the Stoics the Logos was God, a living all-ruling intelligence, and this process of cosmic regularity was therefore not automatic and impersonal: 'It was not natural to Greek thinkers of any period to suppose that what was self-moving and law-like in behaviour was dead or mindless', [9] and cosmic intelligence is basically the same intelligence that exists in men. While Aristotle did not make explicit use of the Logos doctrine, he nevertheless treated the world as inherently rational and knowable, and the human mind as an integral part of it, not as something standing mysteriously apart from an unconscious and mindless process. [10] For the Greeks, especially after Plato, the soul was the rational element of man, through which alone it was possible to attain true knowledge. [11]

In India the concept of Bráhman was evolved, which was very similar to that of Logos – an eternal foundation of cosmic order and law. The original meaning of the word is much disputed; one of the most favoured is 'the sacred utterance of the priest', embodied in the Vedas, and this would of course be very similar to Logos, and to the creative Word of God in Genesis. Another related explanation is that it developed in from 'the notion of "ceremonial form" in the behaviour of the priest who makes the offering and in the operations of the sacrifice'. [12] What might seem the very limited notion of ritual order actually had a universal significance in Indian thought, because it was believed that the rituals of sacrifice could themselves force the gods to do the will of the priests: '. . .the growth of sacrifices helped to establish the unalterable nature of the law by which the (sacrificial) actions produced their effects of themselves'. [13] Ritual law thus became Cosmic Law in the hands of the philosophers:

> Looking at the advancement of thought in the Rg-Veda we find first that a fabric of thought was gradually growing which not only looked upon the universe as a correlation of parts or a construction made of them, but sought to explain it as having emanated from one great being [Brahman] who is sometimes described as one with the universe and surpassing it, and at other times as being separate from it. [14]

In the Upanishads, the final form is given to the idea that the *atman*, the soul, self, or essence of man, is identical with Bráhman: 'The fundamental idea which runs through the early Upanishads is that underlying the exterior world of change there is an unchangeable reality which is identical with that which underlies the essence in man'. [15]

In China the equivalent idea was the Tao, which literally means 'way' or 'course of things', and this idea was extended to the cosmos, becoming an extremely general principle of universal order, transcending even Heaven, which had a personal quality of moral concern entirely lacking in that of Tao. Joseph Needham notes the similarity of the Tao to the Logos of Heraclitus 'controlling the orderly process of change', [16] and the Tao, like Logos and Bráhman, is one: it is 'the Supreme Oneness' and all things meet in it. The Tao was not a Creator standing outside the cosmos creating it out of nothing, but spontaneously generated the order of the cosmos. In this process *li*, meaning pattern/order/organization and therefore something like reason, interacted with *ch'i*, matter/energy. The expansion of *ch'i*, *yang*, and its contraction, *yin*, had brought light out of darkness, and order out of chaos, and generated all the complexities of the universe in accordance with the Tao. 'Under', as it were, the Tao was Heaven, which noted the deeds of men, and came closer to our idea of God; the Mohists, indeed, believed that it was not only benevolent but willed that men should love one another without regard for social status. This was an extreme view, and while Confucius believed that Heaven was benevolent and concerned with human affairs, other philosophers denied even that.

China, like other societies, from an early date had thought of Heaven as giving commands to men. But by our period the dominant notion of the cosmos was, in Joseph Needham's term, 'organic': 'The harmonious co-operation of all beings arose, not from the orders of a superior authority external to themselves, but from the fact that they were all parts in a hierarchy of wholes forming a cosmic pattern, and what they obeyed were the internal dictates of their own nature'. [17] While there was therefore a conviction that there was an order in nature, this did not *necessarily* lead to the belief that human beings were therefore capable of explaining what that order was in their limited languages. 'The Taoists, indeed, would have scorned such an idea as being too naïve for the subtlety and complexity of the universe as they intuited it. Human rational personal beings had another faith: the universal order was intelligible because they themselves had been produced by it. . . "Heaven, Earth, and man have the same Li" '. [18] The Tao, like Bráhman and Logos, therefore manifests itself in the *te*, 'virtue' of each individual, and for the Taoists and Confucians, like the Stoics, 'There seems to have been a common belief that acting naturally is concurrently a means for uniting the self with some greater or comprehensive unity; in fact the good life involved a conscious recognition of the need for this union'. [19]

While the Old Testament contains archaic material going back to the second millennium BC, its finished form was the result of a radical rethinking of the religion of Israel, especially from the sixth century BC, during the exile in Babylon under the Babylonians and Persians. Its most notable achievement was the vision of the transcendent Creator from which, of course, Christianity and Islam both derive; a vision that was prophetic and poetic, not philosophical at all, but which could nevertheless be related to the Logos through the Hebrew notion of Hokhma, wisdom or reason. This, like Logos, Bráhman, and Tao, was common both to the One and to man, and 'Wisdom in the religious sense also is an intellectual quality which produces the key to happiness and success, to "life" in its widest sense'. [20] Isaiah 31:2 is the earliest dateable statement in the Old Testament that Yahweh is wise, and makes it clear that 'the kind of wisdom which is attributed to Him is essentially the same as human wisdom in its quite general sense' [21], except that God, of course, has wisdom in an infinitely greater measure.

> I [wisdom] was set up from everlasting, from the beginning,
> Before the earth was . . .
> When he established the heavens I was there . . .
> When he marked out the foundations of the earth;
> There I was by him as a master-workman
> (Proverbs 8:22-31)

Philo, a Hellenistic Jewish philosopher of Alexandria, and who wrote in the first century AD, was a significant figure in the history of Western thought, because he produced a synthesis of Hebrew ideas, especially on Hokhma, with the Greek philosophy of the Logos as a system of unchanging causes, that later had a profound effect on Christian thought:

> Philo used the term *logos* for the principles on which God modelled his creation, like a city fashioned within the mind of an architect, from which followed with invariable regularity all the operations of this universe. But God could overrule these regularities just as he could have created another kind of universe had he so chosen. . .The most pervasive route through which these ideas passed into Latin medieval thought was Augustine [354-430 AD], for whom the *naturales leges* which God had ordained were the laws of measures, numbers and weights. He applied the concept of natural laws, or laws of nature, to the motions of the heavenly bodies, the generation of living things, and the development of the world itself pregnant with things to come. God could then be discovered in the great open book of nature, as well as in the revealed book of Holy Scripture. These ideas were to become fundamental principles of Western medieval and early modern natural philosophy. [22]

*3. Religion and science.*

This therefore challenges the prevalent modern view that throughout history there has been an inherent conflict between religion and science, between 'superstition' and 'reason', as very simplistic. There are in fact many strands in religious thought; some of these are clearly hostile to the scientific study of nature, but others are favourable, and it is also essential to distinguish between religion as an official Church, or equivalent body, and as the personal faith of individuals.

But before we look further at the range of possible relations between religion and the scientific study of nature, it is very important to note that there has also been what we can call a 'humanistic' contempt for science in most civilisations. According to this, man and how he should

*The New Religions*

live, literature, beauty and the arts are all far more important and interesting than pedantic and useless curiosity about the natural world. This attitude can be found in Socrates, Confucius, and the humanists of the Renaissance, for instance, and Swift satirised the Royal Society as the crazed Academy of Lagado in Gulliver's Travels, trying to extract sunbeams from cucumbers, for example. Indeed, the mad scientist in his laboratory is still ridiculed, and only tolerated and given funds because his research is thought to be useful in improving our standard of living.

Religious explanations of nature are most obviously irrational and anti-scientific when they simply appeal to the nature or will of a deity, when, for example, people 'explain' an eclipse of the moon by saying that it is being eaten by a demon. 'But why does the demon eat the moon?' 'Because that's what demons do'. 'Why does water expand when it freezes?' 'Because that is God's will.' No sense can be made of statements like this, which simply 'explain' one unknown by another. Religious traditions that emphasise the omnipotence of God at the expense of His rationality clearly fall into this category.

Fear of offending the gods could also discourage scientific speculation about the cosmos. Even in relatively free-thinking Athens, philosophers were not immune to prosecution on religious grounds, if their theories were thought disrespectful to the traditional gods of the city-state: 'About 432 BC or a year or two later, disbelief in the supernatural and the teaching of astronomy [Anaxagoras, who said the sun was a flaming rock] were made indictable offences. The next thirty-odd years witnessed a series of heresy trials which is unique in Athenian history.' [23] A major factor here was wartime hysteria, when Athens was losing against Sparta.

Some Indian thinkers, especially the Buddhists, thought that the picture of the physical world given by our senses is an illusion, *maya*, so that studying it could only be a waste of time:

> how could a mere phantasmagoria invite serious scientific study? How could the mentality which averted the eyes from it, and which sought salvation in eternal release from it, encourage the investigation of it? . . . Alas, the 'World', for the Buddhists, was not only 'the world, the flesh, and the devil', but the world of Nature itself. . . Buddhism was not interested in co-ordinating and interpreting experience, or finding reality in the fullest and most harmonious statement of the facts of experience, but in seeking some kind of 'reality' behind the phenomenal world, and then brushing the latter away as a useless curtain. [24]

This profound devaluation of the whole of material existence, by comparison with the spiritual, could produce in all religions a 'holy ignorance' that stifled intellectual enquiry into nature. Some took the view that, while the natural world is not an illusion, it is merely the temporary abode of the soul, a place of trial and temptation to be endured for a brief period, so that by comparison with eternity it is trivial and not worth serious attention. A more hostile view of the study of nature was that trying to understand its mysteries was not just idle curiosity that led to the sin of pride, but positively impious. 'To pry into the mysteries of nature that God chose not to reveal . . . was to transgress the boundary of legitimate intellectual inquiry, to challenge God's majesty, and to enter into the territory of forbidden knowledge.' [25]

The written word of scripture has presented its own problems, too. Lloyd George once said that to the military mind, thinking is a form of mutiny, and this is often true of the religious mind when it interprets sacred texts literally. Lactantius, tutor of the Emperor Constantine's son, ridiculed the Greek notion that the earth is round because the Bible clearly says it is flat, while Protestant fundamentalists from the

Reformation to the present day have demanded a literal interpretation of the Creation in the Book of Genesis. In the nineteenth century, Victorian fundamentalists objected to umbrellas because God said that the rain should fall on the just and the unjust, and opposed anaesthesia for women in childbirth because God had said 'In sorrow shall you bring forth children'.

There was also the tendency for religious leaders to regard secular explanations of the natural world as a challenge to their own intellectual authority. This raises the distinction between religion as a set of ideas, and religion as a social institution, and we shall see that until the nineteenth century European scientists themselves were mostly devout Christians; the essential struggle was not so much between science and *religion* as between scientists and the authority of the Church.

Again, once an organized religion supports the legitimacy of the ruler, and becomes a pillar of the state, the ruler himself may become a defender of religious orthodoxy, as when, for example, the Christian Emperor Justinian closed the philosophical schools at Athens for heresy in 529 AD. The Jesuit missionary Matteo Ricci, in early seventeenth century China, had his books on mathematics confiscated by the authorities, because mathematics and astronomy were imperial monopolies, centred on the Emperor's religious responsibilities for rectifying the calendar.

Even if there was religious interest in nature, as in the early Middle Ages, this might only consist in finding symbolic meanings for the divine. For example,

> The pelican, which was believed to nourish its young with its own blood, was the analogue of Christ who feeds mankind with his own blood. In such a world there was no thought of hiding behind a clump of reeds actually to observe the habits of a pelican. There would have been no point in it. Once one had grasped the spiritual meaning of the pelican, one lost interest in individual pelicans. [26]

On the other hand, there were a number of reasons why religion could foster serious scientific enquiry. In the first place, the study of the heavenly bodies and the calendar, as we saw, had been an integral aspect of religion from very early times, and laid the essential foundations of astronomy. More generally, ancient religions were very interested in how the cosmos was formed by the gods. How things began, the emergence of the first humans, and so on, are standard themes in the myths of tribal societies and the ancient literate civilisations. These creation myths were therefore important sources from which the earliest rational speculation about the nature of things could develop.

But undoubtedly, the most important stimulus here came from those notions of Logos, Bráhman, or Tao, the whole idea that the universe makes sense at some deep level, and that it is governed by a unified body of rational laws given by a divine Creator. This has been an essential belief for the development of natural science, and unless the Greeks, in particular, had been convinced of this they would never have persevered in the serious investigations of nature that they did, and the same is true of medieval and Renaissance science. As Joseph Needham says, '. . . historically the question remains whether natural science could ever have reached its present stage of development without passing through a "theological stage" ', [27] that is, of a rational Creator giving laws to the natural world as well as to man, and which man could understand. The idea that the scientific study of the natural world is to study the mind of God was an extremely important motivation for genuinely scientific studies until well into the nineteenth century, and still survives.

*4. Conclusions.*

Many of the ethical, political, and religious systems of thought produced in the ancient world achieved some powerful insights into the human condition that are still valid today. But the political and moral sciences, like logic and mathematics, were relatively easy applications of formal operational reasoning. It would prove far more difficult to apply it to the natural world and this brings us to the final phase of the book, in which we will see what were the necessary conditions for the development of modern experimental science, and how it took place.

# CHAPTER XII
## Natural Philosophy

These final chapters will be especially concerned with the intellectual and social roots of modern science, because this was essential to every major feature of the industrial revolution – steam power, electricity, the chemical industry, and ultimately atomic energy. Explaining how this kind of mastery over nature developed should therefore be an excellent opportunity for materialist theories to prove their worth. If they are true, we shall find that natural science is linked to the modes of production and technology, and develops in close relation to them. The evolution of the natural sciences is also a test case for the Darwinian model of variation and selection. We should find that in the process of competition between different theories, these are constantly being tested against the facts, so that false theories are steadily eliminated, and true theories selected in their place.

In this chapter we shall see how some highly ingenious thinkers attempted to apply philosophical, formal operational, reasoning to the physical world. This 'natural philosophy' was the earliest science, but talking of ancient science will strike some readers as a contradiction in terms. For them, real science is modern science, which only began when Galileo and his contemporaries discovered the power of experimental method at the beginning of the seventeenth century. But ancient science was the attempt to give rational and systematic explanations for why the physical world is as it is, that was quite self-consciously different from traditional mythical explanations, and a fundamental turning point in the history of human thought. To say that water expands when it freezes because that is God's will is no explanation at all, because God, by definition, is unknowable, and His will is a mystery. A rational explanation has to be something knowable to us, and that means in terms of our own reason and experience, not the arbitrary will of a deity. Natural philosophy, therefore, from classical times until the seventeenth century, was always the search for 'the reason why', some rational explanation in terms of cause and effect by which we attain true knowledge, *scientia*, about the cosmos and the natural world.

## 1. Why were ancient philosophers interested in nature?

We know that understanding how nature works allows us to control it more effectively through technology, and improve our material conditions of life, but this is a fairly modern idea which can tell us nothing about the origins of natural philosophy. In the ancient world, unlike our own, theories about nature were irrelevant to technology, [1] and Greek and Roman natural philosophers had not the slightest intention of being useful. Indeed, a strong belief running from Plato to Seneca held that the search for Truth was a noble quest which should not be tainted by sordid materialism. The high cultures of antiquity, in any case, were overwhelmingly literary and moralistic in their interests: most thinkers were far more concerned about man and society, and with how we should live, than with the nature of light, or how the lever works, or why things float. [2] In later antiquity, however, the development of astrology, alchemy and other arcane or magical sciences encouraged the belief that penetrating the secrets of nature could be the source of extraordinary powers.

In this chapter, and the next, it will be necessary to concentrate mainly on the Greeks. Although I shall occasionally refer to significant developments in India [3] and China, the majority of the important contributions came from the Greeks, and as a matter of history it was Greek science that was the main intellectual stimulus of modern science.

The earliest natural philosophy in Greece, as elsewhere, developed out of an ancient interest in how things began, the birth of the world, the separation of earth and sky, and of day and night, and the fate of the soul after death. Religion, philosophy, and the earliest science were mingled together, and the first natural philosophers were often sage-like figures around whom legends tended to accumulate, and who expressed themselves poetically, not in the cold, analytical prose of Aristotle. For the early Greek philosophers who were interested in how the universe worked, there were two main strands of thought, one of which focused on what things were made of, their matter, the other on their mathematical form or structure. (We shall see what this involved in section 3.)

*2. Matter: elements and atoms.*

The 'Pre-Socratic' philosophers in the sixth century BC, such as Thales of Miletus, wanted to find a simple, general theory to explain what the world is made of, and how it works, but they had nothing to base it on except familiar experience. So they began by applying imagination and some extremely ingenious arguments to the simplest features of daily life in a farming society, such as earth, water, fire, wood, and air or wind. These 'elements', not surprisingly, occur in India and China, as well as Greece, [4] but in the course of philosophical speculation they lost their simple physical meanings, and became rather more like what we now think of as the different states of matter – solid, liquid, and gas. So 'water' came to include metals because they could be melted; 'air' meant any sort of vapour or gas; and 'earth' included a wide range of solids.

But how could only four or five elements explain all the variety of nature? Some proposed that one element, such as water or air, might be fundamental and then become transformed into the others, while another theory was that there was an even more basic matter, the Unbounded, which could be transformed into the elements. Again, the elements might form all the different objects by combining together in different proportions. [5] Secondly, and just as important, how was change brought about? One theory was that the principles of the hot, the cold, the wet, and the dry could operate on the Unbounded, and these opposites were derived from ordinary experience of the seasons and the weather. Like the seasons, no opposite could dominate the others permanently, but operated cyclically in a balanced way, [6] and the relation of opposites was a basic feature of Greek natural philosophy that we shall encounter again and again. Since all change is cyclical, matter can never be destroyed, and must therefore always have existed (another fundamental idea), and has always been in motion.

To the hot and the cold, the wet and the dry as agents of change, could be added the factors of condensation and rarefaction, and this idea, too, was based on simple, natural observations of rainfall. 'Air' – the vapour of clouds – condenses to form water, which in turn condenses to form the solid, ice; and conversely, when water is boiled or evaporated, it turns back into 'air'. [7] Fire, then, is very rarefied air, water is in the middle, while very condensed 'air' forms the solids or 'earth'. But while these theories were inspired by basic features of ordinary experience like farming and the weather, they were essentially metaphorical and imaginative, and not supported by any clear evidence. Instead, the Greeks took the path of logical argument in which they were highly skilled.

So Parmenides (born c.515 BC) of Elea challenged the whole idea of change itself, claiming it was an illusion of the senses. How, he argued, could any real change occur if there was only a single basic kind of matter such as water or air, or the Unbounded, that underlay all apparent change? He was the first philosopher really to defend his theories by powerful logical arguments, and to raise the great question of whether reason or the senses is the better guide to truth. So he used logic in a new argument for the permanence of matter. Reason showed us, he said, that Being necessarily has to exist, while non-Being by its very nature cannot exist; therefore nothing can come into being from non-being or pass away into non-being, and not only is change (and motion [8]) illusory, but the cosmos is also one, continuous and indivisible. He defended this by arguing that once one accepts the possibility of division, there is no logical reason why it should ever stop: the result would therefore be an infinite number of bodies which either still had *some* mass, so that the cosmos would be infinitely large; or, that their mass would have been divided away into nothing, so that the cosmos would vanish. The only alternative to these absurdities, then, is to accept that the cosmos is one.

The challenge of Parmenides was taken up by the Atomists, such as Democritus (c.450-400 BC. [9] They argued that while a line, which is imaginary, can be divided into an infinite number of sections that have no mass, it is impossible to continue the division of physical objects indefinitely because matter cannot be destroyed. Matter must therefore be made up of tiny particles that are themselves indivisible, the atoms (*atomos* means 'indivisible'). [10] The doctrine that 'Nothing can be created out of nothing, nor can it be destroyed and returned to nothing' also meant that the atoms must always have existed, and it was assumed that they had always been in motion. Because the atoms were separated from one another and also in constant movement, logically they therefore needed the Void, a region of non-Being, in which they could move, and it was therefore claimed that non-Being is as real as Being.

The Void, and therefore the cosmos, is infinite in extent, and the proof of this was that geometrically, one cannot have a boundary, a line, without a space on *both* its sides, so the idea of a boundary with space only on *one* of its sides is logically impossible. [11] Correspondingly, matter must also be infinite, since a finite quantity of matter would disperse itself in infinite space and become invisible. There are therefore infinitely many bodies like the earth, and the earth cannot be at the centre of the cosmos, because infinite space can have no centre, nor can it have directions like up and down. Atomic theory also, of course, rejected the four elements, since they themselves had to be composed of atoms.

It was assumed that the atoms constantly collide with one another; and while some rebound others lock together to produce solid objects. They did not do this to fulfil some cosmic plan, so the order of the physical world was the result of purely mechanical processes without any ultimate purpose. [12] In order to explain the different appearances of things they said that atoms differed to some extent in size, in shape (some, for example, had hooks), and in order and position. They also developed the idea that the different combinations of atoms, rather like molecules, could explain many of the changes that we observe, such as the ability of the sea to change colour from green to white foam.

But atoms are not completely locked into the various bodies, because every object gives off a 'film' of atoms that are the origin of visual images and our sensations of colour, taste, smell, sound and so on. On this basis they distinguished between these so-called secondary qualities, and the primary properties of size, shape, weight and motion. [13] The distinction was extremely important because it is these properties that can be measured and which proved to be the central concern of physics.

While atoms are infinite in quantity, they are, then, limited in type, and the various combinations of these types determine the forms of different things. They were specifically compared with letters, because in both cases a small range of basic building blocks can be combined, at different levels of complexity, to produce a limitless variety of words and then sentences, or 'molecules' and things. [14]

Although atomism was very logical and ingenious, the problem was to find some definite evidence for it, and at best atomic theory could only give plausible explanations for some of the features of the physical world. Differences in the density of materials were easily explained as the result of the distance between the atoms in a body, so bodies with more void were lighter than those with less. Iron had more void than lead, so lead was heavier, but iron was harder because its atoms were grouped together in tighter clusters. In the same way, one can only cut an apple with a knife, and light can only pass through air and water because of the spaces between the atoms, and there are many examples of changes in the appearances of things that occur in ways we cannot detect because they are too small to be seen. The wearing away of a stone by water dropping on it, and garments becoming moist when hung out near the sea, are cases in point.

But in many ways the Atomists seem remarkably modern: their atomic theory of matter, and their attack on the four elements, were taken up again in the seventeenth century, as was their distinction between the primary and secondary properties, and eventually found to be basically correct. So, too, was the notion of infinite space with no centre, in which the earth is only one of a myriad of similar bodies all moving in the vacuum of space, while their generally mechanistic theory as a whole is fundamental to modern science. The Greeks (and Indians, who also had a well developed atomic theory) were perfectly capable, then, of producing a mechanistic explanation of the universe. The rejection of it by Plato, Aristotle and other philosophers was not because they couldn't understand mechanistic reasoning, and had nothing to do with the absence of machinery in their day.

Nor did most philosophers reject atomism because they were locked into some sort of 'religious' thought by their culture. To be sure, many regarded their denial that the gods had any influence as shocking and impious, but more significantly, mechanistic reasoning seemed quite unable to account for those aspects of mathematical form and structure in the cosmos that were becoming increasingly obvious, especially in astronomy, and which we shall consider shortly. [15] In particular, they could give no coherent explanation of the orderly motions of the heavenly bodies, and so tried to reject the findings of the astronomers on some very feeble grounds that we will note in the next chapter.

Ancient atomism is therefore an excellent example of what can be achieved by imaginative theory and minimal evidence. As Chomsky has very well said:

> Time after time, people have been able to construct remarkable explanatory theories on the basis of very limited evidence, often rejecting much of the available evidence on obscure intuitive grounds as they sought to construct theories that are deep and intelligible. . . . The theories that have been constructed, regarded as intelligible, and generally accepted as science has progressed have been vastly underdetermined by evidence. Intellectual structures of vast scope have been developed on the basis of limited and (until recently) fairly degenerate evidence. [16]

As we shall see, the ability to ignore inconvenient evidence and to follow the light of intuition and the creative imagination were essential for the progress of science. The atomists were free to speculate precisely because there *was* little evidence for their theories, and because they sidestepped the really difficult problems of motion and force (dynamics) by simply taking the motion of the atoms for granted. If selection had been operating vigorously, however, then a theory with great evolutionary potential would have been prematurely discarded because the actual evidence for it was very weak.

On the other hand, the atomists also shared some primitive ideas about nature with the other philosophers. While the Greeks had developed formal operational thinking in their reflections on the structure of society, language, logic, and mathematics (especially geometry), they did not, and could not, have the necessary concepts for applying this formal operational thought directly to the natural world. Learning to integrate theory and facts in an effective experimental method was an immense task that was only achieved in the seventeenth century. What they fell back on, then, were the primitive, intuitive ideas embedded in the poetry and proverbial folk wisdom of their own culture, which they elaborated in formal operational ways. What else could they have done?

So while atomist theory might look purely mechanistic, it was in fact permeated, like the whole of Greek philosophy, by the popular belief that like attracts like. Democritus, for example, said that 'All animals alike herd together with their own kind – doves with doves and cranes with cranes. And so it is with inanimate things': on a beach, for example, 'the motion of the waves rolls all the long-shaped pebbles into one place, all the round ones to another, showing that the likeness of things tends to draw them together'. [17] As it has been said, if he had looked at a real beach he would have seen that pebbles are sorted by size rather than by shape, which might have suggested a more mechanical explanation in which weight was involved.

The belief that like attracts like became, in China and India as well as in Greece, the basis of increasingly vast systems of symbolic associations between man and nature that were based on affinity of colour, shape, number, and many other qualities. They were believed to provide deep insights into the nature of the cosmos, and we shall see that they were essential factors in the development of astrology, alchemy, and medical theory. In Greek medicine, for example, basic temperaments were linked to four 'humours' or liquids secreted by the body, and in turn to the cosmos as a whole:

| Temperament | Humour | Organ | Quality | Season | Element | Planet |
|---|---|---|---|---|---|---|
| Melancholic* | black bile | spleen | cold & dry | autumn | earth | Saturn |
| Phlegmatic | Phlegm | brain | cold & moist | winter | water | Venus & Moon |
| Sanguine | Blood | liver | hot & moist | spring | air | Mercury & Jupiter? |
| Choleric | yellow bile | gall bladder | hot & dry | summer | fire | Mars & Sun |

*Melancholy: excessively contemplative, brooding, gluttonous*
*Phlegmatic: cowardly, unresponsive, without intellectual vitality*
*Sanguine: kindly, joyful, amorous*
*Choleric: obstinate, vengeful, impatient, easily aroused to anger [18]*

This view, however, could co-exist with the opposite view that like *repels* like, and that there is a natural attraction between opposites of every kind – human, animal, and inanimate, because opposites satisfy each other's deficiencies. Here we are moving beyond magical associations into that general Greek disposition to reason by analogy with easily observable things and events, such as sexual attraction. This was often used to 'establish' important scientific propositions without trying to support them by precise observation or argument. [19]

But despite such primitive features, speculation about matter had achieved some extraordinary triumphs: the idea that all the appearances of things can be reduced to different combinations of basic elements, the conservation of matter, the argument that the earth floats in space because it has no reason to move in any direction, the distinction between matter and force, and, of course, atomic theory, primary and secondary properties, and the notion of the void.

## 3. Mathematics and form.

The idea that the cosmos was based on a rational design of order, and proportion, and form, on Logos, was accepted by all except the atomists. The study of form for the Greeks meant, above all things, geometry and number. There is a plausible tradition that Thales visited Egypt and brought back their (fairly rudimentary) geometrical knowledge to Miletus, where he began transforming it from a set of practical rules about how to measure areas and volumes, construct right-angles, and so on that we noted in Chapter IX into a theoretical science. But the development of rigorous geometrical proofs needed, in particular, an input from the systematic logic being developed by Aristotle. It reached its peak in around 300 BC when Euclid, in his *Elements*, showed that a vast system of proofs could be constructed on the basis of a few apparently self-evident truths. [20]

This type of deductive geometry was a unique achievement of the Greeks; the Romans, by contrast, never had any interest in this sort of thing, and only valued geometry where it could be useful in practical activities like surveying. But for Greek philosophers, deductive geometry was clearly an intoxicating and thrilling achievement that fundamentally affected their outlook on the world. For them, geometry became the supreme expression of the union of reason, truth, and beauty, and the power of the human mind to penetrate the secrets of nature, especially in astronomy. (While the Indians and Chinese had considerable *knowledge* of geometry it remained at the level of techniques of figure construction, such as doubling the square, and the idea of proof never developed.)

Slightly later than Thales was the enigmatic figure of Pythagoras (c.580-c.500 BC). [21] A sage who famously taught the doctrine of reincarnation, he wrote nothing, and his followers were a quasi-religious cult who also became involved in city-state politics. But he was also credited with teaching the fundamental importance of number for understanding the cosmos, and we can get some idea of how the Pythagorean tradition developed from what is known of later members of the school. [22]

The Pythagoreans seem to have been the first to claim that numbers are the basis of everything, indeed that numbers really exist in things. They did not, however, mean number in the sense of the measurements and equations of modern science, which would have been quite impossible at that stage. Theirs was the much easier 'number' of primitive number symbolism. Like the Konso, for example, they regarded 2 as female, 3 as male, and marriage as 5 – the union of 2 and 3. Pythagorean number symbolism was therefore steeped in those magical associations we noted in the previous section, and according to Aristotle they thought that the cosmos is governed by an elaborate system of 10 dualistic oppositions:

| Limited | Unlimited |
|---|---|
| Odd | Even |
| One | Many |
| Right | Left |
| Male | Female |
| Resting | Moving |
| Straight | Curved |
| Light | Darkness |
| Good | Bad |
| Square | Oblong |

(These are also very similar to the properties of Yang and Yin in Chinese thought.)

This magically inspired interest in number was not only a vital contribution to the arcane science mentioned earlier, but also had great evolutionary potential, because 2 and 3, for example, are also even and odd numbers, and this led the Pythagoreans to investigate all sorts of puzzles and relationships between them, such as the multiplication of even by even, odd by odd, and even by odd, and those numbers such as 24, which is the product of even by even (6 x 4) and also even by odd (8 x 3). They also laid the foundations of the study of such fundamental topics as prime and irrational numbers. [23]

Although there is no evidence that Pythagoras could have proved the theorem associated with his name, [24] the Pythagoreans had a particular interest in geometry. For example, they began the investigation of symmetrical, or regular, solids, and by the time of Plato it was known that only five are possible; all their faces are identical, and all the corners can be fitted inside the sphere:

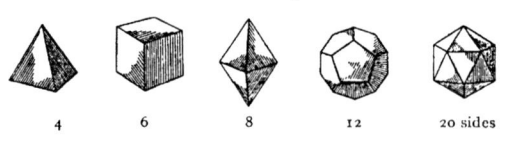

4   6   8   12   20 sides

These became known as the Five Platonic Solids because Plato linked them with the four elements and the cosmos. [25] But the perfect solid for the Pythagoreans was the sphere itself, just as the perfect figure was the circle, and they had even proposed that all the heavenly bodies, including the earth, revolve around the Central Fire, and that the earth itself is a sphere. As we shall see, the idea of circular motion had a profound influence on Greek astronomy, not just because it was eternal like the celestial bodies, but because, being predictable, it was also considered as rational, whereas motion in straight lines was considered irrational, and typical of matter.

Geometry itself is also based on number: 1 is the point, 2 is the line, 3 the triangle, and 4 the simplest solid, the tetrahedron, made up of four equilateral triangles. 1, 2, 3, and 4 also add up to 10, the base of the Greek number system, and of all the other peoples the Greeks knew, and considered the perfect number. The Pythagoreans are also supposed to have made the fundamental discovery that the musical scale is based on these same numbers 1, 2, 3, and 4, so that a single string could be divided into different lengths to produce the basic intervals of 2 : 1 (octave); 3 : 2 (fifth); and 4 : 3 (fourth). This was the first case in history when physical properties could be expressed in terms of number, and must have caused great excitement, and reinforced the belief that number is a fundamental principle of the universe.

The application of numbers to music and harmony (developed eventually into a very complex system of modes or scales) was the most striking example of the Pythagorean interest in proportion and ratios, which became fundamental, together with geometry, in the way the Greeks applied mathematics to nature. [26] All these ideas were highly relevant to the fundamental idea that the universe is regulated by reason, and strongly influenced Plato's (c.429–347 BC) belief that behind appearances there is an eternal reality expressible only in terms of number, geometry, proportion, and harmony. The actual triangles and circles that we draw in the sand are, of course, only material, but behind them are the triangles and circles of thought, and behind these again are the pure Forms of the triangle or the circle themselves. It was these Forms that the Divine Craftsman used, beginning with the right-angled triangle, to generate the Solids, and thus the elements and the rest of the physical world. Indeed, in Plato's view, there are Forms of everything – colours, sounds, the elements, the artefacts we make – in a hierarchy from the Form of the Good downwards, but knowledge of these cannot be obtained through our senses because these only give us a fleeting and illusory impression of reality. The rational soul of man is the only possible source of knowledge of the Forms, and it acquires this knowledge by recollection of its earlier existence in the transcendent realm of pure Form, [27] so that for Plato intellectual and spiritual illumination are almost one and the same.

Plato's universe is good as well as rational, because the Divine

Craftsman is good, and man as a rational being is an integral part of its design, related to it as microcosm to macrocosm. As we have seen, his ideas were therefore very congenial to Christianity, and his belief that the cosmos has a deep mathematical structure has also had a profound influence on Western scientific thought, especially during the Renaissance. But he had no interest in the actual study of nature because he believed that its mathematical structure could only be grasped by pure reason, rather than observation and measurement. [28]

## 4. Aristotle's natural philosophy.

His pupil Aristotle, however, had a much greater interest in the study of nature, and combined many of the ideas that we have been considering in a monumental theory of the physical world whose influence survived into the eighteenth century. Atomism was basically right, but Aristotle seemed more plausible in terms of ordinary experience, and we must now look at his theory in some detail, not only because of its central historical importance, but to see how he combined Fact and theory.

Aristotle followed Plato in holding that the universe is rational, and constructed on intelligible, mathematical principles, and everything in it is for the best, but he rejected Plato's theory that the Forms exist in a transcendental world of their own, and maintained that they are embodied in the material world. He therefore attributed much more value than Plato to observation and the senses as sources of truth, and if Plato's model of knowledge was mathematics, then Aristotle's was biology. He eventually founded his own school of philosophy, the Lyceum, which, unlike the Academy of Plato, made the study of nature an integral part of its research, which was very substantial. [29]

But, as I have said, he completely agreed with Plato that the universe is fundamentally rational, and on this assumption he was justified in asking, not just *how* something came to be what it is, but *why* it did, what purpose it has in the scheme of things. He said that there are four sorts of 'causes', or explanations, of how things come to be as they are. The first is the matter from which they are made, but he regarded Matter as having the *potential* of becoming many things; in order to realise its potential to become some particular thing, there had to be a second (Formal) cause. By Form he meant not just a thing's shape but its essential nature and properties. But imposing form on something required a third cause, the Efficient cause, which is its physical origin, what made it happen, and we cannot explain the whole process unless we also understand the fourth cause, which is tells us why it exists, what its purpose is in the larger scheme of things (the Final 'cause'). So, the matter of a house is the stuff out of which it is built; the form or structure of the house is its design, which distinguishes it from a table or a chair; its efficient cause is its builder; and its purpose, its reason why, is to keep its owner and his property secure from the weather, theft, etc. (Purpose and form are obviously very closely connected.) Aristotle says that natural processes are to be explained in exactly the same way, and he not only includes plants and animals, but the motions of inanimate objects. (We therefore *understand* something when we grasp its Form, its essential nature. This is a basic feature of his philosophy to which we shall return later.)

He therefore totally rejected the mechanistic theories of the Atomists, because they could not explain why things are as they are. So, he says, they argue that nature does not work for the sake of something, or because it is better so, but because things just have to happen in a certain way: in their view it does not rain, for example, *in order to* make the crops grow but because water is drawn up into the clouds and then cools and falls back to earth again as rain. And if crops grow as the result of rain, it had no purpose in doing so but just happened.

But, he replies, would we seriously follow this line of argument in explaining why our front teeth are sharp and appropriate for tearing, whereas our molars are broad and suitable for grinding, and say that this just happened because it had to, that it was a coincidence? So, he concludes, if we regularly have rain in winter, and heat in summer, and if our teeth are clearly organized in a purposeful way, this cannot be the result of coincidence but must be for the sake of something.

Nature, then, always acts for some end, and does nothing in vain; this approach was fruitful in his biology, because it was the basis of a functional analysis of organic design, but it also led him to accept the old idea that Nature constantly strives towards perfection, as Form struggles with Matter to raise it to increasingly higher types. Base metals, for example, 'try' to become gold, while Man (and specifically the male) is, of all living things, the most in accord with nature, so that only in him are all the organs in their proper places. So he maintained that upper, right, and front are not only superior to lower, left, and back, but are also the origins (*archai*) of the three dimensions. The form of man is not just taken as the ideal, but is also supposed to explain the fact, for example, that food is distributed to the other organs of a living thing from the upper region of its body. This theory is obviously derived from the position of the mouth in man, but 'When faced with the fact that in plants it is clearly the roots at the *bottom* of the plant that perform this function, Aristotle concludes that plants are simply upside down', [30] one of the many cases where he subordinates facts to theory.

Now there is nothing inherently absurd in the view that the universe is arranged for some purpose. It has been known for some time that it is 'fine-tuned' in a remarkable way that allows life to develop, and that if, say, the force of gravity, or the relative weights of protons and neutrons were even slightly different, life could not have evolved. But interesting though this is, modern science does not ask why it should be so, or use it as a guide for research, because we have learned that while we can discover what the laws of physics and chemistry are, we cannot go behind them to ask *why* they are as they are. These ultimate 'why' questions are now regarded as subjects for religion and philosophy, while science concentrates simply on the what and the how. But Aristotle's attempts to answer them nevertheless produced a highly integrated and comprehensive theory.

He considered that the theory of the elements, plus the qualities of the hot, cold, moist, and dry provided a more plausible explanation of physical transformations than anything atomism could offer. They were also central not only to his ideas about the structure of the cosmos, but about motion. Here it is important to note that he did not think that motion is the natural state of matter; on the contrary, he assumed from common sense that all matter that is not self-moving is naturally at rest. The Pre-Socratics, the atomists (and the Chinese) [31] had assumed the opposite, and therefore did not think that motion was a problem, but because of Aristotle's false belief that it has to be explained, he was the first philosopher to produce a comprehensive theory of motion.

The origin of motion is the Prime Mover, which itself remains unmoved, but imparts motion to the stars, which in turn transmit motion to each of the planets in turn, ending with the moon. The motion of these heavenly bodies is naturally circular and eternal, because they are made of the element aether, whereas the four earthly elements naturally tend to move in straight lines unless some other motion is imposed on them, because they try to get back to their natural places by the shortest route, which is a straight line. Fire and air naturally move upwards because they are light, while water and earth naturally move downwards because they are heavy – conclusions that again seem obvious to common sense. Motion in a straight line, however, cannot continue for ever in a finite universe but has to be stopped and reversed; circular motion, on the other hand, can go on for ever, and it was this perfect and ceaseless motion of the heavenly bodies, and its complete regularity, which proved their divine nature, and their complete difference from earthly elements. (While he was right to think that linear and circular motions are basically different, it was for the wrong reasons. Motion in a straight line needs one force, whereas circular motion needs two, but this was first realised only in the seventeenth century.) He also made a very important distinction between the *natural* motions of the elements, and motion that is enforced or *unnatural*, as in throwing a stone upwards.

Since each element has its natural form of motion, then the heavenly bodies must be composed of some other element, which is why Aristotle used a fifth element, the 'quintessence' or aether, to account for their circular motion. Aristotle's aether was a weightless, transparent kind of crystalline substance that formed a nested set of hollow spheres one within another, with the stars embedded in the outermost sphere and the planets, sun, and moon in the intervening spheres. The sphere of the moon, as it revolved, in turn stirred up the other four elements below it.

This model of the spheres not only explained planetary motion, but created a fundamental and qualitative distinction between the earthly sphere of fire, air, earth, and water which, in the form of the atmosphere, extended as far as the moon (the 'sublunary' sphere), and the celestial sphere of the aether and the heavens. The sublunary world was one of change and decay, whereas the perfect celestial world was pure and unchanging. Aristotle's universe was therefore filled with matter of different kinds, and motion could only be transmitted by direct contact, with no action at a distance by forces like gravity.

To explain why Aristotle thought that the earth should be at the centre of the cosmos, and why the heavy elements should fall towards it, we must examine his theory of the cosmos in more detail. While Aristotle's cosmos was entirely filled with some kind of matter, outside the sphere of the stars was nothing at all. He refuted the atomists' argument that a bounded cosmos was logically impossible by redefining space itself. The atomists thought of space as an infinite empty box with objects in it, whereas Aristotle defined it in terms of the *relations between physical objects*: no physical objects therefore implied *no space at all* and the boundary argument simply didn't apply: Aristotle's cosmos, therefore, could not possibly float in the atomists' Void. Aristotle was rather obsessed by the Void, and had many arguments for rejecting it. At the level of observable fact he denied the possibility of a vacuum, and this seemed to be supported by the way water can be sucked up through a tube – it flows up the tube to prevent a vacuum being formed. (Nature's abhorrence of the vacuum, referred to in the Middle Ages as the *horror vacui*, was standard doctrine until the seventeenth century.)

But the Void also contradicted his whole theory of motion. In a bounded cosmos, upwards means in the direction of the stars, while everything that naturally moves *downwards* must stop at the centre (going beyond the centre means going up again). So by placing the earth at the centre of the cosmos, Aristotle could then explain why objects always fell towards it. 'Up' and 'down' were also the basis on which the elements could be arranged in their natural positions: it is obvious that earth is at the bottom, then rivers and oceans, then air, with fire going upwards. (We think of this order in operational terms, as simply based on *relative* density: a stone, for example, sinks in water, whereas air rises as bubbles, but for Aristotle the order expressed their *essential* natures.) Naturally, then, he rejected the infinite Void, because without periphery or centre there could be no up or down, and therefore no reason why objects should fall. [32]

By this time philosophers and astronomers could provide a number of sensible reasons for thinking that the earth is round, stationary, and at the centre of the cosmos, and we will come to them in the next chapter when considering Greek astronomy. We must now examine Aristotle's theory of motion in more detail because it has some primitive features that need explaining. An obvious example is the way that he tried to explain the motion of the elements by their Form, their essential nature. While it was quite appropriate to ask what is the essential nature of a triangle or a circle, asking what is the essential nature or Form of a stone was a fatal obstacle to any real understanding of the laws of physics, of explaining what things did, especially the behaviour of moving objects like heavenly bodies and projectiles. This primitive 'essentialism' meant that what we would treat as causal relations between things, like the elements, were treated as expressions of the innate powers of the things, which were awakened in truly intuitive style. It was supposed that each substance 'carried about with it this battery of powers, like a warship bristling with guns ready to discharge a projectile whenever a suitable target is sighted'. [33] So, while Aristotle knew that a lunar eclipse is caused by the shadow of the earth when it comes between the moon and the sun, he still thought of 'being eclipsed', like 'being moved' as a *property* of the moon, like its size, not simply the result of its *relationships* with the earth and sun. [34]

And while the different elements each have their own natural motions, they also tend to occupy their natural places because, as Aristotle explicitly says, place itself exerts an influence on the different elements: the downward tendency of earth is awakened by the centre, just as the upward tendency of fire is elicited by the above. Motion, he says, is in the movable, by which he meant that being moved is a potential of *the thing being moved*, as well as of the mover, just as teaching and being taught are potentials in the tutor and in the pupil. Just as the tutor transmits a Form of some piece of knowledge to the pupil because the pupil is able to receive it, in the same way the centre actuates the potential of earthy things to move downward towards it. We saw in Chapter VI that, for intuitive thought, there is no real transmission of force, but rather that the essential nature of one object is awakened by something else outside it, like obedience to a command. Aristotle's 'bipolar' theory of motion, with place awakening the natural disposition of the elements to move in a certain way, is a classic example of this. [35] So, then, the Formal 'cause' of a stone's motion is its earthiness, its tendency to move downwards in response to the centre, while the Final 'cause' is the power of the centre to attract the element of earth to where it should be in the scheme of things. We then have to explain what *actually* makes it move, the Efficient 'cause', and this has to be some force, because Aristotle assumed that matter is at rest until some force moves it.

He believed that weight itself is a force, (the force is really gravity, acting on the mass of the object), but for him that would only produce a *constant* motion downwards. He knew, however, that objects accelerate as they fall – objects falling from high places have greater impact when they land than objects falling from low down – and therefore had to explain this acceleration. His solution was, that as objects get closer to the centre of the cosmos they become heavier, so more force acts upon them, and therefore they must accelerate. (In fact, it is the constant force of gravity that produces the acceleration, but in the absence of any idea of gravity, Aristotle's theory was ingenious.) Since he thought that weight is a force, and the greater the force, the greater the speed, he was logically bound to maintain that heavy objects must also fall more quickly than light objects, in proportion to their weight. (There are plenty of unscientific people in our society who still believe this.)

An object falling to earth is a case of natural motion, but a man throwing a spear enforces an unnatural motion upon it for a certain time. Aristotle thought that a mover must always be in contact with the thing moved, but he could not see how the force of the thrower's hand could be transmitted into the spear itself. The only alternative was to believe that it was the air which, in various implausible ways, somehow continued to transmit the force of the thrower to the spear after it had left the thrower's hand, with diminishing effect, until it resumed its natural motion and fell to earth. Because the natural motion of falling was different in kind from the enforced motion of being thrown, Aristotle also did not think that both kinds of motion could operate simultaneously: so the enforced motion would move the spear in a straight line, and when this was exhausted, the spear would then fall vertically downwards. (If anyone had objected that javelins and discuses seem to follow a curved path when thrown, he would perhaps have replied that this is an illusion of the senses, like the apparent merging of parallel lines with distance.)

Aristotle could understand quite well that the speed of something is proportional to the time that it takes to travel a certain distance. He says, for example, that what is quicker will travel over a given distance in less time than something slower will take to travel the same distance, or that in an equal time, the quicker will go farther than the slower. But speed *itself* could not be a quantity, a magnitude, because a magnitude was always a ratio between two similar quantities, like two distances, or two times, whereas speed is the ratio between a time and a distance. So Aristotle's notion of speed was limited to those comparisons between the distance covered and the time taken. Not only was there no idea of measuring speed in units equivalent to 'miles per hour', but there was even less idea of measuring the speed of an object that was actually moving, as when we look at our speedometer to find out how fast we are travelling 'now'. Our 'mph', and feet per second, involve the idea of 'instantaneous speed', the distance that *would* be covered if the object *were to* travel from A to B at the same speed as it has at the instant we measure it. The medieval philosophers were the first to grasp this idea, because they discovered how to treat speed as a magnitude, but without this concept, it is understandable that Aristotle could not even attempt an analysis of acceleration. [36] But his was the first comprehensive theory of force and motion, and many aspects of it, such as projectile motion and falling objects, were easily testable. Yet no effective experiments were carried out on these problems in the ancient world, and this brings us to the broader question of scientific method and how fact and theory were related.

*5. The absence of experiment and measurement.*

Aristotle's system was very convincing, not only because it was logically so coherent, but because it was so closely related to ordinary experience. The Forms or essences of the different elements, for example, could be directly observed from their different sorts of movement, unlike the hypothetical motions of the atoms. It was not until the seventeenth century that people realised that this notion of the essences of things was nothing more than a description of what they actually did, and so could have no explanatory value at all. Aristotle and others of his school also made a large number of observations, especially of animals and plants. No science can advance without this preliminary sort of fact-gathering and its systematic classification. But while in this chapter we have seen examples of facts, or alleged facts, being used as arguments for or against particular theories, there was no idea that theories should constantly be tested against facts by any sort of experiment. This was because the evidence of the senses, 'the Facts', were regarded as inherently unreliable and prone to error, and it was the function of theory, of reason, to correct the Facts, not to be corrected by them. The *Meteorologica*, for example, contains many facts relating to the theory of how the hot, cold, wet and dry transform the elements. 'We are given an analysis . . . of which substances are combustible, which incombustible, which can be melted, which solidified under the influence of either cold or heat, which are soluble in water or other liquids.' But, 'the role of the data . . . was to illustrate and support the theory, not to put it to serious risk'. [37]

Not only were theories not systematically tested against evidence, but Aristotle and the other ancient scientists hardly ever measured anything. Indeed, the whole idea of measuring many physical properties such as heat and cold presented enormous intellectual problems. Aristotle considered heat and cold as qualities, and for him, quality and quantity were quite different sorts of things or 'categories'. Quantity can be of two sorts, what is countable – separate things like coins – or what is measurable, like the length and weight of a block of stone. One of the hallmarks of quantities is that they can be added together, such as number to number, length to length, or weight to weight. A quality, on the other hand, means a characteristic feature of something, as having two feet is characteristic of a man, whereas having four feet is a characteristic of an animal. Black and white, hard and soft, hot and cold, heavy and light, moist and dry, and bitter and sweet are all examples of qualities. First of all, they can't be added together like quantities. Adding one bowl of warm soup to another bowl of warm soup will not produce a bowl of hot soup, and while we can add one white tile to another white tile, this will not increase their whiteness, but only their area. Secondly, every quality has a contrary or opposite – black/white, heavy/light, and so on – whereas quantities do not: what is the opposite of four, for example? [38]

Now the problem with thinking of qualities in terms of opposites is that it is then impossible, nonsensical in fact, to think that they could be arranged on a single scale of variation, as we now do with hot and cold on a scale of temperature, or with heavy and light on a scale of relative density – specific weight or gravity. Indeed, qualities can't be measured at all, because they have no size, and do not appear as separate things that can be counted. The Greeks knew, of course, that there are *gradations* between black and white, or between hot and cold, or heavy and light, but these were seen as mixtures between qualitatively different sorts of thing, of Forms, just as, to take a crude analogy, we think of tea and milk as different kinds of things mixed in different proportions in our cup. But we cannot have a scale that includes both tea and milk. All that we can say is the Form of a quality such as black may be more intense in some objects than in others because there is a greater proportion of it. This distinction between quality and quantity, together with the belief that speed is not a magnitude, and the idea of natural and violent motion, was one of the major obstacles to the scientific analysis of nature. We can now see why the Greeks had to apply mathematics to nature by taking the easy gateways of number symbolism, proportion, and geometry, because these largely avoid measurements, and they are also static, avoiding the problems of motion and change.

*6. The lack of the idea of probability in the ancient world.*

Another area of modern science that was conspicuously lacking in Greece, and the ancient world generally, was anything like our notion of probability and statistics. In sixteenth and seventeenth century Europe problems about gambling odds were important in developing the theory of probability, so we might expect that, since dice and similar forms of gambling were extremely popular in the ancient world, these would have provided them, too, with the same sort of laboratory conditions in which to discover the basic laws of probability. But this did not happen because ancient ideas about probability in general were very different from ours.

When we think of a die, we expect that the more often we throw it, the more likely it is that, unless it is biased, each face will turn up an equal number of times – the Law of Large Numbers. But to the ancients this would have seemed ridiculous. Obviously, a Roman would have retorted, increasing the number of throws must make the results *more* unpredictable, not less, and the whole idea that the unpredictable and the irregular could produce order and regularity would have been unthinkable.

There is good evidence that no attempt was made to work out the probabilities involved in games of chance, [39] nor was there any clear idea about how to calculate the odds of different combinations of throws. [40] It was generally believed that in gambling there are hidden forces at work that affect the sequence of throws: these may be our personal luck – they would have loved the case of Lord Yarborough, who was so permanently unlucky at whist that he has given his name to a hand with no points – or it might be a matter of some subtle skill: Plato thought that one could only be good at dice if one had begun as a child. But dice and astragals (ankle-bones) were also used in temples for divination where they were thought to respond to different mystical influences and reveal our future. (This combination of gambling and divination is absolutely typical of pre-modern societies.)

We saw in Chapter VI that while in primitive thought there is a notion of the meaningless accident, no one accepts the possibility of the *meaningful* accident. If an event has some personal significance there must be a reason for it, an explanation for 'why me?', and in the ancient world, of course, many people accepted explanations based on the gods, fate, and so on. Even philosophers such as the Stoics, who believed that all events were rigidly determined and that chance was an illusion, accepted divination in principle, because it was simply making clear the effects of the hidden causes at work behind events, that we could not understand in any other way.

But they would never have accepted that these hidden causes could be understood mathematically, because there was an absolute antithesis between mathematics and chance. Mathematics, notably geometry, was based on proof, and so was the only possible basis of certain knowledge. Below this level of absolute certainty the Greeks were well aware, of course, that we can predict some things with a fair degree of confidence, such as a competent doctor predicting the likely outcome of a disease. This kind of knowledge could be the basis of elementary statistical generalisations, such as 'Why are boys and women less liable to white leprosy than men, and middle aged women more than young?' [41] In law, too, one could estimate degrees of probability or likelihood based on evidence, and the same applied in the natural sciences, where reason and evidence could establish conclusions that were probable in the sense of the likely. [42]

But chance was another thing again, because it was inherently inscrutable. In Aristotle's view of cause and effect there were some things that occurred either necessarily or usually, like the motions of the celestial bodies, or the course of a disease, because they were an essential part of the activity. An accident, however, was something which had no necessary or inherent relation to that process or activity. So if a man were digging a hole in the ground to plant a tree and found some buried treasure (rather than worms or stones), this would be an accident because there was no inherent connection between his aim of planting a tree and finding the treasure, which had been buried there by someone else for entirely unrelated reasons. Or if one goes to see someone to collect a debt, and meets an old friend that, too, is an accident because it has nothing to do with collecting the debt. [43] The possible causes of accidents in this sense are innumerable, and therefore it seemed to Aristotle quite absurd to suppose that there could be a science of the accidental. Science was concerned with the knowable, with what displayed regularity, with the normal, but a science of the accidental would be a science based on exceptions, on the irregular, and would therefore be nonsensical. [44]

It is therefore obvious why neither Aristotle nor other philosophers would have regarded gambling as of any scientific interest. But some people in the ancient world had a more professional interest in gambling: the Emperor Marcus Aurelius, for example, took his personal croupier with him on campaigns, and in China shop-keepers would gamble with their customers in coin-tossing games, giving them back their stake plus their goods free of charge if they won. [45] In order to stay in business the shopkeepers must have had some advantage over their customers, but how did they gain this? To rely on long experience alone would have been very expensive, so they must have had some additional source of knowledge about odds, and the obvious one is a table of possibilities, like this one for the possible permutations of three coins:

| Result | 1st coin | 2nd coin | 3rd coin |
|---|---|---|---|
| 3 Heads | H | H | H |
| 2 Heads | H | H | T |
|  | H | T | H |
|  | T | H | H |
| 2 Tails | T | T | H |
|  | T | H | T |
|  | H | T | T |
| 3 Tails | T | T | T |

This requires no knowledge of mathematics at all, but once it has been set up it becomes obvious that some outcomes have many pathways, and these are the ones to bet on, while others only have one. It seems possible that Chinese shop-keepers, or Roman croupiers, who depended on a better knowledge of gambling odds than that of their customers, would have been familiar with tables of this sort, but of course would have kept such trade secrets to themselves. The customers of the shop-keepers and croupiers would have believed that their success depended on their personal fate or lucky charms, while the shop-keepers would have relied on what they knew about the odds of coin-tossing without bothering with what that implied at some deeper philosophical level.

Again, we have seen that people were perfectly capable of making simple statistical generalisations about, say, the likelihood of different diseases affecting different types or classes of people. But these were never measured or put into actual numbers, and any kind of probabilistic thinking based on relative frequencies is therefore very rare in antiquity. [46] What did not exist in the ancient world were commercial insurance, life assurance, or annuities, in which probabilities have to be given a numerical value because they is the basis on which particular sums of money have to be calculated. Least of all was there any idea of what we know as the Law of Large Numbers, the idea that hidden patterns will emerge from a sufficiently large number of cases. This was first understood in the seventeenth century from the statistics of mortality tables. So it was only when the whole idea of a numerical, quantitative representation of the world had developed, that the theory of games could then take on a completely different significance, and become the basic laboratory of probability theory.

## 7. Conclusions.

Materialist and Darwinian theories, then, have obviously not done well in explaining the emergence of natural philosophy: natural philosophy was driven by the intellectual curiosity of a scholarly elite, and any attempt to derive it from material needs or technology has no foundation. Our senses cannot tell us that the sun is not really moving when it rises and sets, that the air around us is actually pressing down on us with a weight of many tons, and that everything is composed of atoms. Nature does not force these facts upon our minds, but has to be interrogated. To do this, and apply formal operational thought to nature, science has had to develop an effective method of integrating reason and the evidence of our senses by measurement and experiment.

Nor has the Darwinian model fared any better: ancient science was rather like primitive warfare, in the sense that while there was a great deal of conflict, there were no decisive Darwinian battles, because measurement and experiment were too poorly developed to put rival theories to any conclusive test. But despite this, '…had the Greeks confined their attention to those problems where specific issues could be settled definitively by observation and experiment, what we know as Greek natural science would simply never have been brought into existence.' [47] They had to begin by asking grand philosophical questions about what the cosmos was made of, whether it was possible to create or destroy matter, could it be divided up indefinitely or did one have to come to a stop with the atom, what was the nature of motion, and so on. The apparent discoveries that we have looked at in this chapter gave them the confidence that reason could enable them 'to discover the enduring and intelligible reality behind the constant changes perceived by the senses', [48] while the belief that the cosmos was a beautiful harmony was aesthetically and spiritually uplifting. On this basis, and enormously assisted by their deductive geometry, the

Greeks were able to lay some essential foundations of modern science that we shall look at in the next chapter.

# CHAPTER XIII
## Ancient Sciences

For the Greeks, philosophy explained the underlying, essential nature of things, and why the cosmos works as it does. They assumed that it was rational and governed by law, which also meant that everything had a purpose and that nature did nothing in vain, and in the most economical way possible. Nature was also thought to be governed by those principles of affinity we noted in the last chapter, and everything could be graded in terms of relative perfection. The heavens were superior to earth, man was superior to animals, and precious metals superior to base metals, so that astrology and alchemy had philosophical support. The Greeks also assumed that the senses are unreliable by comparison with reason, so it is understandable that scientists seldom used evidence to disprove theories, but instead used theories to 'save the appearances', that is, to show how the illusions of our senses can be given rational explanations. The planets may seem to revolve in irregular orbits, just as sticks appear to bend in water, or parallel lines converge with distance, but these deceptive appearances of things can be corrected by geometry.

Philosophy provided the intellectual foundations for a number of specialized sciences, and in this chapter we shall examine some of these to see how far the Greeks were able to adjust theory to observation, and to what extent they made their observations fit with what they had already decided nature must be like. The first will be astronomy, the most important science, which not only supplied their basic model of the cosmos, but combined a great deal of observation with mathematics. We shall move on to astrology and alchemy because we need to understand how Greek scientists could believe what to us were clearly contrary to the facts, and because these branches of ancient science were in fact of great importance in the development of modern science.

Optics, like astronomy, was an ideal opportunity for using mathematics to solve problems of nature, and had a great influence on later science in Islam and the West. But we shall also try to understand why its practical applications were virtually non-existent. Finally, we shall come to mechanics and pneumatics, which will show us how Greek and Roman scientists interacted with the technology of their day, and the limitations of their theories when applied to force and motion. It will also reveal, again, how little interaction there was between science and practical problems of technology.

The schools of philosophy continued at Athens, where those of Aristotle and the Stoics, [1] rather than the atomists, were the dominant influences, and the Lyceum was a major centre of research. The Museum [2] at Alexandria also came to play a major part in the development of Greek science. Funded by the Ptolemies, and later by the Caesars, it supported three or four dozen researchers in a wide range of subjects, rather in the manner of an Oxford or Cambridge College, with access to the great libraries of Alexandria.

*1. Astronomy and astrology.*

Astronomy has remained at the core of science, and we cannot hope to understand why Copernicus and Newton were so significant, without a good grasp of ancient astronomy, and the problems it was trying to solve. As usual, we must start by explaining its origins.

By about 700 BC, the court astronomers of Babylonia, under the Assyrians, had collected a large mass of observations in the process of recording and predicting celestial omens, including eclipses, relating to the prosperity of the king and the state. [3] The planets, too, especially

Venus and Mars, were important omens that had originally encouraged astronomers to observe their motions. Like eclipses, these had no practical importance, but the mathematical problems in explaining these became the intellectual driving force in Greek astronomy, and laid the foundations for Copernicus and the modern understanding of the universe.

The Babylonians recorded their observations in the form of mathematical tables, but the Greeks used these to create a geometrical model of the universe, to show how the heavenly bodies actually moved around the earth. We are so used to this sort of model of the solar system that it seems obvious, even banal, but in fact it was a revolutionary leap of the imagination. None of the other astronomical civilisations – the Babylonians, the Chinese, the Maya, the Aztecs, or the Indians – thought of converting what they knew about the movements of the planets into geometrical form. (Though the Indians, like the Arabs, later adopted the Greek model.) But the completely false idea that circular motion is natural to the celestial bodies allowed the Greeks to avoid the major problems of gravity, and the real causes of celestial motion. They could not have solved these, so the false belief in natural circular motion was the basis of increasingly ingenious solutions to the problems of planetary motion, based on geometry alone.

In c.450 BC Anaxagoras had argued that the moon gets its light from the sun, and causes solar eclipses by coming between the earth and the sun, just as lunar eclipses are caused by the earth coming between sun and moon. A hundred years later Aristotle could give some standard arguments to show the earth is round: the earth's shadow on the moon in an eclipse is curved; as we travel north or south different constellations rise above the horizon or sink below it; and ships sailing out to sea gradually disappear below the horizon when seen from land.

The belief that the earth is a sphere allowed the use of geometry to estimate its size, and the first attempt about which we know the details was by Eratosthenes (c.275–195 BC) at Alexandria. His estimate of its circumference as 24,389 miles was within 2% of the true figure, if he was using the unit of length we suppose. [4] Establishing the size of the earth was the essential foundation for a number of further calculations, such as the diameter of the moon and its distance from the earth. [5] These results were not accurate, but their technique was entirely correct, and showed what could be achieved by geometry and basic measurements. They were improved by later astronomers, and by the time of Ptolemy in the second century AD it was considered that the earth was of negligible size in relation to the distance of the stars.

But while they knew that the earth is round, they also thought it was stationary. This not only seemed obvious to common sense, but was supported by some compelling arguments. Aristotle, and most of the leading astronomers down to Ptolemy, argued that the earth cannot rotate on its axis because objects thrown vertically into the air come down again to their point of origin. Another argument was that if it rotated, then the clouds and birds would obviously be left behind and all would move towards the west.

It was also believed that the earth was at the centre of the universe, and that all the heavenly bodies revolved around it (the 'geocentric' theory). Aristarchus of Samos (c.275 BC) and Heraclides (who both inspired Copernicus) proposed that the earth might orbit around the sun, but this 'heliocentric' theory, like their suggestion that the earth rotates on its axis, was dismissed for two fundamental reasons. If we walk around our garden, the relative positions of trees and buildings to each other seem to change, and this is known as parallax. Now the ancient astronomers knew very well that, in the course of the year, the relative positions of the fixed stars do not display any parallax at all so how, then, could the earth revolve around the sun? Aristarchus said that we could not see any parallax because the stars were too far away, but Hipparchus and Ptolemy, the greatest astronomers of antiquity, dismissed the idea. Aristarchus was right, in fact, but the parallax is so small that it took the telescopes of the nineteenth century to detect it.

Secondly, according to Aristotle, the earth had to be at the centre of the universe because, since there was no idea of gravity, this was the only way to explain why objects fell towards it. It is important to realise that all the astronomical observations in the ancient world were quite compatible with this belief, and even today navigators and surveyors find it more convenient in practice to treat the earth as if it were really at the centre of the universe. [6]

Eudoxus, a pupil of Plato, is the first Greek known to have worked out a geometrical model for the orbits of all the planets around the earth, which he based on Plato's belief that the planetary orbits should be those of uniform, circular motion, while his observational material for calculating these came from the Babylonians.

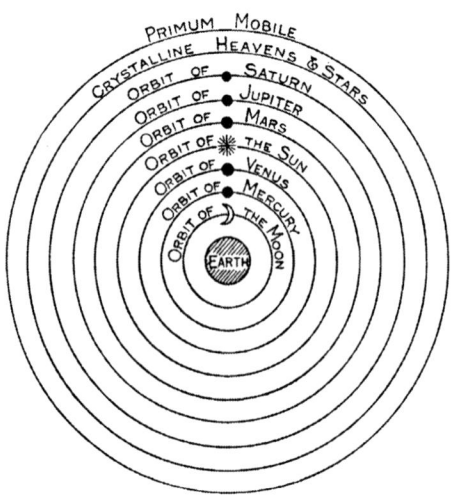

The first task was to arrange the heavenly bodies in an order, below the Prime Mover, with the sphere of the stars as the outer shell of the universe, then the planets with the slowest orbits on the outside, closest to the stars, and those with the fastest orbits closest to the earth. The result is shown in the diagram above.

But although this model explains solar and lunar eclipses, it is far too simple, because it doesn't take account of three major features of planetary motion: (1) the planets vary in brightness during the year; (2) they appear to move backwards ('retrograde motion') at certain times of the year; and (3) the speed of their orbits seems to vary. [7] (1 and 3 also apply to the sun and moon, but to a much smaller degree.) There was, however, a basic philosophical assumption that, whatever the planets might appear to do, because they were heavenly bodies their motion had to be rational, and therefore circular and at a constant speed. To solve this problem, Eudoxus suggested that each planet was moved not just by one, but several crystalline spheres, all centred on the earth, but with different axes and motions, and in this way he was able to explain their varying speeds, and their retrograde motion, by assuming that each sphere could move in a different direction at a different speed. The model needed no less than 55 crystalline spheres and was basically adopted by Aristotle. But it still had one great drawback, which was that since all the spheres were centred on the earth, the distance of the planets from the earth had to remain the same, and so it was hard to see how they could then vary in brightness during the year.

It was probably the great mathematician Apollonius (c. 250–200 BC) who explained this and the other problems of planetary motion by the brilliant idea of the epicycle. This involved two circular orbits, as shown:

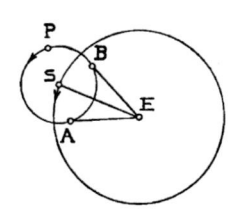

The planet P is on the small circle, the epicycle, and revolves around its centre at S; S is located on a second circle, the deferent, which has its centre at E, the earth. So the point S revolves around the earth, while the planet itself revolves around S. This epicyclic model can explain retrograde motion quite easily: to an observer on earth E, P will first seem to be moving in one direction (anti-clockwise in the diagram), but when it comes towards A it will slow down and then seem to stop, and from A to B will seem to move in the opposite direction. The retrograde motion will also take place while the planet is closest to the earth, and therefore when it is brightest. This model of planetary motion is extremely flexible, because an endless range of different sizes of epicyle and deferent, and different relative speeds, can be devised to take account of all the variations in planetary motion.

The epicycle could also be used to explain a number of other irregularities, such as the fact that that the sun takes six days longer to move between the spring and autumnal equinoxes than between the autumnal and spring equinoxes. [8] But these geometrical models by themselves were not enough to account for the actual positions of the planets during the year, and this was the achievement of Hipparchus (c.190–c.120 BC) of Rhodes, who 'transformed Greek astronomy from a purely theoretical into a practical, predictive science'. [9]

He used the Babylonian and his own observations to compile a star catalogue, graded the stars in classes of brightness (magnitude), mapped them onto a celestial globe, and devised a sort of celestial latitude and longitude system for precisely recording the positions of stars and planets. He introduced the Babylonian circle of 360° into Greek mathematics, developed trigonometry and tables to allow it to be applied to astronomy for the first time in calculating the positions of celestial bodies, and improved astronomical instruments. Astronomy was able to become the most successful of the ancient sciences because the extreme orderliness, and in a sense the simplicity, of the heavens made them ideal for precise observations, while the lack of experimental method was not a disadvantage because experiments are impossible in astronomy anyway.

The atomists, however, always rejected astronomy, not because they had any good evidence for doing so, but because they were hostile to the whole idea of a rationally ordered cosmos. Lucretius, for example, in the first century BC, was still claiming that the sun and moon could be about the same size as they appear, and only a short distance above the earth, and that the planets might be blown around in their orbits by the wind.

The culmination of Greek astronomy was the *Almagest* of Claudius Ptolemy of Alexandria, in the second century AD. Ptolemy made considerable use of Hipparchus' ideas and data, and the *Almagest*

> was the first systematic mathematical treatise to give a complete, detailed, and quantitative account of all the celestial motions. Its results were so good and its methods so powerful that after Ptolemy's death the problem of the planets took a new form... What particular combination of deferents, eccentrics, equants, [10] and epicycles would account for the planetary motions with the greatest simplicity and precision? [11]

The traditional stereotype of Ptolemy in the history of science pictures him desperately trying to maintain the superstition that the earth, as the home of Man, must be the centre of the universe, against all evidence and common sense, by fabricating an ever more elaborate system of epicycles, eccentrics, [12] and equants, until all this nonsense was swept away by the simple solution of making the sun the centre of the solar system instead of the earth. Ptolemy agreed that there was no way of telling if the heliocentric or the geocentric theory was correct by astronomical observations alone (at least in his day). But, like all the leading astronomers, he rejected the heliocentric theory not on mathematical grounds but because, for the reasons we have already discussed, it seemed a practical absurdity. Several times in the *Almagest* Ptolemy says, with Aristotle, that we should always choose the simplest hypothesis that can account for all the facts, so the enormous complexity of his astronomical model was in no way because he did not value simplicity; on the contrary, it resembled Sherlock Holmes' principle that when you have eliminated the impossible, whatever remains, however improbable, must be the truth.

Ptolemy was not only a great astronomer, but used the *Almagest* as the basis of an elaborate system of astrology which he set out in his *Tetrabiblos*. We have seen that Greek astronomers did take evidence seriously, as for example, in their arguments that the earth is round, stationary, and at the centre of the cosmos, and in their efforts to square their geometrical model of the solar system with observations. How, then it might be asked, in this intellectual environment, could astrology survive the selective process? To explain this, we must examine the philosophical background.

In the general sense that the heavenly bodies have power over the world and man, astrology goes back to the beliefs of tribal society but, especially through the many centuries of Babylonian observations, it had become much more precise. They developed the system of the twelve constellations of the zodiac, that spread not only to Greece but, with variations, to India as well. But it was the revolution in scientific astronomy brought about by Hipparchus [13] that gave astrology its mathematical teeth, and allowed astrologers to make precise calculations about the positions of the heavenly bodies. Hipparchus was a firm supporter of astrology, and the discovery around 100 BC, by the leading Stoic philosopher Posidonius, that the moon controls the tides gave astrology a useful boost.

For the Stoics, especially, the idea of a fundamental sympathy between celestial events and those on earth followed inevitably from the idea of Nature as God: 'as above, so below', as it was said. And since Nature is also inherently rational, and everything happens from necessity and not from chance, there is nothing absurd in supposing that man can use his reason to discover how the different celestial bodies and their relationships govern the destiny of earth and the human race. Cicero's objection that the planets are too far away to influence the earth makes sense to us because we think in terms of their gravitational attraction. But since there was then no idea of gravity, the great distances of the stars and planets were irrelevant, and philosophers saw nothing strange in the idea of their influences operating throughout a cosmos that was permeated by Logos and *pneuma*.

The seasons and the weather, argues Ptolemy, are obviously linked with the annual path of the sun through the zodiac, while the moon affects the growth of plants and animals, and the tides, but storms and winds are also affected by the positions of other stars and planets. Planetary influences must also be responsible for the differences of climate from one part of the world to another, and for the different bodies and characters of the various races that live in these regions. [14] Given all these correspondences between 'the above and the below', astrology could also be applied to the fortunes of individuals by casting their horoscopes, particularly because man is united to the cosmos by his reason. (To Cicero's sneer that all the Romans who fell at Cannae must have had different horoscopes, yet still suffered the same fate, Ptolemy would have replied that general calamities can always override individual fortunes.) He also argued that, just as not all diseases are fatal and can be cured by a good physician, so too not all the predictions of horoscopes are inevitable, and that by being forewarned we can avoid some of them. His basic justification for studying astrology was the

Stoic view, that by being in tune with Nature we can best live a rational life, and obtain tranquillity of mind.

The result was an immense system of correlations between the positions of the planets in relation to the stars, and to each other, affecting male and female, young and old, night and day, the regions of the world, races, nations, and individuals. There were also correlations between the planets and plants, animals, organs of the body, bodily humours, metals, colours, and all aspects of the physical world. Astrology thus became closely linked, in particular, with medicine, and with alchemy, in both of which it remained a major driving force until the seventeenth century. Simply to dismiss astrology as 'superstition', however, which has been the standard attitude of historians of science, misses the essential point: that it was an attempt to reduce nature to a systematic and rational order, and therefore played an important part in the development of modern science. As the distinguished authority on ancient science, Otto Neugebauer, has said,

> To a modern scientist an astrological treatise appears as mere nonsense. But we should not forget that we must evaluate such doctrines against the contemporary background. To Greek philosophers and astronomers, the universe was a well defined structure of directly related bodies. The concept of predictable influence between these bodies is in principle not at all different from any modern mechanistic theory. And it stands in sharpest contrast to the ideas of either arbitrary rulership of deities or of the possibility of influencing events by magical operations. [15]

*2. Alchemy and the origins of chemistry.*

Alchemy was the supposed science of transforming base metals like lead, tin, copper, and iron into gold and silver, and Alexandria was the main centre for alchemy in the West, where it seems to have begun developing from around the second century BC onwards. [16] Many of the ores of metals and their compounds are striking and brightly coloured – the gold of iron pyrites, the red ochre of iron oxide, the green of malachite, and so on. The colours of gold and silver had a special fascination for ancient man, but these colours can be produced in a number of ways. Arsenic and mercury can give a silver colour to copper, while sulphur can also produce yellow metallic compounds.

The jewellers of Egypt, and other civilisations, were thoroughly familiar with using these and other materials to make base metals look like silver and gold. [17] But there were also well-known tests for real gold: the touch-stone, a black quartz on which gold leaves marks depending on its purity, and the ability of gold to retain its colour when heated in the furnace; Archimedes had also shown that the gold wreath of King Hiero had been alloyed with silver by his test for their different densities, [18] and kings and emperors knew quite well what they were doing when they debased their coinage.

It might seem, then, that in this practical technological climate, intelligent people would not have believed that it was possible to turn base metals into real gold, and in order to understand how, like astrology, it survived the selective process we have to think ourselves back into their mental world. In the first place, ancient metallurgy was never a purely materialistic craft, like the modern blacksmith's shop where we get our wrought-iron gates. As we saw in Chapter VIII, smelting itself was often seen as a form of sexual procreation, metals were associated with the gods, and the whole of nature was a vast organism: veins of minerals were supposed to grow in the earth, and metals were believed to perfect themselves by slowly developing from the more base forms to the more precious, ending with gold. The miner interrupted this process by extracting the metals from their womb before they were ready, and the alchemist could also artificially speed up the process in his laboratory.

Greek philosophy also made an essential contribution here. Aristotle not only supported the ancient idea that metals developed in the ground, where they were formed out of sulphur and mercury, but maintained that underlying all appearances, including the elements, was basic, primitive matter, the *materia prima*, on which the specific Forms were then imposed, giving metals, say, their different properties. When alloys such as bronze were made the Forms of tin and copper disappeared, and the entirely new Form of bronze came into being, with its own properties owing nothing to those of tin and copper. So against the background of this world-view it was possible for the alchemists to accept that some 'gold' was indeed just fake gold, but also to believe that they could produce a synthetic gold in the laboratory that was as good as, if not better, than natural gold. In a similar way, we too might accept that synthetic rubber made in a factory was more essentially 'rubbery' than rubber from a tree, even though it might be different in some respects from natural rubber.

If, then, one could strip metals of their Forms, and reduce them to basic matter, it should be possible to project a new Form onto them. The Stoic *pneuma*, that permeated all physical things, acted like a seed, and when planted in the *materia prima* it could be nourished with warmth and moisture to grow into gold or silver. These beliefs were the basis of a very concrete research programme that involved using the three main agents of sulphur, arsenic, and mercury on a range of metals to produce a series of colour changes. The base metal was first 'killed', which meant removing its Form and so reducing it to the *materia prima* which, because all colour had been removed with the Form, was black. This was then whitened by mercury or arsenic compounds, and then turned yellow, for example, by the seed of gold, or sulphur water – a solution of hydrogen sulphide. (The 'divine' sulphur water, which was credited with remarkable powers, was one of the roots of the idea of the medieval philosophers' stone.)

The colours of these metals were of special importance to the idea that transmutation was possible. While the atomists believed that colour is a secondary quality of things, the product of our senses, Aristotelian and Stoic philosophers denied this and thought that colour is a fundamental property of matter. This supported the alchemists' belief that changes in colour were the best indicators of the true state of the *pneuma* and the success of their operations.

All this involved very elaborate and prolonged processes, particularly distillation which was often repeated hundreds of times; heating a substance so that its vapours would then condense as a solid deposit, or until it eventually became a powder; heating two or more substances together so that they fused; and the evaporation of liquids to produce crystallization: a whole series of transformations between solids, liquids, and gases, in other words. It is worth noting that a vast range of new apparatus was designed to perform all these experiments, [19] but no practical jeweller or metal-worker would ever have been motivated to conduct these obscure, expensive, and enormously elaborate procedures. Arab and medieval alchemists, however, used them to make a series of profound chemical discoveries that laid some of the essential foundations of modern chemistry. But alchemy also played a more general part in the development of science through its wider associations with astrology and medicine.

Gold and silver had been associated with the sun and moon from very ancient times, and the planets had also become associated with the metals in the new form of astrology, which was an integral part of alchemy from the first, as were a variety of mystical beliefs. The incorruptibility of gold had close associations with immortality, and alchemy always implied more than the mere transmutation of metals. It also signified a process of gradual perfection, from sickness to health (and the base metals, too, were regarded as sick), of moral impurity into spiritual perfection, and even from mortality to eternal life. (In China where minerals had long been used as medicines, there was a search for the elixir of Life itself, immortality, but this idea only reached the West in the Middle Ages, via the Arabs.) [20]

> Alchemy therefore existed in a cosmos governed by all those affinities between things that we have already encountered, of like with like, between metals, planets, colours, bodily humours, moral qualities of perfection and debasement, and so on.
>
> Alchemists saw such echoes and resonances as very serious. They were what deep understanding was about and revealed the intricate interweaving of everything. *The test of a scientific idea was its resonance rather than its testability, which might seem shallow, trivial and boring.* [my emphasis] Through suffering in the furnace, the lead or iron lost its baseness and became eternal gold; no material explanation was adequate for this, and we should similarly be prepared for sacrifice and pain on our way to perfection. Alchemy resembles ritual. [21]

The belief had developed in Alexandria that God had revealed to the ancient Hebrew prophets, even to Adam, the source of all the arts and sciences, and of untold powers over nature. This knowledge, although corrupted, had passed to the Egyptians, in particular the sage Hermes Trismegistus, and then to the Greeks, and these ideas were then taken up by the Arabs in Baghdad who wrote many hundreds of books on these 'Hermetic' secrets. (They were secrets because their practitioners strongly believed that most people were morally and spiritually unworthy of receiving such knowledge.} This natural magic obviously resembles the primitive magic of tribal societies in so far as it accepts a whole series of affinities between things, but, like astrology, it has now become systematized and rationalized at a far more sophisticated level that has some of the features of formal operational thought. In the

Middle Ages and the Renaissance it was to play an extremely important part in the development of experimental science, particularly in those areas that lay outside Aristotelian physics, and mechanics. A good example is magnetism.

## 3. *Magnetism and electricity.*

Knowledge of magnetism and electricity in the ancient world was largely a blank: nature gives us few clues that they even exist, and in any case the two seem very different. Lightning is the most obvious example of electricity, but in antiquity was regarded as a form of fire; the torpedo fish was known to cause numbness to humans as well as killing small fish, but the nature of this power was unknown, as, of course, was the electrical basis of nervous stimuli and muscular activity in the body. Magnetism was originally discovered through the ability of the ore magnetite to attract iron, which was well known in the ancient world. The power of amber, when rubbed, to attract chaff was assumed to be similar, although amber actually attracts by static electricity, but neither of these cases of attraction seems related to lightning or torpedo fish.

The only practical route to the discovery of electricity therefore lay through the study of attraction, and the properties of magnetite, or lodestone, to attract iron. Given the fundamental importance of cosmic sympathy in ancient thought, the lack of scientific interest in magnetism might seem remarkable. [22] While they knew that pieces of iron can be magnetized, no Greek or Roman even seems to have realised that a lodestone has two poles, so that like poles repel each other and opposite poles attract, let alone that a freely suspended lodestone or one floating on a piece of wood will point in a north/south direction. Ptolemy even says that if garlic is rubbed on lodestone it will no longer attract iron. Pliny knew that it can repel iron as well as attract it, and Lucretius describes iron filings dancing in a bronze bowl, obviously repelled by a lodestone, but while these philosophers theorized about its nature they never seem to have made any systematic investigations of lodestones, although such experiments would have been extremely easy. The puzzle is to some extent removed, however, when we realise that magnetism was the prime example of those occult forces that lay outside the scope of the mathematical sciences, and was more appropriate to the Hermetic tradition.

Indeed, the discovery that magnetite orientates itself in a north/south direction seems to have been made by the Chinese through the use of the divination-board in the first millennium AD. This was a map of the heavens at the centre of which was a spoon-shaped object made out of magnetite to imitate The Great Bear, or Big Dipper. This could freely rotate, and the tail of the spoon was used to show the southerly direction and correctly orientate the divining board. It was also used in the orientation of buildings according to the mystic art of geomancy, but the predominance of river and canal traffic in China delayed its use by seafarers until the eleventh century, and the magnetic compass seems to have reached Europe by the end of the twelfth century. We will take up the story of magnetism and electricity again in Chapter XVI.

## 4. Optics.

The microscope and the telescope are probably the two most important instruments in the history of science, while the technique of perspective drawing created enormous excitement when it was first described in 1436, and has been fundamental in Western art and science ever since. But in the ancient world magnification and perspective were only the concerns of craftsmen, and not of the slightest interest to scientists. Rock crystal, which can be remarkably transparent, was known to ornament-makers and gem-cutters of Egypt and Mesopotamia from a very early date, and it seems that while making ornaments out of it they discovered the first magnifying lenses (which only later were made out of glass). These would have been invaluable for engraving fine details on gems and seals, many of which could not have been produced without them, and there seems to have been a close association between lenses and engravers and jewellers through the Minoan, Greek and Etruscan periods down to the Romans. (A plano-convex glass lens, for example, was discovered in the House of the Engraver in Pompeii.) [23]

The earliest reference in Greek literature is not to their ability to magnify at all, but to their use as burning glasses for cauterising wounds. Glass globes filled with water were also used for this, and it is in this connection that Seneca, for example, remarks that 'Letters, however small and indistinct, appear larger and clearer through a glass globe full of water', but the idea of using this for scientific investigation or as an aid to weak vision is not mentioned. [24] Ancient scientists were also very familiar with the magnifying powers of concave mirrors, but here again, they were only interested in their powers of burning and their geometry. [25] While Seneca refers indignantly to the magnifying mirrors which the wealthy debauchee, Hostius Quadra, installed in his house the better to view his sexual orgies, he does not discuss the possible uses of magnification in the study of natural objects, which seems to have remained firmly at the craft level. We have to remember that magnification was also thought of as being essentially an optical illusion, and this view was still current in the seventeenth century, when it was deployed as an objection to Galileo's use of the telescope in astronomy. Like magnetism, magnification was one of those occult phenomena that lay outside normal science, and is another example of how cultural assumptions about 'science' dictated what was going to be studied and what was not.

Greek scientists were primarily interested in optics because it provided excellent opportunities, like astronomy, for using geometry to 'save the appearances'. Parallel lines, for example, appear to merge, and reflection and refraction provide many more examples of illusion: the image in a mirror is not really behind it; the stick seen beneath the water is not really bent, and the letters on a document seen through a globe of water, or in a magnifying mirror, are not really as large as they appear. So what we regard as the power of lenses and mirrors to reveal the truth by assisting the senses, to the Greeks, then, were optical illusions, and geometry is particularly well suited to analysing how these are produced.

But before we come to optical theory, we have to understand their ideas of vision itself. The relation between light and vision is difficult to grasp because seeing does not seem to be just a matter of the eye receiving rays of light: [26] we have to look *at* things, as well as merely receiving visual impressions of them. We now know that when we look at things we focus their image through the lens of the eye, but the Greeks did not understand the physiology of the eye, and the prevailing view was that the eye itself emitted a kind of visual ray or beam as it looked at things, which were illuminated by light. While the idea of the eye emitting rays may seem very strange to us it is actually one of the ideas that occur to children: Piaget records a child asking, "Papa, why don't our looks mix when they meet?" And an adult recalls that "When I was a little girl I used to wonder how it was that when two looks met they did not somehow hit one another." Children also find it hard to distinguish between light and vision: so an adult asks a child "Are to see and to give light the same thing?" "Yes." "Tell me the things that give light." "The sun, the moon, the stars, the clouds and God." "Can you give light?" "No…Yes." "How?" "With the eyes". [27] Here we meet another example of an intuitive mode of thought among the Greek philosophers that they nevertheless developed with deductive geometry in fundamental studies of reflection and refraction.

The eye beams, like rays of light, were thought to radiate from the eye in straight lines because straight lines are the shortest distances between two points, [28] and Euclid held that the visual rays actually form a cone:

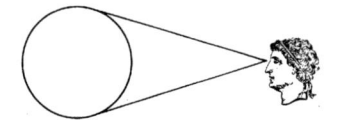

On this fundamental geometric assumption, he proved that the apparent size of an object was proportional to the size of the angle at the apex of this cone, so that a mountain a long distance away would subtend the same angle as a much smaller object close at hand. It seems clear that without the false notion of visual rays it would have been difficult, if not impossible, for the idea of the cone of vision to have developed, and so for Euclid to have applied geometry to optics. The non-existence of visual rays made no practical difference to the study of reflection and refraction, and the imagined visual cone actually replicates the way in which the lens of the eye focuses light rays on the retina. This is another example, like circular motion, of a false idea actually providing the essential basis for scientific advance because the true facts of the matter were too difficult to grasp.

Euclid in his *Optics* discusses a whole series of problems in which the way objects appear to be is not how they really are, and uses geometry to 'save the appearances'. For example, to explain the illusion that parallel lines seem to converge, he shows that because the apparent size of objects depends on the size of the angle that they subtend at the eye, the farther the lines become from the eye, the smaller the angle that separates them becomes, so that they seem to get shorter with increasing distance.

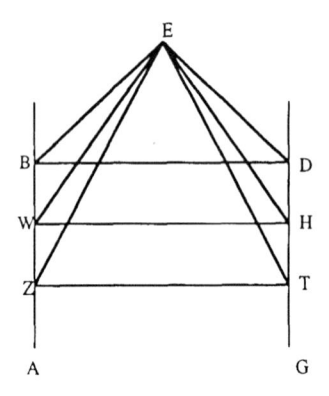

In the diagram, AB, GD are the parallel lines, and E is the eye. Because ∠ZET is smaller than ∠WEH, which is smaller than ∠BED, line ZT would seem shorter than WH, which would seem shorter than BD, so that BA and DG would therefore seem to converge. He then applies this reasoning to planes above the eye (as in a ceiling) and shows that the greater the distance, the lower it seems to become; with planes below the eye (as in the floor) showing that the farther it is the higher it seems to become; and that planes to the left of the eye seem to incline to the right, while planes to the right of the eye seem to incline toward the left. If we were to put these propositions (nos. 11-13 of the *Optics*) together, the result would look as follows, with four planes receding to a vanishing point:

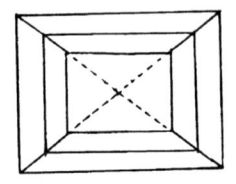

But Euclid's whole purpose was to *explain* an illusion, not to teach painters how to create one, and there is no evidence at all that he had any interest in perspective drawing or in the vanishing point.

Perspective drawing, in so far as it existed in the ancient world, was primarily a craft interest and of no concern to scientists in their study of optics (with one exception that we will come to). The painting of pictures on vases, theatre scenery, and the walls of domestic houses was seen as a challenge to represent three-dimensional reality. The practical study of foreshortening made considerable progress, and a side-view of solid objects such as a throne or tomb, 'three-surface' perspective, which is relatively easy, was fairly common even if it was often incompetent. [29] But to organize all the objects in a scene as though observed from a single point of view requires an understanding of the vanishing point, on which all lines converge, as illustrated above. This is a far more difficult technique, as Piaget notes, and there are in fact only two brief and casual references, by Lucretius and Vitruvius, to the vanishing point in classical literature. [30] Vitruvius refers to the fifth century Athenian Agatharcus, suggesting that he applied geometry to theatrical scene-painting. [31] But whatever he may have achieved, it seems that there was no generally recognized theory of perspective in antiquity, and that any successes in this were due rather to the craft skills of individual artists, than to academic theory. To judge from its rarity, even in villas at Pompeii, only a few painters managed to achieve it with real success, and the impression is that painters had to master these problems without any assistance from science, just as the scientists remained aloof from the magnifying glasses of the jewellers and engravers.

The only significant case of science being applied to the problem of perspective was that of Ptolemy, who showed how to project the earth's sphere on to a map. [32] A crucial idea was that there is a central ray in the cone of vision, which became the focus of what we now know as the vanishing point. Ptolemy's work on map projection was basic to the Renaissance theory of perspective, but he would have had no interest in teaching painters their craft, nor would they have thought of asking him. Ptolemy also carried out some genuine experiments in optics, for which he designed some simple apparatus. [33] Although he could not base any general laws on these, they are one of the very few examples from ancient science of correlations of quantitative measurements obtained by experiment.

We now come to the nature of light itself. To us it seems an obvious Fact that when a sunbeam strikes a piece of glass, and throws a rainbow image on a wall, it is the glass that breaks up the light into its component colours. But to the Greeks it was an equally obvious Fact that colours are in the things we see, and that light simply reveals these and contains no colours itself. So when the colours of the spectrum were seen in sunlight passing through a prism, in the rainbow, or the spray from burst water pipes, they had to be explained as a special kind of illusion produced by reflection. Aristotle, for example, said that the colours of the rainbow are the result of the reflection of the sun in the tiny water droplets in the black cloud opposite the sun; they are individually too small to reflect the image of the sun, so they can only reflect its colour. The different colours are produced by the weakening of the visual rays with distance in the process of reflection: the appearance of red is produced by the strongest rays because those droplets are closest to the sun, green by weaker rays, and violet by the weakest of all, but the matter was debated for centuries. While in a sense the ancients had the evidence [34] at least to question the traditional theory of light, and understood something of refraction, their philosophical assumptions made it very difficult for them to take the evidence seriously.

## 5. Mechanics and engineering.

For the Greeks and Romans, machines had four basic applications: military engineering; the lifting of heavy weights in the construction

Ancient Sciences

of monumental public works; raising water; and gadgetry for theatres and temples. (The water-mill for grinding corn also appeared in the first century BC.) The writings on mechanics that have come down to us seem to have been mainly handbooks for students of military engineering and architecture. (The architect did not just draw up plans but was responsible for the whole building project, and needed to understand the mechanics of heavy lifting.) Most of the ancient engineers wrote on all these topics, particularly catapults, [35] and they all worked at royal courts or the Museum in Alexandria. Archimedes (287–212 BC) is the most outstanding, and we can also include the Alexandrians – Ctesibius, Philo, (both also third century BC), and Hero (first century AD) – and the Roman architect Vitruvius (first century BC) and his contemporary Frontinus, [36] an expert on water supply. Ancient mechanics were therefore very much within the domain of the state, rather than of private commercial interests.

One might think that all these practical applications of force would have led to advances on Aristotle's ideas about force and motion (dynamics). But when Greek and Roman engineers were confronted with difficult problems in dynamics, such as the range and trajectories of missiles, rates of water flow and pressure, or the transmission of forces in buildings, they did not use these to advance the theory of dynamics. They either ignored such problems altogether, or reduced them to a static form, in which they could best use their powerful mathematical tools of geometry, proportion, and number.

For example, the engineers of King Philip II of Macedon, father of Alexander, had invented an entirely new type of catapult which replaced the earlier bow by two separate wooden arms; these were inserted into two vertical skeins of sinew rope mounted in frames, as in the diagram:

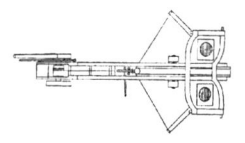

Clearly, the larger these sinew torsion springs were, the heavier the projectile they could fire, but deciding on the correct proportions of catapults of different power was a major problem. The Ptolemies of Egypt employed engineers to carry out extensive and methodical tests with a range of catapults, and after a series of experiments discovered that the fundamental relationship was between the weight of the projectile (whether a bolt or a stone), and the diameter of the torsion spring. Once the diameter of the spring had been decided, a fixed set of proportions then dictated the dimensions of the other parts of the catapult. [37] This is, to my knowledge, the best example of the (fairly elementary) use of mathematics in the ancient world to solve an engineering problem, and it was about proportion, not dynamics.

It is most interesting that there is no attempt in the catapult literature to relate the weight of the projectile with its range – will half the weight travel twice as far from the same catapult, for example? – or to establish what elevation achieves maximum range. Indeed, the whole relation between range and elevation was left to the soldiers to work out by experience. [38] In contrast, Renaissance writers on the ballistics of cannon treated these as central problems, and also applied ballistics to Aristotle's ideas about projectile motion, whereas ancient engineers never used their knowledge of catapults to test Aristotle's theories.

Greek science achieved most in the branch of mechanics known as 'statics', where bodies are stationary. This was essentially founded by Archimedes, the greatest mathematician of antiquity who also made extraordinary practical discoveries in engineering; [39] he is the closest we get to an ancient Galileo, who was inspired by his scientific example.

He proved the basic mathematical laws of the lever, established the centre of gravity in a number of different shapes, and also discovered the principle of buoyancy while working on the problem of King Hiero's wreath [40] to see if its gold had been mixed with silver. An object floats if it weighs less than the volume of water it displaces, i.e. is less dense than water, which is why stones sink and air rises in water. [41] (This insight, however, was never assimilated into theories of the four elements.) Archimedes used his invention of the compound pulley, together with the windlass and the endless screw, to demonstrate how a large ship could be hauled along a beach. [42] The principles of mechanical advantage were therefore well understood because they could be solved geometrically, and Hero later wrote extensively on the principles of the lever, the windlass, the pulley, the wedge, the screw and gear ratios.

Forces also occur in buildings. Vitruvius wrote the classic text on architecture in the Roman world, but he had been in the Imperial service as superintendent of catapults and an administrator of the water system, so had wide mechanical knowledge. His book has many chapters on the proportions and harmony of buildings, on the use of vases of different sizes to improve the acoustics of theatres, and on the minutiae of building construction and the manufacture of concrete. But he has almost nothing to say about the weight of a building's components, and how architects can control the forces they produce. While he refers briefly to the relieving arch in a wall, and describes how it transfers the weight above it outwards to the foot of the arch, he does not discuss the general principles of arches and explain how they work. Nor is there any mention of the forces acting on stone lintels, and the limits beyond which they may crack, or the maximum angle at which a wall can lean before it falls over. Nor does he discuss how a pitched roof can be prevented from pushing out the walls underneath it, although roof trusses or their equivalents were certainly used. [43]

Galileo, by contrast, begins *Two New Sciences* with a discussion of why structures above a certain size collapse under their own weight, and why beams and lintels break. Yet he uses only Euclidean geometry and the work of Archimedes on the lever, all of which were available to Vitruvius. Vitruvius' book also contains a great deal of practical information on machines including cranes, catapults, and water-pumps. So again, it is striking that he takes no more than two short paragraphs to describe the water-wheel and the water-mill, which had appeared in his own lifetime, and does not discuss the differences in power generated by overshot and undershot wheels, although both were known to the Romans.

Water-flow and pressure were obviously basic problems for the designers of aqueducts and piped supplies, and Vitruvius and Frontinus give calculations of the cross-sectional areas of water-nozzles with different diameters, because the rates at which customers were charged were based on these nozzle-sizes. But since they did not attempt to calculate pressure, nozzle size alone could only give the crudest indication of the rate of flow and the amount of water their customers were getting. Again, bursting pipes were a general problem for the Roman water supply, but although it was known from experience of water-clocks that the rate of flow depends on the head of water in the tank, and runs more slowly as the water level goes down, this knowledge was never part of any experimental investigation into water pressure, as we shall see in the next section.

So far, we have not touched on those machines and devices that had no practical purpose, and were for the entertainment of the wealthy and for use in temples. In Alexandria, there was a passion for automata that imitated the motions of the celestial bodies, such as elaborate water clocks and planetaria, and the movements of men, birds and animals. These developed together, and used many of the same principles, especially those based on gear-wheels, siphons, compressed air, and steam. The mathematician and engineer Hero wrote extensively on them, as well as on theoretical mechanics, and historians of science have predictably been outraged by the spectacle of science debased by inventing frivolous conjuring tricks for the rich (singing and drinking birds, satyrs pouring wine, magic jugs that will only pour when one's thumb is removed from the handle, or goblets that can never be drained); and the superstition of temple gadgetry such as slot machines for holy water, temple doors opened by a fire on the altar, or a shrine over which birds can be made to revolve and sing. But it was these attempts by men to play at being gods that led to important advances in mechanics, hydraulics, and pneumatics. 'The most ingenious mechanical devices of antiquity were not useful machines but trivial toys. Only slowly do the machines of everyday life take up the scientific advances and basic principles used long before in the despicable playthings and overly-ingenious, impracticable scientific models and instruments.' [44] This applies particularly to pneumatics, which was mainly used for what we would consider frivolous reasons and led to very important discoveries.

*6. Pneumatics and steam-power.*

Pneumatics began with Ctesibius c.270 BC when he was installing a movable mirror in his father's barbershop. [45] To keep the mirror in place he had devised some counter-weights which ran inside tubes, and compressed the air within them. Ctesibius was a mechanical genius who later worked at the Museum, and this discovery led him to invent the piston and cylinder that, with the addition of valves, could be used as a pump for air and also for raising water and draining mines. He was a rare example of a person in the Graeco-Roman world who rose from the artisan class to the ranks of the intellectual elite.

Researchers at the Museum were also familiar with the siphon, and with using steam and compressed air in a wide range of automata, and so had the opportunity to experiment with the vacuum, steam power, and atmospheric pressure, which is what particularly concern us here. Hero, especially, was not only an impressive practical researcher, but an expert mathematician, and here, as in the case of Archimedes, one might expect to find, if anywhere in the ancient world, some understanding of the power of experimental method.

He begins his *Pneumatics* with a discussion of Aristotle's theory of the vacuum. Philo had already shown that when a vessel is heated with its neck under water, it will expel some of the air within, and when cooled it will suck the water back again, and this and experience of other devices led Hero to conclude that while a continuous vacuum does not occur in nature, 'it is to be found distributed in minute proportions through air, water, fire and all other substances'. This is the 'porosity of matter' of later Aristotelians such as Strato (in his treatise *On the Void*), not really true atomism, and Hero uses it to explain why, when an artificial vacuum is produced by suction or by the cupping glass, the air will immediately rush back again when the vacuum is released. This was because there is a natural elasticity of the air particles like that of a sponge when compressed, which will resume its former shape when released. It is also because the particles 'will have a rapid motion through a vacuum, where there is nothing to obstruct or repel them, until they are in contact [again].' So an artificial vacuum collapses because of the natural tendency of air particles to fill it, and there is no idea that atmospheric pressure could be involved in this.

The idea that suction is a positive force, in a way like magnetism, that pulls up the liquid, is extremely seductive to adults, as well as to children. There is, however, a very curious passage where Hero tries to explain how wine flows up a tube when we suck it. He argues that as we suck the air out of the tube our body becomes proportionally bigger and so presses on the surrounding air, which in turn presses on the surface of the wine. The vacuum we have created in the tube then allows the wine to flow into it. Although he does not offer the slightest experimental evidence in support of this theory it does, of course, contain the germ of the idea that it is the pressure of the atmosphere on the surface of the wine that forces it up the tube to fill the vacuum.

But Hero, like every one else in the ancient world, could not have imagined that the atmosphere exerted any pressure at all on the earth's surface: air is lighter than water, and he regarded its natural movement as upwards, so that the notion of the atmosphere as a whole exerting a downward pressure on a water surface would have seemed absurd. It was thought of as extending up to the moon's sphere, and it was also supposed to be much thinner close to the moon and much denser closer to the earth. No one, of course, had measured this, and the reason for the difference in density was that the atmosphere was thought to be affected by the different qualities of the earth and the heavenly regions. [46]

The depth of air or water *itself* therefore produced no increase in their pressure. Hero's main piece of 'evidence' for this is the 'fact' that 'those who dive deep supporting on their bodies an immense weight of water are not crushed', and he supports this by the claim that air is not forced out of their lungs. At the level of theory he argues as follows: imagine the column of water directly above the diver and resting on him; then imagine that this is replaced by a similar column of some other material than water, but of the same density. Now, he says, Archimedes has shown in his book *On Floating Bodies* that bodies of the same density as water, when immersed in water will neither protrude above it, nor sink in it, so this body could not therefore exert any weight on the diver beneath it, and therefore the water cannot do so either – QED. Although this is an extremely clever argument, it is striking that Hero apparently made no attempt to test it, which he could easily have done by taking an inverted glass down to the bottom of a public bath, and observing that the water level inside clearly rises as pressure increases with depth. The decrease in the flow of water clocks as the level of water in the tank goes down is also a fairly obvious indication that water pressure varies with depth, but which, again, he seems to have ignored.

He does, however, provide one excellent experiment, which is designed to disprove the theory that water flows down the long arm of a siphon because it weighs more than the water in the arm inside the tank. He made a siphon tube in which the arm outside the vessel was shorter but thicker than the arm inside the vessel, so that the water in it must weigh more, but showed that it did not flow, thus proving that it is not the weight of water that produces the flow, but the exit of the external tube being below the level of the water inside the vessel.

Hero and others at the Museum were thoroughly familiar with the expansive power of steam: Hero designed an elaborate boiler to blow steam and hot air on a fire and also his famous type of steam turbine, in which a hollow ball is made to rotate by two jets of steam from nozzles mounted on the ball. But practical tests [47] have shown that it would have had an efficiency of about 1%, so that enormous amounts of fuel would have been needed even for a machine of about $1/10^{th}$ horsepower, essentially that of one man, whom it would have been much simpler and cheaper to employ in the first place. In theory, a jet of steam could have been used to drive turbine blades of windmill form, but a useful steam turbine was technologically impossible at that time.

The only practical gateway to steam power was by use of atmospheric pressure and the vacuum; but while Hero knew that when a vessel is heated the air in it will drive water out, and that when it cools it will suck it in again, (the basis of the thermometer in the 17$^{th}$ century) he does not seem to have realised that the condensation of steam will itself create a vacuum, and of course knew nothing of atmospheric pressure. Pistons, cylinders and valves were well known, as was the basic idea of the beam to transmit motion in the same way as Newcomen's beam engine. Technically, then, a small version of the Newcomen engine could have been built, but it would have seemed pointless since there was no idea of atmospheric pressure, or the power of steam to produce a vacuum. In any case, even if a useful steam engine could have been built, it would have been enormously expensive, and a far cheaper and simpler way of harnessing natural power was the water-wheel. But Hero does not even refer to this, and while he does mention the windmill (the sole reference in antiquity), [48] it is only for driving the bellows of an organ (he is far more interested in describing the piston-bellows than the wind-mill). Like Vitruvius, he seems to have had no interest in harnessing steam, water, or wind power for any serious productive purposes.

*7. Conclusions.*

The particular sciences we have been examining in this chapter were dominated by the immense intellectual and moral prestige of philosophy, of reason and mathematics, whose role was to be the tutor of the senses, not to be taught by them. (A good example is Hero's use of Archimedes to prove that water pressure does not increase with depth.) One of our most important themes is that philosophy is capable of extraordinary insights, but unless it becomes involved in the close investigation of nature, it can only go so far, and ultimately dies out in sterility. This happened to ancient natural philosophy, and again in Islamic civilisation and in medieval Europe. Greek science, as we have seen, made some investigations of nature, but apart from medicine, biology, and music, which we have not had space to consider, it was at its most impressive in the narrow area where mathematics, particularly geometry, could be applied and from which forces, motion, and change were excluded. While their astronomy was based on premises that were fundamentally wrong, these were essential conditions for its development. It was a remarkable intellectual achievement, and like the geometrical optics of Euclid and his successors, formed the basis for subsequent developments by Islamic, Medieval, and Renaissance scientists. Archimedes' work on the lever and buoyancy founded the science of statics, and the work of Ctesibius, Hero, and others at Alexandria on basic machines such as gears, inclined planes, and pulleys, and on pneumatics, were of permanent value, and were also taken up by later scientists.

What is striking, however, is that once forces and motion became involved, and any measurement other than weighing was required, what we would consider problems were either ignored, or converted into static forms that could handled by geometry. Just as knowledge of military catapults did not stimulate any challenges to Aristotle's theories of motion, the knowledge of building techniques did not generate any theoretical interest in the forces generated within structures; water-clocks, pumps, and siphons did not prompt any experimental investigations of air and water pressure; and steam and water power were of no scientific interest at all. More widely, neither lenses nor the problems of perspective painting stimulated any real scientific investigation; awareness that the lodestone attracts iron did not lead to any studies of magnetism; and dice games did not lead to any investigation of probability. To be sure, some measurements were made, [49] and a few experiments were performed, by Ptolemy in optics, Hero on the siphon, and in acoustics with the monochord and similar devices. But these were very rare, and it is more striking that so often they did not perform experiments that would have been easy to set up. [50]

'The facts', then, may only have a very limited impact on theory, and technological experience does not automatically suggest the value of testing theories by experiment as a means of discovering scientific truths. While Hero and Archimedes were obviously familiar with all sorts of mechanical devices, and it is highly likely that it stimulated their thinking on, say, the lever, they would not have thought this workshop experience had any place in a scientific treatise, which was about proof. As we have seen, most ancient scientists were entirely aloof from the technical world of the crafts, and did not imagine that they could learn anything from it, nor did they have any association with the sordid world of commerce and manufacture. Researchers were gentlemen scholars, of independent means or funded by the state, and the state's main interests were in public works and entertaining gadgetry for a wealthy elite, not in serving the interests of businessmen by encouraging useful inventions that could have made a profit. In these respects the Greeks and Romans were typical of all the ancient civilisations, so in the next chapter we will have to see why it was only in the exceptional social conditions of Western Europe in the Middle Ages that science and modern technology could emerge, and why in the normal course of events this would not have happened.

# CHAPTER XIV
## The Uniqueness of Western Society

*1. The limits of the ancient civilisations.*

If 'Science's flourishing periods are found to coincide with economic activity and technical advance', [1] these early developments in science should have continued, either under the Romans and Byzantines, or else in India, the Islamic world, or China, and at least one of these empires should have produced modern experimental science. Yet despite periods of great economic prosperity, with major flowerings of culture, technology, and the arts in the first millennium AD, science never developed beyond a certain point in any of these civilisations. This chapter will therefore be a further test of the materialist theory that science simply emerges as a response to economic and technological conditions, and will show instead that the whole organization of society, and its values and beliefs, also have a profound effect on the development of science and on technology itself. We shall pay particular attention here to the extent that capitalist enterprise was able to develop, and the possibility for craftsmen to associate with scholars, because both these factors were crucial for the development of modern science. They were closely associated with the organization and status of towns, which will also be of special importance in this chapter.

*The Roman Empire.*

The Roman Empire was the only one in which the Emperor had to consult an assembly, the Senate. But its political power had passed to the Emperor, who had sole control of legislation and the powerful bureaucracy, and of the army which was the foundation of the Roman state and of political success; the authority of the Senate was merely moral and cultural, representing as it did the educated upper classes. The Senate certainly had no power over taxes, which were decided by the central government, and collected by tax-farmers or local town councils. The city-state had ceased to exist, and while the Empire was based on towns, they were designed on a standard pattern to spread '*Romanitas*' to the provinces. Town finances were subject to particularly rigorous control by the Imperial bureaucracy, and their charters were all based on a fixed model issued from Rome. This allowed them the privilege of their own elected councils, but while freemen could elect the councillors, these had to be from the decurion order, basically well-educated landed gentry, who alone were supposed to have developed the ability to engage in rational debate, which effectively excluded merchants and artisans from political office.

By the time of Augustus, (63 BC–14 AD) civil law was well adjusted to the realities of commercial life, the Romans had a relaxed view about charging interest, and large-scale banking and credit were well developed, and highly respectable. Money-making by trade, particularly retail trade, still had a disreputable image, however. Lyons (Lugdunum), for example, was one of the great trading and manufacturing centres of the Roman Empire, with many wealthy merchants. But they were all freedmen and foreigners, and none was even a citizen of Lyons, much less a member of the local or imperial aristocracy. [2] Artisans and manual skills generally were despised, not least because of their associations with slavery: Cicero, for example, referred to shopkeepers and artisans as 'the sewage of the state', and the fundamental distinction between the liberal and mechanical arts continued to make any collaboration between educated scholars and craftsmen socially impossible. Despite the great importance of commerce, this was a fundamentally agrarian civilisation, and the basic sources of wealth and prestige were land. 'Power, honour and status were the most potent forces in the ancient world' and while obviously business-men wanted to become rich, 'When a man had made a fortune from other sources, commerce or more rarely manufacturing, his surpluses were more often devoted to estates than ploughed back…into the business. Wealth was, in fact, more often treated as a means to an end – a means of entrée into the circle of landed aristocrats – rather than as an end in itself'. [3]

The way to become rich by manufacturing was to produce fine metal-work and jewellery, ceramics and textiles for the aristocracy, not by industrialised production for the masses. 'When they could, in short, the ancients turned their crafts into arts: they did not, with few exceptions, attempt to convert them into industries.' [4] There was obviously, in addition, a plentiful supply of slaves and cheap labour that would also have made expensive investment in water-wheels and other sorts of machinery uneconomic. There was not, then, much incentive for the merchant class to invest in technology. Those who owned workshops thought of their markets as more or less fixed, often governed mostly by local needs because of transport limitations, and certainly not as indefinitely expandable if only they could turn out enough goods at a competitive price. While *we* can imagine water-wheels driving banks of potters' wheels, or powering blast-furnaces and trip-hammers in iron foundries, although such things would have been technically feasible they would have been very expensive to develop, and Roman business-men did not invest capital in that way.

In the Graeco-Roman world the state was the only important source of investment in large-scale technology, but was basically only interested in military technology and ships, public works and water supplies, and entertaining gadgets, while the water-mill was introduced to grind flour, a monopoly of the Roman state.

> the Ptolemies founded and financed the Museum at Alexandria, for two centuries the main western centre of scientific research and

invention. Great things emerged from the Museum, in military technology and in ingenious mechanical toys. But no one, not even the Ptolemies themselves, who would have profited directly and handsomely, thought to turn the energy and inventiveness of a Ctesibius to agricultural and industrial technology. The contrast with the Royal Society in England [1660] is inescapable. [5]

While the mechanicians at the Museum produced a wide range of remarkable inventions, these were not taken up by the wider society because the Museum and its patrons were an isolated and elite class, with no effective links with the merchant class and the artisans: the rapid spread of the mechanical clock in the towns of medieval Europe, for example, is a striking contrast to this.

Nor was there any general belief in material progress. Aristotle, for example, had maintained that 'the advances made by the arts and sciences in each civilization were the fulfilment of the potentialities of their natural form, beyond which they could not go.' [6] While Seneca said that 'One day our posterity will marvel about our ignorance of causes that are so clear to them', he was thinking only of intellectual, not technical advance. People assumed that the technology of their day was about as highly developed as it could be, and in any case there was no reason to think that it could be improved by pure science, which even most of the educated found pointless and boring, and much less interesting than Man. Roman education was based on literature and grammar and, at the highest level, on acquiring the arts of rhetoric to be an effective speaker in the courts and the Senate. There were many public libraries, and some centres of learning such as the philosophy schools of Athens, and the Museum of Alexandria, but there were no equivalents to the universities that developed in medieval Europe, and higher studies in the sciences were mainly the private pursuits of a few wealthy men.

The Romans, in particular, valued practicality, not theory, and preferred books that dealt with useful subjects and were morally edifying. [7] Even to Greek intellectuals, science was 'a generally unattractive field, not to be compared with the delights of literature and philosophy', [8] and by the third century AD it was intellectually pretty well exhausted: 'The main effort tends more and more to be spent on preserving knowledge rather than on attempting to increase it'. [9] Although they were the heirs of Greek science, the actual demand for science among educated Greeks and Romans was therefore at an elementary level, much titillated by curiosities, and was catered for by hack writers who compiled popular text-books and encyclopaedias: the Imperial court certainly had no interest in fostering the natural sciences.

Developments in science had come to a halt well before Christianity became the official state religion under Constantine in about 312 AD. But while Christianity did not kill science off, it was certainly not interested in reviving it. Whereas for the Stoics, in particular, knowledge of how nature works was part of the intellectual equipment of the rational man, as one of the means by which he attained tranquillity of mind, for the Christian all that was necessary for salvation were the essential truths of religion. So the more extreme advocates of 'holy ignorance', such as Tertullian, could say 'We have no need of curiosity after Jesus Christ, nor of research after the Gospel', [10] and even St Augustine, who was primarily responsible for adopting Plato into Christian thought, and personally admired Greek science, did not think it necessary for the Christian to know anything about it.

It was probably more important for the history of philosophy and science in the Roman and Byzantine Empires that Constantine made Christianity the *official* religion: 'What marked Christianity out [from paganism] is not the particular doctrines associated with it, so much as the fact that those doctrines eventually received unprecedented state approval and support'. [11] Once the Emperor was legitimated by an official Church, theological disputes inevitably became matters of imperial concern, and Constantine, for example, convened the Council of Nicaea in 325 to settle a controversy about the Trinity which seemed to threaten the unity of the Church. The pagan Emperor Diocletian had already ordered all alchemical books burned in 292 AD, and banned all divination, including astrology, as subversive to the state. (In the Middle Ages, alchemy and astrology were to be very important in promoting the idea that science could be useful.) Constantine reinforced these prohibitions – the Church was particularly hostile to astrology and divination in general because they seemed to deny Free Will – and Justinian closed the philosophical schools at Athens in 529 AD on the grounds of heresy. The Western Empire had already collapsed by 476 AD, as the result of the barbarian invasions, but even if it had survived, it would have been as unfavourable to the advancement of science as the Eastern Empire proved to be.

*Islamic Society.*

Unlike Rome and China, Islamic society was only briefly a united empire, and for most of its history has been a civilisation based on the religion and culture of Islam, but split up into a number of empires and states. This Islamic civilisation passed through several phases. During the century of its initial expansion, after the death of Muhammad in 632 AD, into Syria, Egypt, Iraq, and Persia, it was essentially Arab in religion, language, and culture. Though its disparate tribes were held together by adherence to Islam, their primary aim was conquest and booty, not the conversion of unbelievers, and initially the desire of some of their new non-Arab subjects to convert to Islam was unexpected and even unwelcome. The Umayyad Caliphs, who had inherited the authority of the Prophet, and moved their capital from Arabia to Damascus, found themselves masters of an expanding empire that comprised a variety of peoples, notably Egyptians, Greeks and Persians. While Christians, Jews, and Zoroastrians were second-class citizens in relation to Arabs, as 'peoples of the book', they were given protected, *dhimmi*, status and often employed by their new rulers regardless of their faith.

The basic weakness of Arab rule was the endemic tribal and clan rivalries of their society, together with competing hereditary claims to the Caliphate itself, and in 750 AD a new dynasty, the Abbasids, overthrew the Umayyads and soon shifted their capital to the new city of Baghdad, in what had been part of the Persian Empire. This marked the beginning of a very different epoch, in which the emphasis shifted from conquest to the assimilation of Greek and Persian culture, and Arab descent became less and less important. Converts to Islam were now fully accepted, and Persians in particular played a much more important part in the court and in the intellectual life of Islam, although Arabic remained the official language. They had a tendency to belong to the Shia sect, which had an interest in natural science, as opposed to the Sunni, who were more interested in law and theology, and there was a replay of many of those Axial Age developments we examined earlier, that from the eighth to the twelfth centuries produced a golden age of philosophy, science, and technology.

Although Arabs were literate and had cultivated the art of poetry, their traditional culture could not compare in sophistication with that of the Greeks and Persians, and the Arabs had to defend Islam in debate at an intellectual level for which they found themselves unprepared, particularly against Greek philosophy. The Abbasid court at Baghdad not only patronised the arts and technology, in which great developments were made, but encouraged vigorous discussion about philosophy and religion, and a House of Wisdom was established to translate Persian, and then Greek works into Arabic. Especially under the Caliphs Harun al-Rashid (786-809), and his son al-Mamun (813-833), many works of Greek philosophy and science were translated into Arabic; Aristotle was particularly admired, becoming known as 'The Philosopher'.

The state encouraged a number of sciences because they were considered useful: medicine, obviously; mechanics; alchemy; mathematics and astronomy for calculating celestial events in relation to religious observances, and for astrology; and geography and cartography in the interests of long-distance trade. Many distinguished scholars used the library and other facilities of the House of Wisdom to translate and study Greek works on these and other sciences, especially statics and optics, and develop them further. (For the details, see note 12.)

Greek philosophy also had a great impact on religious thought, especially on the theological school known as the M'utazillites who, in a similar rationalist tradition to Philo, Augustine, and Aquinas, attempted to show that the claims of reason and revelation were compatible. For a time they enjoyed the favour of the court, but Caliph al-Mutawakkil (847–61) turned against them and restored the influence of the orthodox theologians. These had always been hostile to Greek philosophy, and their antagonism culminated in the work of al-Ghazzali (1058-1111) who, in particular, helped to discourage the study of the natural sciences as well as philosophy in the Muslim world. He had been brought up as a Sufi, an Islamic sect that favoured a mystical approach to God, and when he came to Baghdad and encountered philosophy he experienced a period of intellectual turmoil. As a result, he became an especially determined and influential opponent of Aristotle and philosophy, particularly in his book *The Incoherence of the Philosophers*. One of his most important teachings was that what appear to our reason as natural causes and effects, are in fact directly created by God, who constantly intervenes in all aspects of His creation. There are thus no real laws of nature at all, but only the arbitrary workings of God's omnipotent will. If this is the case, then theoretical natural science becomes rather pointless [13] and after al-Ghazzali, during the twelfth century, the natural sciences gradually lost their hold on the interest of learned men. So in the next phase of its development, after the absorption of new ideas under the early Abbasids, Islamic civilisation proved too conservative to allow the sort of scientific development that occurred in Europe.

> The symbol of Islamic civilization is not a flowing river, but the cube of the Kaaba, the stability of which symbolizes the permanent and immutable character of Islam. Once the spirit of the Islamic revelation had brought into being, out of the heritage of previous civilizations and through its own genius, the civilization whose manifestations can be called distinctly Islamic, the main interest turned away from change and 'adaptation'. [14]

Science remained subordinated to the world-view of Islam, dominated by Shari'a law which, based on the Qur'an and the traditions of the Prophet, pervaded every aspect of Islamic life: 'The *Shari'a* was not only a normative code of law but also, in its social and political aspects, a pattern of conduct, an ideal towards which people and society must strive.' [15]

> In the early days of Islamic society
> there had been a rule called *ijtihad*, the exercise of independent judgement, whereby Muslim scholars, theologians, and jurists were able to resolve problems of theology and law for which scripture and tradition provided no explicit answer. A large part of the corpus of Muslim theology and jurisprudence came into being in this way. In due course the process came to an end when all the questions had

been answered: in the traditional formulation 'the gate of *ijtihad* was closed' and henceforth no further exercise of independent judgement was required or permitted. All the answers were already there, and all that was needed was to follow and obey. One is tempted to seek a parallel in the development of Muslim science, when the exercise of independent judgement in early days provided a rich flowering of scientific discovery but where, too, the gate of *ijtihad* was subsequently closed and a long period followed during which Muslim science consisted almost entirely of compilation and repetition. [16]

So in astronomy the Ptolemaic model, for example, could not be challenged because the framework of Islamic cosmology was derived from the Qur'an: none [of the astronomers] did, nor could they, take the step to break with the traditional world view, as was to happen during the Renaissance in the West – because that would have meant not only a revolution in astronomy, but also an upheaval in the religious, philosophical and social domain. No one can overestimate the influence of the astronomical revolution upon the minds of men. [17]

Beyond the natural sciences, any departure from established tradition also came to be problematic in the Islamic world:

> In the Muslim tradition, innovation is generally assumed to be bad unless it can be shown to be good. The word *bid'a*, innovation or novelty, denotes a departure from the sacred precept and practice communicated to mankind by the Prophet, his disciples, and the early Muslims. Departure from tradition is therefore bad, and in time *bid'a*, among Muslims, came to have approximately the same connotation as heresy in Christendom. [18]

Imitating the infidels by borrowing their customs and inventions became a particularly bad form of *bid'a*: as the Prophet said, 'whoever imitates a people becomes one of them'. This was a powerful argument used by religious conservatives to block Western technological innovations, and even printing was forbidden because it would defile the name of God, and spread subversive ideas. (An exception was made for weapons and techniques of warfare, because in *jihad* it was permissible to borrow these from the infidel to use against him.)

Quite apart from these religious and cultural attitudes that ultimately discouraged the development of science, the wider society outside the high culture of the court did not allow the development of capitalism in the manner of medieval Europe, or that *rapprochement* between scholars and craftsmen that was so essential to the development of modern science. The Bedouin warriors who had been the basis of the first Islamic armies had been increasingly displaced by non-Arab recruits, many of whom were slaves. The slave-soldier, the *ghulam*, had the major advantages that he had no tribal or clan loyalties, but in theory was loyal only to the Caliph, and even if he were promoted to the highest rank could still be executed at the whim of the ruler.

The Abbasids based their armies on these *ghulam*, particularly the Turks, who were now entering the Islamic world from the steppes and, unlike the Arabs, had abandoned their clans and tribes. Their power over the Caliph grew, especially during the reign of al-Muqtadir, and by 935 AD the new office of the Amir, military governor, reduced the Caliph to a puppet. In 1055 the Seljuk Turks captured Baghdad, and from then onwards Turks dominated the central region of the Islamic world, as the Mamluks of Egypt, and eventually as the Ottoman empire. Their rule developed the *ghulam* principle, and the Safavids of Persia and the Moghuls of India, also with Turkic roots like the Ottoman empire, have all been described as 'military patronage states'. [19] In these, an essentially military government, oriented to conquest, provided security for an ethnically diverse population, and its administrative organization, based on an elaborate bureaucracy, was geared to the efficient extraction of troops and taxes from the towns and villages.

The towns were the focal points of government and of commercial life, and their ruling class was formed by the governor and his military officers and civil officials, religious scholars and lawyers, and rich merchants. Below these were the tradesmen and artisans, and then labourers and the poor. But although Islam was an essentially urban civilisation, supported by its rural peasantry, even the greatest cities, such as Baghdad, Damascus, and Cairo, were part of central state administration, and therefore could never develop into anything like the city-states and semi-autonomous towns of medieval Europe.

Long-distance trade was vital to Islamic civilisation and was regarded as a blessing; Mecca had been a merchant city, the Prophet himself had engaged in trade, and it was said that the honest merchant would enter Paradise. But Shari'a law maintained a strong control on commerce, interest was forbidden, and while rich merchants might individually be men of culture, as a class they remained a subordinate element in the 'military patronage state': 'The Muslim merchant class failed to achieve and maintain a bourgeois society, or seriously to challenge the hold of the military, bureaucratic and religious elites on the state and the schools. It was a difference [from the West] the consequences of which can be seen in every aspect of Muslim social and intellectual history'. [20] Nor, within this highly cultured society, was there any chance of that interaction between the learned and the skilled craftsmen which developed in European cities. The basic pattern of production, as it had been in the Roman world, was that of workshops, especially in textiles and carpets, metal-working, and ceramics, and the artisans were also petty traders in the town markets. But although, like the merchants, they had their guilds, these never had any influence on town administration, or raised the status of the artisans.

Muhammad had taught respect for learning – 'Seek for knowledge even from as far as China', for example – and, especially after the discovery of paper-making from Chinese prisoners-of-war, which was far cheaper than papyrus or parchment, book-shops and public libraries proliferated all over the Islamic world, as did the ideal of the *adib*, the man of all-round education. But 'The idea of the university as it first arose in the West is one which is totally alien to classical Islam', [21] because Islamic law had no notion of the legal corporation. The medieval European university, unlike, for example, the House of Wisdom in Baghdad, was an independent corporation of masters and scholars, on the model of a guild, with a curriculum, faculties of divinity, law, medicine, and arts, which conducted examinations, and granted degrees that were basically licences to teach.

The closest that Islam came to this was the charitable trust, *waqf*, whose founders put their private wealth to public and charitable purposes. The idea of obtaining *waqf* status was to protect the endowment from confiscation by the state, and some of the earliest such foundations were mosques. But while they provided free tuition, students had to find their own board and lodging, and in the tenth century there developed the *madrasa*, a type of college, where the endowment made it possible for the students to be full-time residents. The instructors were independent scholars, and their students studied under them on an individual basis, in an apprentice-master relation for many years; law was the principal subject, but some logic, arithmetic, and astronomy (necessary for calculating the hours of prayer and other religious observances) might be taught. They emphasised, however, the learning of authoritative texts and were not centres of scientific research. Students with a deeper interest in the natural sciences had to pursue them privately with an individual teacher and in the public libraries, with no institutional support. We must also note that 'In the Muslim world the supreme goal of philosophy and science, like that of Hermetic wisdom, was to achieve religious understanding, or gnosis. The Muslim scientist (*hakim*) was essentially a sage to whom knowledge was entrusted. To him it was a sacred duty to guard that knowledge against contamination by the unworthy', [22] especially in the case of alchemy and astrology.

While Islamic society therefore played a vital role in preserving, improving, and transmitting Greek science to the West, it did not have the conditions in which modern experimental science could develop.

*The Chinese Empire.*

Unlike the Roman Empire and Islam, with their great diversity of subject peoples and cultures, the Chinese Empire had a far more homogeneous population. It was also the classic example of the bureaucratic state in pre-modern times. The basis of this was laid after the First Unification in 221 BC, especially under the Han dynasty, which combined some Legalist principles with Confucianism. The Confucian classics had in China the same status as the Qura'n and Shari'a law for Islam, and literature, grammar, and rhetoric for the Romans, and exercised a profound influence on Chinese society. China was divided into about forty provinces, administered by civilian governors and officials, to whom military officers were always subordinate – the soldier was regarded as distinctly inferior to the scholar-official. The old landed nobility were largely dispossessed after the Unification, and replaced by the scholar-gentry, the families of officials and retired officials who had purchased land from the rewards of office. Chinese rulers, with a huge labour force of peasants at their command, were exceptionally active in public works, notably irrigation and flood-control, canals, ship-building, roads, bridges and, of course, the Great Wall itself. All this required an efficient system of tax collection, whereby the peasants' grain could be brought into government granaries, some to be kept in reserve against famine, and the rest to be transported to the capital along the canal system. They also took an active part in promoting agriculture and manufactures, such as the silk industry, even establishing monopolies in some such as iron and salt production.

The Imperial Court was not only the administrative centre of China, and the major source of patronage for technology and the arts, but, because Confucianism was a theory of government as well as of ethics, it was also a centre of learning. In 124 BC the Imperial University, T'ai Hsüeh, was established, with a professorial chair for each of the classic Confucian texts, and it functioned rather like a French *Grande École* for training top administrators. Although, as we saw, Confucianism was opposed to the study of nature for its own sake, there was a Bureau of Astronomy because the Emperor had a religious responsibility for the calendar, and the imperial astronomers not only ensured that it remained synchronized with the sun and moon, but also attempted to predict comets and eclipses, and other omens. Officials studied many other subjects that were considered useful to the state, such as medicine, agriculture, and history, and in 754 AD the Imperial Han-Lin Academy was founded, which provided scholarly assistance of all kinds to the Emperor and government ministries, and also compiled encyclopaedias and histories.

China was unique in recruiting its administrators by competitive examination, and since becoming a scholar-official was the pinnacle of social status, the whole educational system of the country was determined by this. The teaching of the Confucian classics, on which the examinations were based, was originally by private tutors and schools attached to Confucian temples, but by the Sung dynasty (960-1279 AD) many academies, *shu-yüan*, began to be founded, and this process was accelerated by the availability of printed books in the following centuries. The *shu-yüan* were originally inspired by the Taoist and Buddhist ideal of groups of monks meeting in idyllic rural settings for meditation and discussion, but came to replace the temple-schools as urban academies for instruction in the Confucian classics, to prepare students for the civil service examinations. Some were founded by the government, and others by local gentry and rich merchants, but they were all subject to official supervision, the curriculum was based on the examination requirements, and the examinations themselves were conducted by officials. [23] The whole educational orientation of the state was therefore towards a humanistic orthodoxy, and in so far as the natural sciences were studied, it was for their use to the state, not for their theoretical value.

In view of its involvement with public works, it is not surprising that the state was a very effective sponsor of technology as well as learning. By, say, 1300 AD the Chinese were far superior to the West in the use of water and wind power and in the sophistication of mechanical devices, many of which were developed centuries before they appeared in Europe. Apart from using water power to grind corn, they also used it to drive trip-hammers for pounding ores and other metallurgical processes, blowing engines for furnaces and forges and winnowing machines, and operating silk-processing machines. While they did not invent the windmill, the Chinese had long used wind for sailing carriages, with which the first really high land speeds were attained. They also developed elaborate kites, helicopter tops spun by cords, and used the hot air from lamps to drive revolving fans. Cast iron in industrial quantities was produced about a thousand years before the West, as well as chevron-cut gears and differential gears, the crank and connecting rod to convert rotary to linear motion, gimbals (the Cardan suspension), the wheel-barrow, the segmental arch bridge, iron-chain suspension bridges, canal lock-gates and chambers, stern-post rudders and water-tight compartments on ships, the magnetic compass, gunpowder and artillery, paper and printing. [24] But if the scholar-officials regarded soldiers as their inferiors, the artisans in the Imperial workshops responsible for these developments were entirely beneath their notice, and pursued their activities in an essentially non-literate world.

Although Confucianism was basically concerned with the problems of society and government, not the intellectual study of nature, we have seen that astronomy was a major exception because of importance of the calendar and heavenly omens to the Emperor's rule. It was therefore a government monopoly whose study was forbidden to unauthorised persons – by the ninth century AD it had even become a state secret. It is not surprising, then, that the officials of the Bureau of Astronomy who were responsible for observing the heavens, and trying to predict eclipses and other omens, like the Babylonians, were extremely good at observation, data recording, and the development of instruments. Imperial sponsorship led to a flowering of astronomical studies under the Sung dynasty, including Su Sung's remarkable mechanical clock (c.1088), [25] but after the collapse of the Sung the astronomical tradition stagnated. This not only illustrates the basic weakness of all learning, in any civilisation, that primarily depends on government sponsorship, but may perhaps help to explain why Chinese astronomy, for all its sophistication in many ways, lacked the theoretical daring and ingenuity of the Greeks.

Astronomers were officials, and it was not their function to speculate on how the cosmos works: how much pure economic theory has ever come out of the British Treasury, for example? Even the doctrine that the earth is a sphere was only finally accepted as official doctrine in the sixth century AD (it had previously been conceived as an inverted bowl), and very little attention was given to celestial mechanics. There was only the simple idea that the earth was related to the heavens as the yoke is to the egg, floating in the centre of the void, and that the stars and planets were carried round it by layers of 'hard wind'. [26] It is most striking that the Chinese never developed a thorough geometrical model of the heavens in the Greek manner. 'In spite of so much accurate observation, Chinese studies of planetary motion remained purely non-representational in character. Unlike that of the Greeks, in which the geometry of circles and curves was so prominent, it perpetuated the algebraic treatment of the Babylonians. . . and never sought a geometrical theory of planetary motion.' [27]

In addition, the Chinese had no theoretical interest in problems of motion as such. After some initial speculations by the Mohists, that were not taken up, 'it seems almost incredible that through the subsequent millennia of Chinese history there are no recorded discussions of the motions of bodies, whether impelled or freely falling'. [28] For Aristotle, of course, the problems of motion were central for his whole cosmic model, and establishing the laws of motion in the seventeenth century was crucial in the emergence of modern experimental science. The lack of any body of theory comparable to Aristotle's therefore seems to have been an important obstacle to the further development of science in China. So, too, was the lack of mechanistic thinking.

We shall see that the basic image of mechanistic thinking in Europe was provided by the clock; but while Su Sung's clock could only have been known to a tiny elite, the Chinese had an abundance of other machines that would have been broadly familiar to the educated classes, yet mechanistic thinking did not develop among them. This was perhaps because, despite their advanced technology, the whole idea of the 'pushes and pulls' of mechanical causality that was an integral part of experimental science was quite contrary to the Chinese organic worldview, which thought in terms of pattern and order: 'Things behaved in particular ways not necessarily because of prior actions or impulsions of other things, but because their position in the ever-moving cyclical universe was such that they were endowed with intrinsic natures which made that behaviour inevitable for them'. [29] Finally, while atomism had a very important part in developing that mechanistic world-view which was a necessary condition of experimental science, we have seen that atomism was also unknown in Chinese thought.

So, although they made important advances in many areas of science such as mathematics, acoustics, optics, magnetism, hydraulics, alchemy and chemistry, medicine, botany, and agriculture, these studies did not produce modern experimental science. A number of works were published by scholars in these fields of study, but their readership was small and, as in the other ancient civilisations, 'The sciences were primarily pursued by learned men whose education was in the classics and whose views of the world stemmed largely from books. The technologies were primarily undertaken by artisans and master-craftsmen who were often illiterate or semiliterate and whose achievements depended more on practical experience than on abstract theory'. [30] For example:

> The Chinese artisans who built rectangular-trough pallet-chain pumps for drainage and irrigation in the lower Yangtze region in late imperial times altered the proportions of troughs and pallets by (one assumes) trial-and-error to optimise the use of energy for different angles of inclination of the trough, presumably knowing but keeping to themselves the appropriate empirical proportions. The 18th-century French hydraulic engineer De Bélidor, using simple Euclidean geometry plus simple mechanics, *calculated* and *published* the specific optimal ratios for the same type of pump – which was clearly, by a nice twist, originally borrowed from China. [31]

There was no practical likelihood that the social gulf between the artisan and the scholarly elite could have been bridged in China, and the Chinese case reinforces the same conclusions that we drew from the Graeco-Roman and Islamic evidence: that technology by itself does not automatically stimulate experimental science when it becomes sufficiently advanced.

Not only did state bureaucracy stifle intellectual originality, but a change of dynasty, or a ruler with new interests, could result in the withdrawal of patronage and a rapid decline; this happened to astronomy after the Sung, and another good example is the imperial sponsorship of naval expeditions of exploration. An expedition had been sent to the South Seas in the eighth century AD to chart the constellations of the southern hemisphere, and by the fifteenth century it might seem that the Chinese were far better placed than Western Europeans to spread their culture, trade, and conquests around the world. The great imperial fleet of Admiral Cheng Ho had, between 1405 and 1433, sailed on seven voyages of discovery and diplomacy to South East Asia, India, and to the east coast of Africa. But a change of government policy prevented any further voyages of exploration, and in 1479 even the records of these voyages were deliberately destroyed by the civil service. [32]

> Chinese governmental monopoly of certain fields of science and technology led to achievements that would have been unattainable by private efforts alone. On the other hand, such a monopoly could raise these achievements only to a certain level and no further. Ultimately they were followed by decline. We may speculate that if, in the West, a single state or church had continued indefinitely to exercise the same control over astronomy, for example, as was taken for granted in China, the Copernican-Galilean revolution would not soon and possibly would not ever have taken place. [33]

During the first millennium of our era at least, the government regarded merchants as a threat to their own relations with the peasants, by diverting their surplus production into their own pockets instead of into taxes, by speculating in food stocks, by promoting materialistic, commercial attitudes, and by giving subversive examples of how self-advancement by wealth was possible. Merchant activity was therefore strictly controlled by the government, markets closely supervised, the wearing of luxurious garments forbidden, and merchant travel restricted, while government control of the distribution of grain and cloth, and its salt monopoly, severely limited the opportunities of merchants to make profits from these enterprises. Chinese law was basically criminal, and no civil law developed, comparable to Roman or Islamic law, by which commercial life could have been stabilized and encouraged.

But especially during the Sung Dynasty (960–1279 AD), there was enormous economic development: the peasants became free to buy and sell land, production and population greatly increased, as did the entrepreneurial activity of the merchants, and the supply of money in circulation. The merchants, who had originally been excluded from the civil service examinations, increasingly used the opportunities of the academies to enter the civil service, but the merchant class as a whole never challenged the dominant Confucian ethic that regarded money-making as sordid materialism. While merchant families could therefore become rich, their aim was to become scholar-officials and bring glory to their ancestors and their clan, rather as Roman business-men had aspired to be landed gentry. With these attitudes, they could never become a powerful political force as they did in Europe. The towns themselves, as in the other pre-modern empires, were not autonomous, but an integral part of central state administration, and were run by a civilian governor and his military subordinate, not by any form of elected council. So while the merchants mainly lived in the towns, since these were dominated by the official class, there was no way in which an independent bourgeoisie could have developed, and stamped its character on urban life.

Getting finance was difficult, especially as private banks that were genuinely independent of the state were not allowed to develop, since the government looked with suspicion on all significant accumulations of wealth in private hands. This also included the ownership of large landed estates, and at the beginning of the Ming dynasty, for example, there was a massive purge of major landowners in which as many as a hundred thousand people are said to have died. There was therefore no possibility of any significant private investment in technology, either by the merchants or the landowners, and the Chinese achievements in this were primarily the result of initiatives from the state.

The honour of the clan and the ancestors, the solidarity of kin groups, and the Confucian emphasis on family loyalty and filial obedience, affected all classes of society and encouraged a spirit of conformity and submission to authority that was very different from that of Medieval Europe, where kin-groups extended little further than the nuclear family. Those core principles of Chinese culture we discussed at the end of Chapter VII were also fostered by the bureaucracy, and increasingly encouraged stability and orthodoxy, while the debate, competition, and willingness to challenge authority that are fundamental to science, and to capitalism, were repugnant to deeply held Chinese attitudes:

> Among them can be mentioned the subordination of the individual to the group in Chinese thinking, the constant preference for a middle way in matters of behaviour and thought, the desire to avoid disagreement in interpersonal relations and intellectual issues at all costs, and the customary readiness to accept the status quo and established authority. [34]

## 2. The uniqueness of Western Europe.

Even if the Western Roman Empire had survived and prospered after 476 AD, it would not have produced modern science and capitalism but, like the Eastern Empire of Byzantium which lasted until 1453, would have continued in the same centralised and authoritarian mode as China, India, and Islam. The revolutionary transformation that sent it off in a totally new direction was the influx of Germanic barbarians from 376–476 AD. In the fourth century, outside Scandinavia, the region east of the Rhine and north of the Danube was occupied by a large number of Germanic peoples, notably the Saxons, Franks, Lombards, Goths, and Vandals. They were driven by pressure from the Asiatic Huns farther east to settle within the Roman Empire, either by permission or increasingly by invasion, in sufficient numbers and military strength to deprive the imperial government of tax revenues from more and more provinces, and so progressively to destroy the military and civil administration that had depended on this revenue, until the Western Roman Empire ceased to exist in 476 AD. [35]

The collapse of centralized political authority took with it not only the imperial tax structure and the army that had relied on it, but the currency, and the bureaucracy responsible for the administration of roads, bridges, and communications, and the institutions of law and justice. Towns were abandoned in many areas, though in some cases their walls served as refuges from the general disorder. Roman law (except in Italy, where a crude version survived) was replaced by the law codes of the different ethnic groups, and was destroyed together with municipal administration and the landed gentry who had sustained it. They had been the bearers of the classical education in Latin grammar and rhetoric and literature that was a key feature of Roman civilisation in the provinces, especially in their role as town councillors.

The invaders, however, were basically illiterate, and when they destroyed the highly cultured way of life of villas and estates, and the towns, higher Roman education disappeared, together with literacy in general, although the Latin language in its popular form survived and evolved in Italy and the Romano-Celtic regions of France and Spain. (The only future for the landed elite itself was to join the freemen as warriors.) Though some of the kingdoms that emerged after the collapse of the Western Empire were large, such as Vizigothic Spain before its conquest by the Arabs, for some centuries the basic pattern was one of political fragmentation into small kingdoms and warring chiefdoms, with a profound localisation of trade and economic life, in which money became increasingly scarce. Rebuilding effective administrative systems that could collect taxes, maintain order and promote economic prosperity was to take many centuries.

The Germanic peoples brought with them many of those features of Indo-European society we noted in Chapter VII. They were still migratory and predatory, a tribal army of freemen, led by god-descended kings and chiefs elected from the royal line, and acclaimed in the popular assembly. Leaders in war were men of noble family, who were supported in their search for plunder by their war bands of professional warriors, who swore undying loyalty to their lord, and who rewarded them with gifts of booty, land, and slaves. The customary law of the tribe, and the whole idea of the rule of law, was venerated, and continued to be so through the Middle Ages:

> The law by which all human affairs were regulated was that of the ancestors preserved in custom, an eternal system of right relations. Good law was ancient law, and new law was bad. The function of the legislator, therefore, was to proclaim or interpret the law, not to change or make it; the laws of King Alfred, which open with the Ten Commandments and continue with a compilation of earlier law, are wholly characteristic. [36]

In pagan society the law was under the control of the priests, because the oath and the contract were sacred to the gods, and the legal ordeals that established guilt or innocence were presided over by the priests.

In those groups that had converted to Christianity, Christian priests supervised the judicial ordeals of red-hot iron and boiling water. Lawsuits were decided, and customary law was declared at the periodic assemblies of the freemen, that also conducted public business such as co-ordinating military actions, or acclaiming a newly elected king. The freemen, a substantial proportion of the total population, therefore had an honourable status, and were the farmer-herdsmen who could be called on to fight, but were not in the front line. Below them were the freedmen, and finally, at the very bottom, the numerous slaves. [37] Social status was more important than kinship, and the kin-groups of Germanic society were small cognatic 'kindreds', descendants of a common great-grandparent, who paid and received blood-money. But since women could inherit property the kindred could not be a corporate land-owning group, and its control over its individual members was much looser than that of the strong clans of Chinese and Arab society.

But when these different peoples finally settled down in the territories they had conquered, a number of profound changes occurred to this traditional order over the ensuing centuries. One of these derived from the changing pattern of warfare and conquest. 'Formerly, each spring, the Frankish kings had led their people into battle and pillage; every autumn, the captives and booty carried back from these seasonal escapades were shared out among the military chiefs and the guardians of the sanctuaries', and to the free men [38] and this applied generally in Europe. But once the Romano-Celts had been subdued, and local kingdoms established, by the ninth century the opportunities for plundering other groups had greatly diminished.

The pattern of authority changed as well, from popular assemblies to power based on land. The traditional warrior leaders of the war-bands became a local landed nobility, who retained their military retinue of freemen by gifts of gold, horses, weapons and land, and these gradually became an armoured cavalry force of knights who plundered the local peasants instead of foreigners, and created conditions of violent anarchy. The free peasantry were as liable to suffer at their hands as were the conquered native population, and while in northern Europe a significant number of freemen retained their status, in Europe generally they became increasingly oppressed by poverty, war, and the burden of taxation, and many were forced down into the status of un-free serfs, bound to their land. Even in England, 'Many an independent freeman may have had to purchase protection and financial help in time of stress and disorder at the cost of relinquishing some of his rights', [39] such as handing over his land to his lord, and receiving it back again as a feudal tenure, for which he had to perform onerous services.

Royal power was weak because the general illiteracy, and the lack of coinage, proper roads, and communications, meant that there could be no effective bureaucracy, centralised administration, or system of tax-collection, and certainly no standing army, and kings had to travel around their domains to enforce their authority, and consume their rents on their estates. Inevitably, they had to make use of great magnates in running their kingdoms, granting them lands in return for homage and services, not only providing men for military service, but in local administration and the enforcement of law and order. These local magnates constantly struggled to assert themselves against the king's authority, and to convert feudal rights to land granted in this way into absolute hereditary rights, so that there grew up a new emphasis on patrilineal descent and on primogeniture, at the expense of the cognatic kin-group, in the inheritance of land and public office. Public authority itself therefore tended to become converted into private rights that, like land, could be inherited.

Just as magnates held their land from the king, these lords in turn also granted lands to their tenants in return for homage and services, diluting central control by a chain of client-patron relationships from the royal down to the local level, in which the number and status of one's followers was a key indication of one's own standing. By the twelfth century feudal tenure was found all over Europe: the free tenant's duty was to attend his lord and give him support and counsel at the great feasts of the year, which also functioned as parliaments and law courts, to pronounce judgements in their lord's court, escort him through his estates, and provide a contingent of men-at-arms in war, and rent-paying tenants in times of peace. The lord in return had to protect his man in war and in the law courts.

But customary feudal law also imposed important restrictions on the demands that lords could make of their tenants: military service became limited to 40-60 days in the field, and aid from the tenant to his lord could only be demanded on specific occasions. For further grants outside these obligations, the lord had to obtain the consent of the tenant, and these restrictions applied to the king as well as to the nobility.

Even in Norman England, which was one of the more effective kingdoms, central government was relatively weak by comparison with the empires we have been considering. It was based on the shires, each presided over by the king's representative, an earl, together with his own vassals, who would be local magnates with their own castles, 'castellans', and with their own military retinues. The court over which the earl presided was the old popular shire assembly, and the law was customary law as it had been codified by Saxon kings. The shire-court's responsibility was to assess revenue, summon warriors for the royal army, and maintain the peace, and all free men continued to bring their disputes before it, while below the shire-court was that of the local district, or hundred. It seems to have taken over the open-air meeting place, as well as the function of the earlier popular assembly, though it was now presided over by the local castellan.

It had long been in the lords' own interests to encourage agricultural prosperity and population growth, and from the seventh and eighth centuries onwards there was an intensification of agriculture – the cultivation of formerly virgin lands, the planting of new vineyards and improvements in farming techniques, such as crop rotation, the planting of roots and legumes, and the increasing use of the heavy plough. The use of water power (there were 5,000 water-mills in England at the time of the Conquest), mining, metallurgy, with water-powered blast furnaces and hammers, bronze founding for bells, glass-making, and the increasing use of stone for cathedrals and castles are all examples of improved technology.

By the eighth century trade, especially for spices via the Mediterranean, had already become considerable, and Charlemagne in 800 AD introduced the first new coinage since the fall of Rome, the silver penny, with 240 to the pound of silver. Coins remained scarce, however, and great magnates also increasingly operated their own mints, as the cities did later on, making coinage very chaotic in France, Germany, and Italy. Lords also had the revenues from tolls on bridges, rivers, roads, and market taxes, and were increasingly able to live in splendour, not like squalid peasants, with luxury goods, wine, spices, and silks and the finest woollen cloth. But these could only be provided by merchants in the local towns that were springing up again, especially outside Italy, by harbours, navigable rivers, the junctions of major routes, and near important royal estates and monasteries. In England, London was by far the most important.

Lords favoured the rise and prosperity of their towns; normally in debt, they were happy to sell charters and exemptions from various feudal dues to their local town. The financial weakness of medieval kings, with increasingly expensive wars to conduct, and facing the threat of rebellious nobles, led them, too, to court the favour of the towns, as essential sources of revenue from their manufactures and trade, of loans from increasingly rich merchants, and also as supplying a popular counterweight to the power of the nobility. It was difficult, however, to apply the feudal relationship to relations between lords and merchants, and the new towns and their inhabitants gradually acquired a special legal status that set them apart from the rural society around them. The towns were able to buy charters of incorporation and other 'liberties' from the king, granting them their own assembly, the town council, whose jurisdiction was independent of the shire.

The town-dwellers in time also became an officially recognised social order, the burgesses, with a distinct status from the peasantry. Burgesses were freemen who held land in the town at a fixed rent, of which they could freely dispose, and to which they were not bound. These fixed obligations and personal freedom were essential to the progress of a borough, just as fixed customs duties and personal security were necessary to foreign merchants. By the twelfth century the burgesses were binding themselves together by an oath in 'communes' of mutual aid and assistance between all fellow burgesses. At the end of the century, the London commune, for example, received royal recognition, and in 1199 they were allowed to elect their own Sheriff, the king's representative.

Essential features of these towns were the guilds of merchants and craftsmen. Although these had existed in all the other civilisations, in Europe they developed much more self-sufficiency and independence from the state. In northern Europe they had their origin in ancient Germanic and Scandinavian society, where non-kin could join themselves together in voluntary associations by an oath, and a religious ceremony involving sacrifice, feasting, and drinking mead or beer from a common cup. Members pledged themselves to mutual assistance, and the principle of contributions by members, initially only to defray the costs of feasts, could easily be applied to business activities, such as insurance against fire, theft, and shipwreck, and become the basis of substantial investment. The guilds had essential religious functions for their members, with their own churches, and the performance of mystery plays and pageants, as well as charitable and educational work.

This type of association between non-kin was ideally suited to the growing towns, where many unrelated individuals congregated together. In contrast to the contempt for trade that was so typical of the ancient civilisations, in northern Europe merchants had long been honoured as heroic sea-voyagers, who braved the elements to bring back precious cargoes, and in Anglo-Saxon England a merchant who crossed the sea three times at his own expense was entitled to noble rank.

The merchant guilds initially dominated the town councils, but in time their power was challenged by the craft guilds, who finally succeeded in claiming their place in the running of the town. The higher artisans, such as goldsmiths, armourers, and master-masons had always been greatly honoured, and they had frequently been given lands. But many blacksmiths, carpenters, weavers, tailors, bakers and so on were also freemen, and the new towns provided far more opportunities for them than the countryside. One of the towns' most significant features was therefore that they provided islands of opportunity for men of initiative to free themselves from the restraints that operated in the feudal world of the countryside. In the urban environment the individual was relatively free both of inherited status – a serf who managed to live for a year and a day in a town could not be reclaimed by his lord – and of the claims of kin beyond his wife and children, and considerable social mobility was possible, especially in large cities like London.

In London the rights of the crafts were decisively marked by the ordinance of King Edward II, which required every citizen to be a member of some trade, and by the ordinance of 1375 which transferred the right of electing corporate officers from the ward representatives to the trading companies, which we know as the Livery Companies, such as Grocers, Drapers, Fishmongers, Goldsmiths, etc. (King Edward III was himself a member of the Merchant Tailors Company.) This participation in municipal government by the craftsmen was of great importance in raising their social status, and urban society was now very different from that of the towns under the Roman, Islamic, and Chinese Empires, which had been organs of the state, dominated by the official classes, and in whose government merchants and craftsmen had had no part.

Medieval Europe is also unique among the pre-modern agrarian civilisations in its development of parliamentary assemblies for the whole nation, and it was at these parliaments that the towns came to be officially represented. It was in the interests of kings to hold periodic courts, to which their chief tenants and the agents of royal government were summoned. They had a feudal duty to give counsel to their lord, and the greater the number of magnates who attended, the greater the proof of the king's authority. 'Since these large sessions of the tenants were also the most solemn courts a king could hold, great issues would be brought to it; since they were usually held at known times and places, those who desired the king's help in seeing justice would seek them out.' [40]

But kings found it increasingly useful to enlarge the functions of these courts into national assemblies, because the complex and archaic system of feudal land tenure no longer provided sufficient revenue for running the state and the ever-increasing expenditure on war. If the king wanted more revenue he could only obtain it, under feudal law, by the consent of the tenant, just as the local lord had to obtain the consent of his free tenants for further aid. The national parliament therefore provided a new form in which the whole community could give its assent to taxes.

> The princes of the late 13th C claimed to be lords of states, not merely of associations of men; in their conflicts with the papacy, in their assertion of legislative authority, in their claims to the financial support of the whole community, kings required a very general assent. Unless qualified representatives could be gathered, the king's will could not be known or the justification of changes publicized; the rise of the representative assembly is parallel to the rise of royal propaganda. [41]

In Europe, the Parliaments consisted of the three estates of nobility, clergy, and burgesses, while uniquely in England the nobility came to have one chamber of Parliament, while the knights of the shire and the burgesses had the other. It is quite possible, though it cannot be definitely established, that the idea of the three estates of 'those who fight, those who pray, and those who work' originated from the warriors, priests, and farmer/herdsmen of Germanic society. [42] But the ancient popular assembly had taken on a new life at the national level, now incorporating the towns with their merchants and craftsmen. It is important to stress that in no ancient empire do we find rulers having to gain the consent of their subjects to taxation, and while the old elective monarchy had generally been replaced by hereditary monarchy (except in Scandinavia and Poland), there was still the idea that in theory the king was chosen by his people.

This, however, was the highest level of *effective* political integration that could be achieved in Western Europe. Charlemagne (c.742-814), it is true, the King of the Franks, was proclaimed Roman Emperor by the Pope in 800 AD, and briefly united France, Germany, and half of Italy, but on his death in 814 this empire rapidly disintegrated, partly because it had to be divided among his sons, but also because of the inherent weakness of the political and administrative institutions. In 962, Otto of Saxony was crowned Holy Roman Emperor by the Pope, but the Emperor's rule over Germany and Italy was weak: the Empire was too large to be effectively governed, the powerful German princes who elected him resisted his authority, while the Italian city-states established their independence, beginning with the defeat of the Emperor's forces in the twelfth century by the cities of the Lombard League. In Germany, many cities established comparable independence from the princes, and also formed alliances, notably the very powerful Hanseatic League in the thirteenth and fourteenth centuries.

At this point we must consider the status of the Church, which was very different from what it had been under the Roman Empire, and from comparable religious bodies in the other ancient empires. While, for example, the Byzantine Emperors always kept the Patriarchs of Constantinople as firmly under their thumbs as Tudor monarchs did their Archbishops of Canterbury, the Roman Church achieved far greater independence, and to understand how this was so we must go back to the barbarian invaders. Like the Roman Empire, the Papacy had great prestige among the barbarians. Some of these, such as the Goths, had already converted to Arian Christianity, but Clovis (481-511), leader of the Salian Franks, the principal Germanic tribe, converted to the Roman Church in 496, and crusaded vigorously against the Arian Visigoths and Burgundians. Making his capital in Paris, he was a notable champion and patron of the Church, endowing it with lands and property. When the Merovingian line founded by Clovis gave place to the Carolingians in the eighth century, Charlemagne was another notable champion of the Church, particularly against the Lombards, and gave the Pope large territories in Italy. He also crusaded for many years against the Saxons and Bavarians, forcibly converting them to Catholic Christianity.

The Church could therefore give very valuable religious legitimacy to rulers, it helped them keep the peace, and for centuries also provided them with literate administrators, while the bishops in their cathedral cities were of great importance in supporting royal authority. Kings repaid the Church with protection and with munificent gifts of land, so that it acquired vast wealth and played a major role in the re-growth of Western civilisation, notably in the spread of education and literacy, and the idea of Christendom as a universal society.

By the middle of the eleventh century, indeed, the Papacy was starting to assert itself in what was to be a long conflict with the Emperor, and the supremacy of the Pope over earthly kings was declared at the Lateran Council of 1215. But while the Pope and canon law claimed absolute authority over substantial aspects of European life, Christianity had established from the beginning that God and Caesar were owed different kinds of allegiance, and the Pope had to deal with a number of rulers, not only the Emperor but other kings, notably those of France and England, all of whom resisted his claims.

The result was a number of conflicting jurisdictions, between the Pope and the various rulers, and also between kings, nobles, bishops, towns, and universities, who collectively lacked the power to impose a single authoritarian rule. One can find no plausible parallel in this society to the Emperor Justinian closing the schools of philosophy in Athens. The Church's power over men's minds was also increasingly limited by the rise of a literate laity in the medieval period, and then by Protestantism in the sixteenth century. This all allowed the same sort of creativity in art, thought, and technology as had appeared in the city-states of ancient Greece:

The history of the Middle Ages leaves us, above all, with a sense of the extraordinary vigour and creativity which derives from the fragmentation of power and wealth into innumerable centres, competing and expanding into different and unexpected directions. The places where political fragmentation was most complete, such as Tuscany, the Low Countries, and the Rhineland, were perhaps the most creative. That division of authority was caused partly by small political units, partly by the overlapping of royal power, independent cities, strong seigneurs, and finally ecclesiastical authority. Hence the multifarious creativity of medieval Europeans. [43]

The Church also played a central part in the revival of higher learning in the Middle Ages. Since the clergy initially provided the great majority of the educated class, they were naturally the primary route by which Greek philosophy and science reached medieval Europe from the Islamic world, both from Spain and the eastern Mediterranean, and from Byzantium. We also need to note the vital part of the Church in establishing the medieval universities in the twelfth and thirteenth centuries, and the religious teaching orders such as the Franciscans and the Dominicans. The universities were corporations of masters and scholars (*universitas* = 'corporation'), modelled on the guilds, and like the chartered boroughs, governed themselves by their own rules. They were granted autonomous status by a series of Popes; masters and students enjoyed the privileged status of clergy, which gave them immunity from the secular courts; and universities were outside the jurisdiction of the city authorities, with whom they were often on thoroughly bad terms.

They were centres not only for the study of canon law and theology, and training administrators for church and state, but also of philosophy, medicine, and the *quadrivium* of geometry, arithmetic, astronomy, and music. This range of subjects was more favourable for the development of science than those taught in comparable Chinese and Islamic institutions, and Western universities also had considerably more freedom from the direct control of Church and State.

> One of the indispensable contributions made by medieval Western Europe was to provide in the universities a secure institutional context for learning and teaching over a wide range of subjects...No such context was established in the ancient or Islamic worlds, and this certainly left the natural sciences in a much weaker position in those societies than was achieved in the West. [44]

While some of Plato's ideas had been assimilated by the early Church, little had been known about the scientific works of Aristotle before the twelfth century, and there was initially strong resistance to his ideas. Innovation was suspect, curiosity about the mysteries of nature was often condemned, there were many theologians who thought like al-Ghazzali, and in the thirteenth century several Popes tried to forbid the study of Aristotle's works in the universities, but without success. By the time of his death in 1274, Aquinas had written his *Summa Theologiae*, in which he claimed to show that divine revelation and human reason are both genuine sources of knowledge, so that Greek philosophy and, for that matter, Islamic science, were legitimate subjects of study for Christians. Aquinas said, for example, 'There is a certain eternal law, to wit, Reason, existing in the mind of God, and governing the whole universe', an entirely different view from that of most Muslim theologians, and one that was very favourable to the scientific study of the natural world as a means of understanding God's handiwork.

The revival of science in the West was originally, therefore, in the hands of the Church and universities, and we shall see in the next chapter that, despite the later repression by the Inquisition, it was the work of the medieval philosophers on Aristotle that provided some important foundations for the scientific developments of the

Renaissance. From the Renaissance onwards, the universities became increasingly important in the development of science.

In the thirteenth century, especially, a stream of inventions, a few indigenous, but many from outside, especially from China, had an enormous impact on European society that was quite different from their effect where they had been invented. The clock, (uninfluenced by China), the distillation of alcohol, and spectacles (which were the basis of the telescope and the microscope) were European in origin, but gunpowder, paper, printing, and the marine compass and improved methods of ship-building were Chinese.

Western Europe had a far greater need for trade, particularly for spices, than had the largely self-sufficient Chinese Empire, so when the compass and the stern-post rudder, and new forms of ship-building and rigging that allowed ships to tack into the wind, reached Western Europe from China, they made it possible for the Portuguese and the Spaniards to circumnavigate Africa and cross the Atlantic in search of new routes to the spice islands. Prince Henry of Portugal ('The Navigator') was a key figure in these developments, since he was not only able to talk with astronomers and cartographers in their own language, but also with shipbuilders and sailors in theirs. Under his inspiration Vasco da Gama discovered the sea route around Africa to India, [45] while Columbus under Spanish patronage was able to cross the Atlantic to the Americas. Although the voyages of the Portuguese and Spaniards was supported by royal patronage, they also attracted substantial private investment so that profit, missionary zeal, and conquest in search of gold and silver were fundamental motives for these voyages, as they subsequently were for the Dutch, French, and English.

These voyages had a significant impact on science and mathematics, because long distance ocean navigation of this sort was highly dependent on astronomy and cartography, and therefore on geometry and mathematics in general. Whereas the officers of the Chinese Imperial fleet were members of the literate class, the mariners who sailed the European ships were not, and this helped spread the need for mathematical knowledge among the artisan class of Western Europe. In the same way clocks, and the need to calculate their intricate gear ratios and the geometry of dials, and spectacles that involved a good knowledge of the optical properties of lenses and how to grind them, also increased the need for mathematical knowledge among the skilled artisans, as did commercial arithmetic.

In an increasingly wealthy and capitalistic urban environment, respected by the educated classes and enjoying the patronage of royal courts and the nobility, the skilled craftsmen such as lens grinders and instrument makers, clockmakers, printers, goldsmiths, artists and engravers had great opportunities for displaying their ability and forging intellectual relationships with men of learning and wealth. Printers, for example, were usually scholars as well as businessmen, and had to be able to bridge the world of their authors and proof-readers on one side, and the technical world of the workshop with its typesetters, draughtsmen, engravers, ink and paper makers, on the other. (Gutenberg, the inventor of European printing, was originally a goldsmith.)

In this environment the man who had learnt his trade as an apprentice, and knew no Latin or Greek, could still develop the kind of knowledge that was worthy of respect from the theoretical scientist. The high status of artisans in Western Europe led, in fact, to a revolutionary transformation of traditional attitudes to the intellectual value of manual skill. This was the completely novel idea that the techniques and skills of the artisan, far from being a sub-rational form of knowledge, as had always been the view of the educated elite, in fact gave them an understanding of matter that was in some ways equal to the *scientia,* the theoretical knowledge, of the philosopher.

The crafts had traditionally protected their 'mysteries' by secrecy, enforced by the guilds, but,

> in the fifteenth century, city governments also began to realize that technical knowledge was valuable intellectual property. In order to ensure that the local economy would incur the benefits of inventions, city governments took measures to protect inventors' rights. Patents emerged in response to a growing awareness that knowledge could be put to practical use, and that as long as new discoveries were kept secret, the advancement of knowledge, and hence of profit, would be retarded. [46]

[The city of Florence issued the first patent in 1421 to the architect Brunelleschi for a design for a cargo ship.]

By the next century the patent had become a standard institution across Europe, and one of the basic means by which inventors could become wealthy.

The idea that theoretical scientific knowledge could and should be useful had also grown steadily stronger, (for reasons we shall see in the next chapter), and by 1700 scientific research and education was supported by royal patronage across Europe, partly out of concern for public welfare, and also for patriotic purposes in military and economic competition with other nation states. In Britain and elsewhere in Europe a number of colleges and learned bodies were established to promote collaboration between craftsmen and scientists, and also, very importantly, to publicise their discoveries in learned journals: in the sixteenth century Gresham College in London, in the seventeenth century the Royal Society, and in the eighteenth century Glasgow University.

A very good illustration of the collaboration that eventually developed between theoretical science, practical craft knowledge, and capitalist enterprise is the Lunar Society of late eighteenth century Birmingham (c.1765–c.1809). Its members were a group of friends who met every month close to the full moon (to see their way home), and included: James Watt and Matthew Boulton, (steam power) William Murdock (gaslight); Joseph Priestley (chemistry); John Baskerville (printing); Josiah Wedgwood (pottery) Erasmus Darwin, (botany and evolutionary theory), James Keir, (chemistry and glass-making), Richard Edgeworth, (electricity and the telegraph), Thomas Day (educational reform), with Benjamin Franklin and John Smeaton, the civil engineer, as corresponding members.

In the nineteenth century the great advances in industrial methods of production were able to combine with, and make ever more effective use of, the advances in pure science that were also occurring, culminating in the full development of chemistry and electricity and, of course, of nuclear power and rocket science in the twentieth century. The end result of the industrial revolution and capitalism, combined with other social traditions that encouraged the forces of political liberty and individual rights, was the total subversion of the traditional forms of state that we have been considering, with its sacred, hierarchical order, and the coming of modern mass society, dominated by commerce and technology.

*3. Conclusions.*

The great civilisations of Rome, Islam, and China followed what we can call the normal pattern, then, but those very conditions that produced their greatness were, paradoxically, those that worked against the emergence of modern science. Powerful military regimes, with elaborate bureaucracies and tax systems, and an official religion or ideology can foster great art, architecture, literature, and learning, and sophisticated science and technology. But science, in particular, in these conditions of top-down state patronage and official orthodoxy, can only go so far. It was the abnormal conditions of Western Europe, and the multiplicity of financially and militarily weak kingdoms, with their relatively primitive bureaucracies, that allowed the rebirth of something like the city-state, or at least fairly independent urban centres largely outside the constraints of feudalism. Unlike Rome, Islam, China, (and India), cities were dominated by an increasingly prosperous merchant class that developed a vibrant capitalism, and a cultural milieu where creativity could flourish and did not depend on the changing whims and interests of an imperial court.

The Papacy could not impose the same degree of orthodoxy over people's beliefs as Confucianism or Islam, and was often in conflict with secular rulers. Contrary to the popular stereotype, the medieval Church did not forbid the scientific study of nature, and in fact provided the scholars who first took up the torch of Greek and Arab science. This was a crucial factor, because it is highly unlikely that the relatively barbaric society of Western Europe would have been able to generate anything comparable from its own resources. But while the Roman, Byzantine, and especially the Islamic worlds had been familiar with Greek science, they could only take it to a certain point. Further developments also needed the unique urban world of Western Europe, where there emerged those unprecedented opportunities for skilled craftsmen to acquire enough social status so that educated men could associate with them, and acquire the craft knowledge that was so important for the development of experimental science.

While technological and economic factors were therefore of great importance in the emergence of modern science, to give them the dominant role in its development is naïve exaggeration, since they were only a part of a much more complex accumulation of necessary conditions.

In the next chapter we will see how some further conditions for the emergence of experimental science were established during the Middle Ages and the Renaissance, laying the foundations for the scientific mastery of nature that has made modern civilisation possible.

# CHAPTER XV
## How We Learned to Experiment

Experimental science, in the sense of the deliberate testing of theories by properly designed experiments, using measurement and mathematics, to solve specific problems, was as great a revolution in human knowledge as the state had been in politics. Not only did it give us a uniquely powerful means of understanding nature, but the control over natural forces that this made possible was one of the essential foundations of the industrial revolution. Like the state, this type of experimental science was not some simple Darwinian mutation. A number of cultural and social conditions had to be in place, and in the previous chapter we saw why the most important of these only occurred in Western Europe, among all the other literate civilisations. Now we need to consider a number of more specific developments from the Middle Ages to Galileo that made experimental science possible.

*1. The secrets of nature.*

From the tenth century onwards, the West was becoming aware of the great superiority of Arab learning and science, and many scholars travelled to Spain and other parts of the Islamic world to translate and bring back works on mathematics, Aristotle's physics and astronomy, optics, and mechanics which were eagerly studied, especially in the universities when these developed. (We shall give them special attention in section 3.) But strikingly, by the thirteenth century we have the Franciscan Friar, Roger Bacon, (c.1219–c.1292) of Oxford and Paris universities, predicting that, by applying philosophy and mathematics to the study of nature it would be possible to produce all sorts of technological marvels, including great ships controlled by a single man, vehicles without horses moving at incredible speeds, flying machines in which one man operated a motor that flapped the wings, submarines, glasses for seeing great distances, or to help the elderly to read, and burning mirrors to destroy enemies in battle. 'No medieval thinker had ever argued the case of utility so forcefully or hammered it home so often. In his repeated insistence on the practical application of scientific knowledge, on its beneficiality for the individual and the state, Bacon is the advocate for a programme that has become our own'. [1]

This attitude, totally different from that of the ancient world, has created the illusion that he was a medieval Jules Verne or H.G. Wells, a prophet of the Industrial Revolution born out of his time, but he was really invoking the powers of what became known in the Renaissance as 'natural magic', distinct both from demonic magic, and from Aristotelian science. In natural magic, 'The entire universe was a vast system of interrelated correspondences, a hierarchy in which everything acts upon everything else', [2] and it was based on those magical affinities between elements, numbers, colours, planets, humours and so on of the Hermetic tradition that we have already discussed.

The most powerful influence promoting natural magic in the Middle Ages, and beyond, was a large number of Arab works purporting to reveal the secrets of nature, the most famous of which was the *Secretum Secretorum*, The Secret of Secrets. [3] Apart from being a manual for princes in the arts of government and healthy living, it was also filled with esoteric learning on how to read character from the face and body, how to tell which princes would be victorious from the numerological significance of their names, astrology, especially medical astrology, the secret qualities of herbs and stones for making talismans, and alchemy, especially the philosopher's 'stone' for making unlimited amounts of gold for the state, the elixir of life, and the cure for all diseases, the panacea. For centuries thereafter, the *Secretum* was one of the most widely read and influential books among the educated classes generally in Europe. [4]

It had a great impact on Bacon, [5] who referred to it in his work more than any other medieval scholar, and it was his belief in the possibilities of natural magic that was primarily behind his demands that science should be put to practical use by harnessing the hidden powers of nature through alchemy and astrology, and also by magnetism, optics, perpetual motion, and amazing machines. The alchemical tradition had already produced many 'recipe' books on metallurgy, glass-making, ceramics, painting and dyeing, and the appearance around this time of spectacles, gunpowder, and the magnetic compass, with all of which Bacon had some association, and the clock, would have seemed good evidence that his claims were not fantasies but might really be achieved.

But this secret knowledge was concerned with the abnormal, the strange, and the wonderful that fell outside the scope of normal academic science, and very importantly, its practitioners could not demonstrate the *propter quid*, the reason why of their results, which was the essence of true, rational *scientia*. So, 'Despite its nearly ubiquitous presence in the West after the twelfth century, the literature of secrets did not find a place among the official sciences of the universities'. [6] Natural magic developed at the hands of learned medical astrologers and alchemists outside the universities, and was greatly boosted by the revival of Platonism and Pythagoreanism in the Renaissance. Its practitioners were always liable, like Doctor Faust, to be suspected of witchcraft and other diabolical practices, to which of course the Church was relentlessly hostile, and like della Porta (1535–1615), they often attracted the attention of the Inquisition.

Della Porta's practical researches did not involve testing hypotheses by experiment like Galileo: he saw natural magic as the empirical study of the clues that God has provided us about the hidden properties of nature, so that we can put them to use. His *Natural Magic* of 1589 was only one of his many books on the physiognomy of the hands of criminals, and of the whole body, the 'signatures' of plants, the mechanics of water and steam, distillation, agriculture, meteorology, metallurgy, military fortification, and optics and the *camera obscura*. At the end of the sixteenth century he was one of the most famous scientists in Europe: 'The contribution of Porta's conception and practice of natural magic to the emerging idea of science is not merely rational or theoretical or contemplative. Rather, science must represent theory and contemplation coming to practical and experimental expression.' [7]

This kind of science was very different from that of the Aristotelian professors in the universities, [8] who regarded the way of reason as far superior to the way of experience. By 1605 Francis Bacon (1561–1626) was furiously denouncing the indifference of philosophers, both ancient and modern, to human welfare, and their frivolous academic speculation about insoluble enigmas. The nineteenth century historian Macaulay presents him as a fellow Victorian, a Mr Gradgrind dedicated to Utility and Progress, for whom science was the study of facts that would bear fruit, in 'the multiplying of human enjoyment and the mitigating of human sufferings'. [9] But 'Recent scholarship has made it abundantly clear that the old view of Bacon as a modern scientific observer and experimentalist emerging out of a superstitious past is no longer valid.' [10] In fact, his world-view was the same Hermetic and natural magic tradition as that of della Porta, and for him, too, scientific progress consisted in recovering by experiment the knowledge that God had originally imparted to Adam before the Fall. [11]

In natural magic, as we have seen, experiment was still basically exploratory, 'to see what happens if. . .', and Bacon died after catching a chill caused by stuffing a chicken with snow, to see if this prevented putrefaction, but he was essentially a philosopher, and not a practical scientist. He developed an elaborate theory of scientific method, and claimed that by applying it, anyone could do science, and he advocated the establishment of research institutes funded by the state, that would study the crafts, and perform useful experiments. He was particularly critical of the secrecy and mystification that had been typical of the Hermetic and natural magic traditions, and insisted that the progress of science required co-operation between scientists, and the wide publication of their results.

These revolutionary ideas had a direct influence on the founding of the Royal Society in 1660, a major example of those 'environments that foster creativity'. It had a vigorous programme of research, inspired especially by the genius of its Curator of Experiments, Robert Hooke (1635–1703), [12] who might be described as 'experimental man' personified. Natural magic was therefore one of the main ingredients in the emergence of modern science: it involved scholars in the physical investigation of nature, and it tried to discover its underlying principles. But other profound changes in society and thought were needed for modern experimental science, one of the most important being the enormous growth in the understanding of mathematics and measurement, at all levels of society.

## 2. *The growth of mathematics and a quantified society.*

Indian astronomers had brought their powerful number system to Baghdad in 773 AD, and the great mathematician al-Khwarazmi (died c.863) in particular encouraged its use. Gerbert of Aurillac (945–1003), a major scholar who had travelled to Spain to study astronomy and mathematics with Arab masters, was the first to spread the new notation in Europe, together with the very important use of the abacus, and he was followed by others such as Gerard of Cremona and Adelard of Bath, and Fibonacci who brought Islamic mathematics to Italy from North Africa. We have to remember that, until the end of the tenth century, and outside calendrical studies by a few experts, the use and understanding of mathematics in Europe was in a very primitive state. Before even the thirteenth century, precise numbers such as dates of birth and death, year numbers, populations, quantities, distances, prices, and ages were normally not given in documents at all. [13] But merchants and craftsmen, rather than the learned, accepted Arabic numerals enthusiastically, and their first recorded use in England is on the roof timbers of Salisbury Cathedral, where they were carved by French carpenters in 1224. [14]

By the end of the twelfth century, the use of numbers in public records had become much commoner, and the ability to count money (whose use had expanded enormously), and measure cloth had become a test of majority for a burgher's son. In the thirteenth century, especially in Italy, the urban commercial boom led to a huge growth in the awareness of mathematics among the middle classes. There seems no doubt at all that the spread of money in medieval society, by which labour, as well as goods, became a commodity with a market price, was a very powerful factor in what we can call the general 'quantification' of society, and which made mathematics increasingly familiar to the ordinary person in the towns. It was generally agreed that merchants were good reckoners, particularly with the rise of double-entry bookkeeping, and this has led to the theory that they were the driving force behind the new interest in mathematics – 'the counting house hypothesis'. [15] But while commercial arithmetic played a vital part in spreading numeracy in medieval society, it was an elementary branch of mathematics, which at the academic level had nothing to do with the counting-house. 'There was a vast gap between the theoretical arithmetic of the scholar and the practical arithmetic of the flourishing trade of the thirteenth century.' [16]

But the growing complexity of money certainly stimulated the learned to think about the problems of measurement. By the fourteenth century there was extreme economic dislocation caused by war, famine, plague, and the need to raise more and more revenue for warfare. In France, in particular, this led to debasement of the coinage, and constantly shifting prices, there were many different currencies in circulation, and uncertainty about what coins were worth. While, obviously, medieval Europe was not the first civilisation to use money, one has the impression that its forms of money were unusually challenging: '[Money's] everyday use was a highly complex affair and required constant computation and activity on the personal level'. [17]

Some of the leading mathematicians of the fourteenth century, such as Nicholas Oresme, Bishop of Lisieux, wrote important books on money as well as on the philosophical problems of measurement. The old Aristotelian theory that only quantities, not qualities, could be measured, had been challenged by philosophers since the thirteenth century, and Joel Kaye has argued that because money was seen as a common measure of value, it also assisted the idea of measuring qualities. 'Scores of subjective, "qualitative" factors such as social position, or a fighting man's "worth", or a working man's time and labour became newly conceivable as measurable, relatable and quantifiable entities.' [18] (We shall come back to this in the next section.)

The general demand for mathematical knowledge was not only stimulated by its value for the merchant in his book-keeping. Agricola (1490–1555) argued, for example, that the miner needed sufficient arithmetic to be able to calculate the cost of his machinery and the working of the mines. Navigation, surveying, architecture, the making of dial-plates for clocks and sundials, the art of perspective painting, and gunnery all required some knowledge of arithmetic and geometry, and by the sixteenth century the urban middle classes had developed something of a craze for mathematics in general. In England, for example, the number of printers had increased from six in 1500 to about twenty by 1550, and popular literacy, and the appetite for books in English rather than Latin, had increased proportionally. (In the sixteenth century about 90% of science books were in English.) Robert Recorde (c.1510–1558), a university mathematician, physician, and astrologer, among other interests, published a series of basic text-books designed to bring mathematics to the urban middle classes, including the artisans. [19]

Until Recorde, and then Viète (1580), even more fundamentally, had introduced a concise algebraic notation to replace words, 'Twenty or thirty lines of Latin were often needed to express an idea that we can express in a single formula.'[20] Even the strongest minds tottered under the strain, which meant that only real experts could solve equations that are quite simple by modern standards. Recorde's text-books were reprinted in numerous editions, and had enormous influence in developing a society where thinking in mathematical and quantitative terms became the norm for artisans as well as the learned: 'Carpenters, carvers, joiners and masons, do willingly acknowledge that they can work nothing without reason of geometry', for example.

Whereas by 1550, therefore, European mathematics had not advanced much beyond what the Arabs had achieved, development then became rapid. After Recorde and Viète had established a satisfactory algebraic notation, Stevin (1585) invented decimals, Napier (1614) logarithms, and Gunter (1620) the slide-rule. Descartes (1637) introduced one of the most fundamental ideas in mathematics: co-ordinate and analytic geometry, which made it possible to describe different curves by algebraic equations. Wallis (1655) developed the theory of infinitesimals, and Newton and Leibnitz later in the century invented calculus, which finally allowed mathematics to handle motion and change.

This quantification of the world also extended to a new, mathematical understanding of probability that developed from the fourteenth century onwards, particularly through association with commerce and insurance, and which I discuss in detail in note 21.

*3. Medieval philosophy and early science.*

In the Middle Ages, most of the surviving texts of philosophy and science were in Arabic and Greek, and only the Church had the intellectual resources to translate them into Latin, and begin their assimilation into Christian thought in the universities. It was inevitable, then, that more or less all medieval science should be conducted by churchmen, and we saw in the last chapter that, despite conservative opposition, Aquinas in particular helped establish that this was not incompatible with religious faith. The Bible and Nature both came from the same rational God, so they could not ultimately contradict one another, and apparent contradictions could always be resolved. The Church did not insist that every word in the Bible had to be taken literally (as the Protestant reformers did later), and Scripture could be reinterpreted if there were conclusive evidence that its actual words could not be true as they stood: the Church had long accepted, for example, that the earth was round and not flat. Moses was thought to have been the author of the first five books of the Bible, and it was held that he had had to adjust his teachings, in some matters, to the limited understanding of the Children of Israel. St Augustine, indeed, had seen the dangers of trying to use religion to settle scientific questions. In his youth he had been a Manichean, and they believed that eclipses were caused by the struggle between Good and Evil. But his studies of Greek astronomy had shown him that this was nonsense, and he wrote a book *On the Literal Interpretation of Genesis* to persuade the Church that the Bible was essentially about faith and morals, not the physical sciences.

Aquinas, in his *Summa Theologiae*, had shown how most of Aristotle's theories could be made compatible with Christian theology. 'By making Aristotle orthodox [Aquinas and his thirteenth-century contemporaries] licensed his cosmology to become a creative element in Christian thought.' [22] But the Church did not swallow him whole, and condemned in particular his beliefs that the world had always existed, and that every event was pre-determined as conflicting with God's omnipotence. This allowed philosophers to begin discussing all sorts of theories that he had dismissed, but could be possible for God, such as atoms, the void, plural worlds, and a moving earth. [23] Aristotle nevertheless provided a uniquely clear and coherent explanation of the natural world, and many of his theories could also be tested. Although the scholastic philosophers of the Middle Ages were justifiably ridiculed for their slavish devotion to everything Aristotle said, it was in fact their *criticisms* of Aristotle that laid some of the essential foundations for modern science:

> The great new scientific theories of the sixteenth and seventeenth centuries all originate from rents torn by scholastic criticism in the fabric of Aristotelian thought. And more important even than these is the attitude that modern scientists inherited from their medieval predecessors: an unbounded faith in the power of human reason to solve the problems of nature. [24]

In the twelfth century, Europe was still culturally backward and first needed to assimilate what the ancient thinkers had had to say about the understanding of nature, before they could make any contributions to science. Here an outstanding figure was Robert Grosseteste (c.1168–1253), Bishop of Lincoln, who had a leading part in introducing the new learning, especially Aristotle, to the universities. He was Chancellor of Oxford University, and was responsible for directing the studies of the Franciscans there, notably Roger Bacon, encouraging them towards languages, mathematics, and natural science in particular. [25] Grosseteste's main interest was in optics, which he regarded as the most important of the sciences, and the Latin version of al-Hazen's *Optics* was the principal basis of medieval scientific research in the subject.

On the first day of Creation God had said, 'Let there be Light', and quite apart from the association of light with Divine reason and the grace of Christ ('I am the Light of the World'), Grosseteste maintained that light was also the basis of the physical universe. At Creation, it had expanded from an original point into the sphere of the whole cosmos, and therefore was the basis of the extension of matter in space, and of all motion. Since straight lines and angles are fundamental to the propagation of light, this means that geometry gives us the mathematical means of understanding not only light, but the structure of the cosmos. He therefore pioneered the study of optics, which was continued by Bacon. [26] It was in his optical researches that Bacon was the first medieval scholar to use the term 'laws of nature' *(leges naturae)*, in relation to reflection and refraction, and he thought of 'universal nature' itself as one in which everything behaved in accordance with such laws. [27]

Laws of nature came to replace the Forms of the Aristotelian world, in which everything behaved according to its own essential nature, with a much more mechanical and quantitative model based on *relationships*, of a universe explicitly compared to the recently invented clock. Establishing mathematical laws of nature involved two related problems. The first was Aristotle's denial that qualities, unlike quantities, can be measured. The need to measure qualities first arose in theological debates about the varying degrees of sin, grace, or charity that might be present in the soul at different times.

This 'latitude of forms', or variability of qualities, generated philosophical problems of mind-boggling subtlety, but it also had profound implications for science because it could be applied to many physical states that were also considered as qualities – the opposites of hot and cold, wet and dry, slow and fast, heavy and light, for example.

The second, related, problem was that Aristotle's physical qualities refer to human *sensations*, whereas quantities, on the other hand, like size and weight, are not affected by our sensations of them, and really refer to the primary properties of things. The development of science therefore involved taking a physical quality like hot/cold and converting it from a sensation into a quantity, by means of an instrument by which it could be measured. The first thermometer, for example, in 1612, converted different amounts of heat into the height of a column of water in a tube, in other words, into the primary property of length that could be measured by a scale. The development of scalar thinking therefore played a fundamental part in the replacement of the Aristotelian essences and Forms, which were basically qualitative, by quantification and measurement.

The latitude of forms was first thought of as a scale of abstract degrees, that expressed the differing 'intensities' of the quality in question in the form of a line, [28] which came to function in the same way as the water in the thermometer was later to do, as a uniform scale of measurement. So it was now possible to construct ratios between two lines of different lengths, even if they represented two different things, like time and distance. This provided a way round another of Aristotle's problems: that speed itself can't be a magnitude, because time and distance are different kinds of magnitude. By the fourteenth century, the so-called 'Oxford Calculators', of Merton College, now regarded speed, in the modern way, as the ratio of distance travelled to the elapsed time. [29]

Since they were not concerned with the causes of motion, they could ignore Aristotle's distinction between natural and violent motion, and so were able to discuss motion in general. They defined uniform motion as that which covers equal distances in equal times, and could therefore define accelerated motion as one in which, in equal times, there are equal increases in speed. They also understood the notion of an instantaneous speed, as measured by the distance that *would* be covered by a body as if it were to move for a given period of time at the same speed as during that instant. Finally, in what is known as the Merton Mean Speed formula, they said that a body that is uniformly accelerating, or decelerating, will cover a distance in a given time equal to that which it *would* cover if it moved at its *mean speed* during the *whole* of that time. All these ideas were fundamental to an understanding of the laws of motion, and were used by Galileo, but, even more importantly, the philosophers were taking physics out of the Aristotelian world of qualities, of different, unrelated categories into that of measurable quantities, that could be related to one another by mathematics.

The most important work on force and motion, dynamics, was done at Paris, especially by Jean Buridan and Nicholas of Oresme, Bishop of Lisieux. Projectile motion had originally been discussed in relation to the sacraments: just as it can be asked if the grace obtained through the sacraments comes from them, or only from God, so, too, it can be asked if a projectile's motion comes from a force within it, or from the thrower.

Aristotle's theory of motion had been challenged long before Galileo by John Philoponus in the early sixth century AD, who had the idea that 'an incorporeal force' was imparted to the projectile when it was thrown. [30] This idea of a force imparted to the projectile itself was independently taken up again in the thirteenth and fourteenth centuries as the idea of impetus, which is the amount of matter or weight in a body multiplied by its speed – an idea that would have been nonsensical to Aristotle. They believed that the air resists both natural and violent motion, and said that the arrow keeps moving because it has received an impetus proportional to its weight, and the force with which it has been fired. This impetus remains for a while, just as a piece of iron removed from the fire stays hot for some time, but eventually dies away, due to air resistance and gravity. Since heavy things like rocks can acquire more impetus than light things like feathers, this explains why they can be thrown farther, and the natural movements of the heavens were also explained by the original impetus given them by God at the Creation.

The impetus theory was also used by Oresme against one of Aristotle's most important arguments that the earth is stationary: that an arrow returns to its place of origin when fired vertically into the air. If the earth is rotating eastward, then this motion will give an impetus to the arrow that will also carry it eastward, and this will explain why, on a rotating earth, it can still land in the place from which it was fired. [31] Impetus theory was also used to explain the acceleration of falling bodies, because the impetus was seen as a force that acted continually on the body while falling due to its natural motion, so that the impetus would continually increase. [32]

All in all, then, it can be said that these fourteenth-century philosophers were 'moving towards a newly fluid, complex and relational conception of the world', [33] an *operational* conception, in contrast to the static world of Aristotle's essences and Forms. But it has often been noted that, despite all their enthusiasm for measurement, none of them actually measured anything at all. They could, however, make all their theoretical points by dealing in ratios, in *relative* quantities, not in absolute quantities, and in any case, despite Roger Bacon, they had no serious interest in actual physical experiments in which measurements could have been used. [34] While they talked a great deal about the need for experiment, this was all in the context of debates about correct scientific method, and the few experiments that were actually done were merely intended to show the superiority of one idea about scientific method over another.

In fact, the small amount of experimental work of the medieval philosophers, beginning with Grossteste, and which centred on optics, had died out by the middle of the fourteenth century, and this is yet another example of the limits of pure philosophy. As in the Hellenistic and Islamic worlds, medieval philosophy could only go so far, and the development of experimental science would not occur until scholars could become familiar with the practical knowledge of commerce, surveying, navigation, and crafts like clock-making and metallurgy, where accurate physical measurement was important and where it first developed. [35] However, most of the significant scientific writings of the medieval period went into print and were available to sixteenth and seventeenth century scientists, on whom they had an important influence. [36]

## 4. *Craftsmen and science.*

As early as the thirteenth century, Bacon had said that the scientist should not despise craft knowledge but try to acquire it, [37] and there was also a whole series of inventions that linked crafts and science, notably printing, clock making, lens-grinding, gunpowder and artillery, and the magnetic compass. The printing press was of special importance here, since the printer's workshop 'brought together scholars, craftsmen, merchants, and humanists engaged in common pursuits', and made it far easier for books to be published in the language of the artisan and produce a popular scientific culture, as distinct from the elite Latin culture of the universities. 'When apothecaries, potters, sailors, distillers, and midwives got into print along with scholars, humanists and clerics, the Republic of Letters was permanently changed.' [38]

The growing participation of artisans in scientific culture, and the importance of astrology, medicine, and alchemy produced such figures as the artist Dürer (1471–1528), the instrument maker and mathematician Jamnitzer (1508–85), the potter Palissy (c.1510–90), the doctor and chemist Paracelsus (1493–1541), and the metallurgist Agricola (1490–1555), all of whom wrote books promoting the idea that doing in the form of craft skills is an essential form of knowledge. By interacting with matter in the performance of a craft one came to understand the hidden structure with which God had designed physical objects and which it is the duty of man to reveal. Jamnitzer, for example, believed that the design of all living creatures was based on the five Platonic Solids and produced a book illustrating 140 of these combinations. [39]

By the time of the Reformation scientific instrument making had become closely linked with printing and books.

> Not only did they share the strategic craft of the engraver – often the same person made the sundials and scales and the engraved plates for printed illustrations – but they interacted also in the evolution of a practitioner literature. Books on surveying and navigation abound, and the instrument makers often became entrepreneurs for the "practical" science of the $15^{th}$–$17^{th}$ centuries. [40]

It is not therefore surprising that the skilled artisans were becoming increasingly valued by scientists.

A good example of the fruitful interaction that could develop between craftsmen and scholars is that of Robert Norman, a compass-maker of London, who could not understand why his compass-needles became unbalanced as soon as he magnetized them, dipping to the north. In his book *The Newe Attractive* (1581) he says that having tried to rectify this without success, he consulted his learned friends, who advised him on how to perform some experiments. He mounted a needle so that it could move vertically, and so discovered the very important phenomenon of magnetic dip. Norman dedicated his book to one of his 'learned friends', William Borough, comptroller of the Navy, whom Norman thanks for his 'encouragement, good counsel, accustomed courtesy, and friendly affection towards me, an unlearned mechanician', which were apparently crucial to Norman's success. [41]

On the other hand, the learned sciences of medicine and medical astrology were also very important in breaking down the gulf between theoretical and practical knowledge. Dissection gave the physician the opportunity to see the construction of the human body for himself ('autopsy' literally means 'seeing for oneself), and this 'autoptic' authority was increasingly regarded as superior to the authority of the classical medical texts. Medical astrology had close associations with mathematics and astronomy, because medical horoscopes needed careful planetary observations and time measurements for the revision of astronomical tables. Physicians began to make their own observations, and soon found the designs of instruments inherited from the Arabs too crude, so began developing improved versions. We have seen that the greatest physicians were usually attached to royal courts, and attended their patrons to the wars, which were almost perpetual. Naturally, they became interested in artillery and siege engines, so that 'Nearly every important treatise on technology produced in Europe during the fourteenth century and the first quarter of the fifteenth century comes to us from the quill of a medical astrologer.' [42]

We can now look at some other inventions and explore this point about the interaction between everyday crafts and scientific theory in more detail.

*The clock.*

There was no use in medieval society for devices that could measure time within minutes, and the origins of the clock lay quite elsewhere. The life of monastic communities was organized around a daily cycle of religious services that extended into the hours of darkness. Matins, the first, was held at midnight and the last, Compline, also after dark in the winter months of northern latitudes. The monks were summoned to each service by a bell rung by the sacristan, and a water-clock operating an alarm was used to wake him up in time to ring the bell for Matins. [43] As far as we can judge, the alarm was set at Compline by filling the container with water so that, at the proper time, a lever released the striking mechanism to ring the sacristan's bell. But water clocks tend to freeze up at night in northern winters, and somewhere between 1250–1300 some unknown genius found that the oscillating mechanism for striking the bell could also work in reverse, and be used as a time-keeping device that eliminated the water-clock. [44] Until the mechanical clock, time had been measured by devices that used continuous motion of some kind: sundials, water clocks, graduated candles and burning fuses. [45] The idea of measuring time by an oscillating escapement, instead of continuous motion, was therefore deeply counter-intuitive, and it produced for the first time a machine whose speed was not regulated by anything external, like a flow of water, but was entirely internal and self-governing.

Apart from ringing bells in monasteries, one of the first applications of clocks was in great astronomical clocks in cathedrals, showing the movements of the sun, moon and other planets [46] (for example Norwich in 1325, Richard of Wallingford's great clock at St Albans in 1330, and Salisbury in 1380). Until the use of the pendulum three centuries later, however, these clocks were not accurate enough for astronomical research. The astronomical clock was a revival of those automata that had been so popular in the ancient world, in which man played at being God, and became a standard image of the universe set in motion by the Creator. Indeed, the clock, and the machine which it typified, rapidly replaced the organism in the scientific imagination as reason incarnate in the physical world. [47]

Clocks were also eagerly adopted by the merchants and manufacturers in the towns, as a means of regulating the hours of the workers. Unlike the natural hours of the day, which were longer in summer than in winter, clock hours had to be the same length, and uniform hours had been adopted by 1330 in Germany, and 1370 in England. Clocks soon appeared in private houses as well, and were one of the first of those measuring instruments which inspired a thriving craft in which both artisans and theorists were closely involved together. The clock required a good deal of mathematical ability to calculate the gear ratios and relate these to the length of the day, as well as the accurate division of dial plates, but it also required a very high level of manual skill to make the components, such as the cutting of fine screw-threads and gears, the manufacture of bearings, and understanding the expansion of different metals, which were all fundamental skills of machine technology. The later discovery of the pendulum by Galileo, and its application to clocks by Huygens, is another good example of the involvement of pure science with manual crafts.

The elaborate clock of the Imperial Astronomer, Su Sung, in China had remained a unique achievement, hidden away in the Bureau of Astronomy and then forgotten, but in Europe, by contrast, there was no such secrecy, and clocks rapidly became part of urban life. Like other machinery such as automata, water and wind mills, clocks also made it much easier to think about cause and effect. 'The test of whether we have acquired a mechanical understanding of something is for us to be able to take it to pieces and put it together again' [48], and Piaget has noted the crucial part that the study of machines plays in the development of causal understanding in children. By the sixteenth and seventeenth centuries, clockwork and machinery had become the models for all the processes in nature, from the motions of the heavenly bodies to the heart as pump, or the eye as a *camera obscura*. Men hoped to be able to explain all these natural phenomena, too, in the same way as one understood how a clock works, by taking it apart and examining the relations between its components.

*The lens.*

In the ancient world, as we saw, lenses were used by seal and gem engravers for very fine work, but lenses of this sort were generally of 3x–10x magnification, and so too high in power and too short in focal length for use as spectacles, which were unknown. The first suggestion that lenses could be used as a reading aid for the long-sighted came from Roger Bacon in the course of his work on optics, where he notes that a plano-convex lens placed on writing will make letters appear much larger, 'And therefore this instrument is useful for the aged and for those with weak eyes. For they can see a letter, no matter how small, at sufficient magnitude.' [49] The totally different attitude towards practical usefulness from that of the ancient world is very obvious here.

Before the telescope could become a practical possibility, it was first necessary to develop lenses for reading, and this occurred within twenty years of Bacon's mention of them, in about 1286; his ideas may have helped stimulate their invention, perhaps by a cleric who had read him and asked a glass-maker to make him up a pair of spectacles. [50]

Lens-grinding became a specialized trade, much assisted by developments in the lathe. The telescope needed the combination of the small concave lens as the eye-piece, (to produce an image that was the right way up), and a larger convex lens for the object glass, and it seems clear that the practical foundations of the telescope were laid by lens-grinders and spectacle-makers, but probably stimulated by the natural magicians, such as della Porta, who were also very interested in the magnifying properties of lenses. But recent research has made it clear that a telescope using a bi-convex lens as eye-piece, and a large concave mirror as objective had actually been constructed by another natural magician, Thomas Digges, in England by 1580. [51] The year 1608 is generally taken as the year of the invention of the telescope, merely because it was then that the spectacle-maker Hans Lipper[s]hey applied in Holland for a patent for a refracting telescope. This was denied because others in Holland also claimed to have invented it, but it was Galileo who made it famous by using it for astronomy, as we shall see in the next chapter. People nevertheless objected to Galileo's telescope because of 'the association of optical devices with "natural magic" aimed at illusion and deceit, and the belief that only direct vision, certified by touch, could show the world as it really was.' [52]

The microscope almost inevitably appeared at about the same time as the telescope, but whereas the telescope had a long association with natural magic, the microscope seems to have emerged from the craft of spectacle-making, perhaps invented by Zacharias Janssen (1580–1638) of Holland. These first microscopes were compound (with two or more lenses) but were feeble devices of no more than about 10x magnification. They were greatly improved by scientists, notably Robert Hooke, who increased their power to about 50x, but really high power, of around 275x, could only be achieved in the seventeenth century by single lenses in the form of tiny glass beads. These may have been invented by Hooke,

[53] and revealed the hitherto unsuspected world of insects' eyes, the stings of nettles, and the cellular structure of plants that Hooke was the first to see. Later in the century, with van Leewenhoek, the existence of bacteria and spermatozoa was also discovered. The superb illustrations of Hooke's *Micrographia* (1665) showed what the microscope could achieve, and had a very powerful impact. Meanwhile, Kepler and then Descartes had provided a full mathematical treatment of magnification, which was the basis of further improvements in the microscope and the telescope, and Kepler had finally separated the optics of the eye from our perceptions of the world, by showing that the image on the retina is reversed and upside down, as in a *camera obscura*, and that it is the brain that then corrects this image. [54]

*The thermometer.*

The thermometer [55] was the first of the modern instruments that replaced sensations by measurement. Physicians had always had a special interest in degrees of heat and cold, and in the ancient world Galen had proposed that, between Aristotle's opposed qualities of hot and cold there was a neutral zone that was neither hot nor cold, but could be produced by mixing boiling water with ice. He also thought that there were degrees of heat and cold, but no real developments in measuring these occurred until 1578, when Hasler proposed a single numerical scale from 1–9. This comprised four degrees of heat from 1–4, and four degrees of cold, with a zero point between them. Each degree was divided into ten parts, making ninety in all, which he attempted to correlate with the prevailing temperatures at the different degrees of latitude. The idea of scales of temperature, including both heat and cold, was in fact becoming common in the second half of the sixteenth century, although the belief that heat and cold were different kinds of thing long survived the introduction of the thermometer.

The invention was based on the rediscovery of the *Pneumatics* of Hero of Alexandria. We saw in Chapter XIII that Hero described how, when vessels containing water were heated, the air would drive the water out, while on cooling the water would be sucked back again. This was the basic idea behind the thermometer, and in the 1589 edition of *Natural Magic*, della Porta describes some of Hero's experiments involving this principle. While della Porta did not use it to measure temperature, the same basic idea of an inverted glass bulb, with the end of its neck in a water-trough, was described by Santorio Santorii of Padua in a book published in 1612, specifically recommended as a device for measuring temperature. Santorio was a physician and physiologist, with good reason for being interested in developing this device, which he explicitly claimed to have adapted from Hero's apparatus.

He also used his thermometer to measure the heat of the sun and moon, claiming that while the sun caused a fall of 120° (remembering that these thermometers were upside down), the moon only caused a fall of 10°. [56] The main problem of the air thermometer was that it was also influenced by barometric changes in air pressure. The reality of atmospheric pressure was only discovered in about 1640 by Torricelli, and by the 1660s thermometers were being made as sealed tubes in which coloured alcohol had replaced air and water. They were to play an extremely important part in the development of chemistry and physics.

### Gunnery and the laws of motion.

It is often believed that the practical need for accurate gunnery drove scientists to investigate the laws of motion, but in fact, of course, medieval philosophers had made substantial progress in understanding motion long before cannon appeared on the scene. Whereas, however, the catapults of the ancient world had been of no interest to philosophers, gunnery fascinated a series of Renaissance and seventeenth century mathematicians, as an opportunity to explore Aristotle's theory of projectile motion, and the relation between range and elevation. In the spirit of the age, the mathematicians also wanted to make an important practical contribution to gunnery, and published many range tables, but these hopes rested on an illusion. In the first place, the actual trajectory of cannon balls was to prove a baffling problem because it is influenced by a number of factors besides the force of the explosion itself: gravity, air resistance, temperature, humidity, wind speed and so on, and the mathematical solution of trajectories proved enormously complex. Secondly, the guns of the day were simply too inaccurate, and differed too widely from one another in their performance, to provide any reliable test of mathematical theories. [57]

'The standard of engineering technology was not merely insufficient to make scientific gunnery possible, it deprived ballistics of all experimental foundation and almost of the status of an applied science, since there was no technique to which it could, in fact, be applied.' [58] It was only in the nineteenth century that a practical ballistic science could be developed, because only then had industrially produced guns become sufficiently precise and standardized. In any case, since field commanders and naval captains knew that the greatest damage was inflicted on the enemy at close quarters, warfare on land and sea ideally consisted of opponents battering each other into submission at point blank range. While gunners needed some knowledge of mathematics, they could manage perfectly well without mathematical ballistics.

The real significance of the gun to scientists was that it acted as a perfect machine in thought experiments about projectile motion, and ultimately Newton was to imagine the moon as like a cannon ball, fired with enough velocity to keep it in permanent orbit around the earth. Niccolo Tartaglia (c.1500–1557) wrote the first book on ballistics, the

*Nova Scientia*, in 1537. There is a materialist myth that he was basically a practical experimentalist, uninterested in theory, [59] but in fact he was a brilliant mathematician who knew nothing about guns, until he was asked in 1531, by a close friend of his who was a master-gunner, at what angle of elevation a gun would shoot the farthest. He deduced that logically this must be 45°, as the mid-point between the horizontal and the vertical, and claimed this had subsequently been verified by gunners.

Tartaglia is a fascinating figure because he is right on the cusp between Aristotle and modern theories of motion. He accepted impetus theory, and the idea that the air resists all motion, but still believed in the Aristotelian distinction between violent and natural motion: violent motion began with a high velocity that steadily diminished until rest, whereas natural motion was the opposite, beginning at rest and then accelerating indefinitely. Obviously, then, as Aristotle had said, the two kinds of motion could not co-exist in the same body simultaneously, and at the time this was a most convincing argument. Violent motion therefore had to precede natural motion, as when a ball fired from a cannon loses its impetus, and then starts to fall vertically. But whereas the Aristotelian trajectory had been based on straight lines, with a sudden change of motion from violent to natural, Tartaglia maintained that part of the trajectory must be curved, because only with a curve could there be perfect continuity of motion: [60]

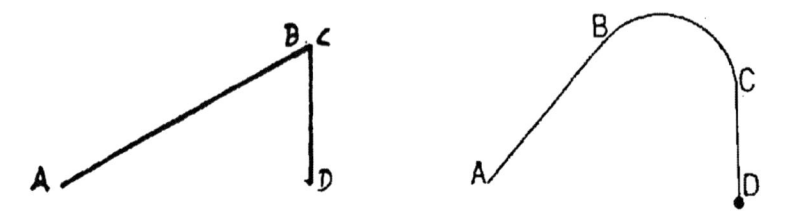

His reasoning was purely theoretical and geometrical, unrelated to any experiments with actual guns, and we shall see in the next chapter how Galileo applied far more rigorous experimental techniques to the problem. Tartaglia did not even doubt Aristotle's law that heavy objects fall faster than light ones, because it seemed self-evident that weight was a force, and that more force must produce greater speed.

Apart from the theory of projectile motion, gunnery also provided considerable opportunities for experiments on the power of gunpowder. Unlike the catapult, it was possible to calculate the force applied to the projectile quite precisely by measuring the amount of gunpowder used, and instruments to test the strength of gunpowder were also developed. Thomas Digges, [61] in the final chapter of his *Stratioticos* (1579) isolates four principal factors in a gun's performance: the quantity and quality of the powder; the length of the gun and the proportions of the bore; the size and weight of the projectile; and the angle of elevation; as well as some minor factors, such as the regularity of the bore, how well the ball fits, and so on. On this basis he then proposes about two dozen experiments, all of which are based on varying one factor and keeping all the others constant. For example, if two similar guns, with the same weight of ball, are loaded with the same type of powder, but of different weights, will the different ranges be directly proportional to the different weights, or to their square or cube roots? Again, if an iron and then a lead ball are fired from the same gun, with the same powder charge, will the difference in range be directly proportional to the difference in their weight, i.e. does a ball of half the weight go twice as far? While these experiments seem only to have been imaginary, they illustrate how it was now possible to think about force and motion in much more precise ways that could be measured. Artillery also led to a new geometric science of fortification, and rational town planning. [62]

## 5. The rational artist and experimental science.

The relation between craft knowledge and theory reached its highest development in the artist-engineer. We saw that while the first writers on topics of military engineering were predominantly physicians in the service of princes, by the fifteenth century artists were becoming increasingly interested in machines of all types, including those for war, and the artist-engineer became a familiar European figure.

> At the end of the fourteenth century the universities went into decline and the leaders in original thought and action became a different group, largely outside them, of. . . artist-engineers. Their expertise lay in the rational control of materials, processes and practices of all kinds, from painting to music, from architecture to machinery, from cartography and navigation to accountancy. They brought about a general transformation of European intellectual life. [63]

The central idea of the rational artist, engineer, and business-man is summed up in the Italian notion of *virtù*:

> A man of *virtù* in Renaissance Italian, coming from the Latin *virtus* meaning power or capability, was a man with active intellectual power to command any situation, to do as he intended, like an architect producing a building according to his design; by contrast with someone at the mercy of *fortuna*, of chance or luck, of the accidents of fortuitous circumstance, unforeseen and hence out of control. [64]

*Virtù* not only required a theoretical understanding of how nature works, grounded especially in mathematics, but also a mastery of practical techniques and materials, typified by those artist-engineers of whom Leonardo da Vinci is the best example [65]. These men

> were a product of Italian urban society and essentially practical in their outlook. Their contribution to the intellectual context of European

science was to add to the logical control of argument achieved by the philosophers, a rational control of materials of many different kinds in painting, sculpture, architecture, canal building, fortification, gunnery, music. The leaders of this group increasingly came to have training in anatomy and in the theoretical sciences of perspective and mechanics as well as in such practical skills as bronze-casting and masonry. . . it shows a characteristic attitude towards the nature that was to be imitated by artifice and mathematical science. It was an active, rationally aggressive attitude to nature. [66]

A good example of the artist-engineer at work is the discoverer of perspective painting, Filippo Brunelleschi (1379–1446), somewhere between 1413 and 1425. He was a Florentine clock-maker, military engineer, fanatical measurer of everything, and the great architect who designed and built the famous dome of the Cathedral in Florence. [67] The modern story of perspective begins with the arrival in Florence, from Byzantium, of a manuscript of Ptolemy's *Geography*, in about 1400, and its translation into Latin by around 1406. [68] We remember that Ptolemy had described how to project an area on a model globe on to the flat surface of a map by using the principle of the visual cone or pyramid, with its central ray. The map forms the base of the visual cone, and the single view-point of the central ray is the key element, because it is the basis on which all the relations between the different points on the map, and the lines of longitude and latitude, are co-ordinated. But this is only half the story, and to understand the rest we need to bring in mirrors. Flat, lead-backed glass mirrors were relatively new in Europe at that time, and were of particular interest to painters because they provided a new way of grappling with the apparent convergence of parallel lines with distance. It was, of course, observed that anything in front of a mirror seemed to be exactly the same distance behind its surface, in the imaginary space 'through the looking glass', as it was in front of the mirror in reality. (For further details see note 69.)

We know that mirrors were fundamental in Brunelleschi's thinking, because of the remarkable experiment establishing the laws of perspective painting that he carried out in about 1425, and very carefully described by Manetti, his biographer. (See note 70.) The first book on perspective, by Alberti, (*De Perspectiva*) was based on Brunelleschi's work, and the practice of perspective drawing became highly technical, involving a variety of instruments to assist the artist. [71] The exact measurement, scale drawing, and three-dimensional effects required meant that pictorial art could become an integral part of scientific research, especially in the drawings of anatomy and machines.

Just as the study of geometric optics had provided a theory linking the objective properties of the physical world with what we see, so, too, theories of acoustics were developed to relate the sensations of our ears with the actual properties of nature. [72] One of the fundamental problems was to explain how our sensations of harmony and discord are produced by the mathematical properties of vibrating strings. The scientific study of music and the effects of sound on the ear progressed considerably in the sixteenth century, and Galileo's father, Vincenzo, made important experimental contributions. [73] The person who did most to set music on a scientific foundation was Marin Mersenne, (1588–1648) Jesuit theologian, mathematician, and musical theorist, who published *L'Harmonie Universelle* in 1636. [74]

With Galileo, whom he greatly admired, he was a pioneer of experimental science, combining what he called 'well arranged and well made experiments' with theory and mathematical analysis. These enabled him to measure the speed of sound, and prove, for example, that harmonious intervals like octaves, fourths, and so on were produced in the ear by two matching sound waves, while dissonance increases as coincidence of the waves decreases. This finally explained why the Pythagorean harmonies had to be whole tones, because these corresponded to whole sound waves. [75] So,

> by discovering, obeying, and manipulating natural laws, with increasing quantification and measurement, art was seen to deprive nature of her mysteries and to achieve its mastery by reasoned foresight, whether in the representation of a visual scene, the design and control of a machine, the composition of music, the navigation of a ship across the ocean. . . or control of a disease or even of the affairs of state. [76]

Brunelleschi and Mersenne illustrate how the rational study of art could become the experimental science of those like Galileo, who followed his father in this tradition. Like the artist-engineer, he took nature to pieces and reassembled it from known theoretical principles, and he saw himself as the heir of Archimedes 'who transformed the questions of the philosophers to make them capable of technical answers'. [77]

## 6. Experimental science.

Galileo and della Porta were both members of the Academy of the Lynxes, a scientific society so named because the gaze of the lynx was supposed to be able to penetrate into the hidden parts of whatever it looked at. We have seen that natural magic made a fundamental contribution to the rise of modern science, and remained an important influence during the seventeenth century, but that its experiments were basically exploratory, and did not test theories or use mathematics in the manner of Galileo or Mersenne. Della Porta had been the star of the Academy, until Galileo showed what the telescope could do in refuting Aristotle, and established the superiority of his type of science.

With Galileo, science was now seen as the art of the soluble, the pursuit of answers to specific questions that could be solved in some clear and definitive way, and this meant by well-designed experiments. To design a good experiment one had first of all to ask the right question, and this depended on having a theory that gave the experiment a specific purpose, and in terms of which the results could be interpreted in mathematical form. This needed efficient measurement, which in turn involved making the results as unambiguous as possible by stripping away all extraneous and confusing factors, above all by distinguishing the primary properties of matter from the secondary qualities of our sensations. Kepler, for example, had solved one of the main problems in the history of optics – the confusion between physics and the human senses – when he realised the different optical functions of the eye and the brain. Experiments therefore naturally concentrated on the primary properties of matter, and this made the design of apparatus and instruments for accurate measurement of the first importance. Instruments were the only means by which one could measure what hitherto had merely been sensed, such as the thermometer to measure heat, or the frequency of a vibrating string to measure the pitch of a sound.

The mechanistic model of the universe, derived from the clock in particular was, of course, very appropriate for the experimental project. The machine, regular and rational, could be taken apart and reassembled to see how it worked, and provided the perfect model of causal explanation. But another essential feature of the mechanical world-view is that it abandons Aristotle's ambition of explaining *why* things are as they are, and concentrates only on his 'efficient' cause, the 'how'. The Greeks would never have accepted this as true science, whose main business was discover the reason why, and many of Galileo's contemporaries rejected his method for this very reason. But,

The great advantage of seizing upon efficient cause as "the" cause of anything happening, and of interpreting it in mechanical terms, the pushes and pulls of clockwork, was that it led to answerable questions in what we have come to call physics. Instead of asking how motion was possible, or why bodies should be attracted towards the centre of the Earth, Galileo asked how fast they fall. [78]

Ancient atomism, which had been rediscovered, also fitted extremely well with the mechanical model, because of its belief that matter in motion explains all forms of change. In *The Assayer* Galileo developed an atomic or 'corpuscular' (*corpusculus* means 'little body') theory of matter. As in the case of the ancient atomists, it made the primary properties of size, shape, weight, and motion very easy to distinguish from the secondary qualities of our sensations. 'I hold that there exists nothing in external bodies for exciting in us tastes, odours, and sounds, except sizes, shapes, numbers and slow or swift motions.' and it was the behaviour of these with which mathematical laws were concerned. Galileo used his notion of atoms to explain heat and light, for example, in terms of very rapid vibrations of corpuscles, and as we shall see, it was the idea of different atomic weights that was to become the theoretical foundation of chemistry by the beginning of the nineteenth century.

The laws of motion were particularly well suited for attack by this experimental, mechanistic science, and provided the easiest gateway into a truly scientific understanding of the physical world. Unlike optics and music they don't involve our sensations, and unlike chemistry, biology, and electricity, they are concerned with very few factors, that can be summed up in the simple question 'What happens when we whirl a stone round on a string and then let it go?' Experiments in motion are fairly simple to set up, and the results easy to measure, and express mathematically, so that the laws of motion as established by Galileo, Kepler, and Newton became the model for the rest of science in the following centuries. 'Natural philosophy was incorporated into the fundamental tone of modern Western civilization, characteristic of its art and its politics as much as its science: its coercive, military approach to nature and mankind, its restless curiosity and striving for progress, in striking contrast with every other civilization.' [79]

Experiment therefore raised the competition between rival theories to a new and much more decisive level than was possible with purely philosophical debate. Philosophical inspiration, of course, still had a fundamental part to play, and scientists were willing to be convinced by theories that seemed beautiful and elegant, even if there was much evidence against them. But, in the last resort, experiment was now the final test of scientific truth.

# CHAPTER XVI

## Modern science and industrialism

Many historians of science have believed that its progress was purely intellectual, the result of men of genius triumphing over ignorance and superstition, and that social conditions were quite irrelevant. Marxists, however, went to the other extreme, [1] and claimed that all sixteenth and seventeenth century science was 'really' driven by the need to solve technological problems that were economically important to the rising bourgeoisie. Since modern experimental science only developed in Western Europe because of the special nature of its society, and especially because craft knowledge and scientific theory could be brought into collaboration, in a capitalist milieu, the Marxists obviously have a point. Problems in mechanics, hydraulics, optics, and acoustics, for example, 'middle-range science', that had been left unresolved by the ancient world, were taken up again by the artist-engineers and the early experimental scientists, such as Simon Stevin (1548–1620), [2] Mersenne, and Galileo. While some problems, such as ballistics, were too difficult for them, they made very important discoveries that were technologically valuable, and so profitable as well. Galileo, for example, offered his improved telescope to Venice for observing enemy ships at sea, and the Republic rewarded him with tenure for his Chair of Mathematics at Padua University and a raise in salary. Yet it is obvious that what really excited Galileo about the telescope was not looking at ships, but discovering the secrets of the Heavens and proving that Copernicus was right.

When trying to explain why seventeenth century scientists pursued such *fundamental* inquiries as astronomy, the laws of motion, the atomic nature of matter, magnetism, and the vacuum, the Marxist argument is therefore rather like claiming that Einstein was really motivated to discover that $e = mc^2$ by the need to give us an alternative to fossil fuels. While the fundamental problems of science turned out to be essential to modern industrial civilisation, solving them had no immediate practical pay-off.

We have, then, to recognize that scientists investigated these problems from intellectual curiosity about the basic secrets of nature; as part of a philosophical battle with Aristotle's ideas on the elements, atomism, the void, and motion; and for religious reasons. As the whole issue of natural magic shows so clearly, seventeenth century scientists inhabited a very different thought-world from our own, which was nevertheless an integral part of the advancement of modern science. The real importance of technology in this process was that it often stimulated the speculations of theoretical scientists on fundamental issues.

*1. The laws of motion and the new astronomy.*

Galileo's work on the laws of motion, in *Two New Sciences* (1638), is a classic demonstration of experimental science in action, expressed in measurements and in mathematical concepts and laws. He regarded Aristotle's search for the essences of things – such as natural and violent motion – as basically misguided, because they were nothing more than superficial descriptions of what things did, or how they looked. He was a Platonist, and thought that the scientist could only understand reality by ignoring the actual appearances of things, and concentrating on their primary qualities, which were essentially mathematical. So, he said, he could not sufficiently admire those like Aristarchus and Copernicus, who had ignored the evidence of their senses, and applied mathematical reasoning to the heavens. But understanding this mathematical structure of things could only be attained by rigorous experiments.

The story that he dropped two cannon balls of different weights from the Leaning Tower of Pisa before an invited audience, to humiliate the Aristotelians, has a large component of myth. [3] All we need to accept is that Galileo, like others, had experimented with falling bodies and found that, making allowance for air resistance, bodies of different weights fall at the same speed. [4] The next problem that Galileo set out to solve was how fast they accelerate. He therefore devised a method of slowing down the whole process of falling, so that he could set up an experimental apparatus to obtain exact measurements. He did this by rolling a ball many times down an incline, and timing the results very accurately. [5]

His results showed that acceleration is constant, which contradicted ideas that the speed of falling bodies might increase very rapidly to begin with, and then stay the same. There is, therefore, an equal increase of speed in equal amounts of *time*, (not of distance, as Galileo and many before him had previously supposed) and in free fall this is 32 feet a second, every second. A stone falling for twice as long as another therefore goes four times as far; for three times as long, nine times as far, and so on, the 'times squared law' of acceleration. He was not concerned with what he saw as the philosophical problem of what *made* the stone fall, but simply with measuring how fast it actually accelerated, but he had also found a precise measurement of the force of gravity as it affects objects on earth.

Since a ball rolling down a slope would accelerate, while a ball rolling up a slope would decelerate, it also followed, logically and inevitably, that a ball rolling along a horizontal path in a straight line must continue to move for ever at the same speed. This is the conservation of motion; it looks like inertia, as defined in Newton's First Law of Motion, but the central idea of inertia is *resistance* to a disturbing force, and Galileo did not consider forces. [6]

Galileo used a similar combination of theory and experiment to discover what the true path of a cannon ball would be. The relevance of these experiments was that, like the motion of the planets, the path of the cannon ball is curved, and Galileo showed that this was the result of two separate motions. One was horizontal motion – provided in this case by the explosion – and the other was the downward, accelerated motion of all falling objects.[7] The resulting path of the cannon ball was in fact a parabola. These ideas that the path of projectiles combined two motions, the conservation of motion, and his law of acceleration were all highly relevant to planetary motion and astronomy, where he was a notable champion of Copernicus.

There were two major reasons for the revival of astronomy in the sixteenth century: the demands of navigation for more accurate charts of the heavens, and the inaccuracies of the Julian calendar, especially in relation to Easter, and there was also the need to reconcile the Arab data with Ptolemy's. But these problems could have been solved without worrying whether the earth or the sun is at the centre of the cosmos, which had no relevance whatever to the calendar or to navigation, or even to tide-tables. Astronomy was about cosmic order, and its revival coincided with a new interest in Platonic and Pythagorean thinking; this approach to nature not only included the idea of number and geometry as the true realities underlying the appearances of things; but also sun worship, and the idea of the Sun as the visible symbol of God.

Nicolaus Copernicus, (1473–1543), a Pole by birth and a Canon of Frauenberg Cathedral in East Prussia, was an expert mathematical astronomer (his calculations were used for the Gregorian calendar reform of 1582). He was much influenced by Neo-Platonism, and it therefore seemed to him that the Sun, the 'mind', the 'ruler' of the universe, which Hermes Trismegistos (the ancient and mythical originator of alchemy) called a living god, should also be at its centre. [8] Since in his view God must have designed a perfect universe, He would have based the movements of the heavenly bodies on perfect circles, so Ptolemy's elaborate systems of epicycles and equants had to be wrong. He therefore revived ancient Greek theories that the sun, not the earth, is the centre of the universe, and this also required the earth, like all spheres, not only to have a circular motion around the sun, but to rotate on its axis. (It is hard to imagine anyone approaching astronomy from a more 'other-worldly' perspective.)

But he then had to meet the ancient objections to the idea that the earth revolves and the heavens are stationary. These were, in particular, that if the earth is rotating rapidly eastward, then objects thrown upwards would not come down in the same place, and the clouds and birds would all drift westwards, and so on. Copernicus replies that rotation is actually natural for spheres, and that the lower air close to the earth is carried along with it as it rotates, which is why none of the predicted effects on clouds and birds, actually occur. He answers the objection that objects thrown upwards return to their point of origin, by using Oresme's argument that they return to the same place because they are carried along by the circular impetus they receive from the earth's motion. And how likely is it, Copernicus asks, that the whole universe could rotate around the earth in twenty-four hours, since the vast distances of the stars would involve their travelling at an unimaginable speed to achieve this? As the shore seems to be moving when seen from a ship in motion, so it is possible that the apparent motion of the heavenly bodies is actually the result of the earth's motion.

Once we also think of the earth as orbiting the sun among the other planets, the major peculiarities of planetary motion – their retrogressions, and their variations in brightness – can easily be explained by the fact that the earth periodically overtakes the outer planets in their orbits round the sun, and is itself overtaken by the inner planets. Retrogression only occurs when this is happening, and it is also at this time that the earth is nearest to the planet in question, which is why it is also at its brightest. This was an enormous intellectual simplification of planetary motion, which had great appeal to many scientists. It is the form in which the Copernican revolution is typically presented in popular histories of science, so readers naturally wonder why every intelligent person wasn't immediately converted to it.

But there were other problems that this simple model could not solve, such as the fact that the sun moves more quickly between the autumnal and vernal equinoxes. To account for this, and for similar irregularities of the planets, Copernicus in fact had to go back again to eccentrics and epicycles; so the earth, for example, revolves on a circle whose centre itself revolves around another point on another circle which itself revolves around the sun. To make his system work as accurately as Ptolemy's, in fact, he needed to use almost as many eccentrics and epicycles, with no significant improvement in accuracy. This was bound to be the case as long as it was believed that the planets revolve in circular, rather than elliptical orbits.

By removing the earth from the centre of the cosmos, Copernicus also had to answer the ancient objection that if the earth moved this would be detectable by stellar parallax. Here he could only argue that the stars must be farther away than previously supposed, but he also had to answer the even harder question of why objects fall towards the centre of the earth, now that it was no longer at the centre of the cosmos. He therefore began thinking of some other principle entirely, and proposed that heavy bodies are drawn to earth by some intrinsic attraction. [9] Here we have the first intimations of a gravitational force between objects, which was taken up by Kepler, whom we shall come to in a moment, but it was vague and very unsatisfactory. Yet in spite of these serious limitations, a number of astronomers found the simplicity of the new model aesthetically convincing. [10]

Actual evidence in its favour, however, remained distinctly thin for a long time, [11] and over the next decades, the main developments in astronomy, as far as actual evidence went, were a series of observations that increasingly discredited Aristotle's distinction between the unchanging, perfect heavens, and the corruptible realm of earthly things.

In 1572 a brilliant new star, which we now know was a supernova, appeared in these 'unchanging heavens', and the great Danish astronomer Tycho Brahe (1546–1601) proved that it was stationary, and therefore a genuine star. A major comet appeared in 1577, and Brahe proved that it could not be in the atmosphere, whereas Aristotle had maintained that comets *are* in the atmosphere, between the earth and the moon. It was, however, the telescope in the hands of Galileo from 1609 that particularly threatened the Aristotelian model. This showed that the surface of the moon was not pure but pitted and cratered, and thoroughly earth-like in appearance, and in 1612 Galileo and a number of astronomers observed spots on the face of the 'incorruptible' sun. Most seriously of all, Galileo's telescope revealed that Venus has phases like the moon, an effective retort to those who argued that this could only occur if the planets revolved around the Sun.

While these discoveries were extremely damaging to the Ptolemaic/Aristotelian model, they still did not, by themselves, prove that the Copernican model must be true. Tycho Brahe, for example, had produced a feasible alternative model in which the moon and sun revolved around the earth, while the planets revolved around the sun. Despite all these defects, Galileo in particular claimed that the Copernican model was right, and his ensuing conflict with the Church about this was a definitive episode in European history, which is discussed in note 12, but astronomy moved on regardless of this. Protestant Europe was beyond the reach of the Inquisition, and the eccentric Emperor Rudolph II (1552-1612) in Prague was a notable patron of astronomers, astrologers, and alchemists, and these included Brahe and Kepler, whom we shall consider in a moment.

Copernicus' book was translated into English by Thomas Digges in 1576, who also reintroduced the ancient atomist conception of the Void in which the stars are scattered through an infinite universe, not all confined to an outer sphere. Atomism and the idea of the void therefore had just as fundamental a place in the new astronomy as they had in the new ideas about matter. [13] Aristotle's universe was also filled with matter, in the form of the five elements, and all motion was transmitted by direct contact. But in the new Void this was impossible, and it raised again the new and fundamental problem for astronomers of how forces like gravity could act at a distance.

Johannes Kepler (1571–1630) was another Neo-Platonist and Pythagorean, and a mystic and astrologer, who had been converted to Copernicanism in his student days. In 1600 he became Brahe's assistant in Prague, and gained access to his vast accumulation of very accurate observations on planetary movements. Brahe had asked him to solve the irregularities of the Martian orbit, which were the most perplexing of all the planets. His work on the Martian orbit convinced Kepler that no combinations of eccentrics and epicycles in the Copernican model, would ever be able to account for these, and he therefore cast about for some simpler alternative. Since circular orbits had to be abandoned, and after experimenting with various egg-shaped orbits, he found that the ellipse would fit his data perfectly. [14] But why should this be so?

As a Platonist he was convinced that the cosmos had to be based on simple mathematical principles, and some years earlier, in the *Mysterium Cosmographicum* of 1596, [15] he had gone back to Copernicus' problem of why the planets move more slowly as their orbits become farther from the sun. This was of the very first importance, since it raised, for the first time, the whole problem of the *forces* involved in planetary motion, as well as their geometry. He had suggested that perhaps the sun had some power by which it could drive the planets round, and that this might become exhausted with distance. [16]

William Gilbert's book on the magnet was published in 1600, in which the earth itself is represented as a giant magnet. Since de Maricourt, (see section 3) magnetism had been associated with the cosmos, and now appeared to explain that mysterious force, acting at a distance, which Kepler had in mind. It was therefore the entirely incorrect idea that magnetism could attract celestial bodies that was so important in persuading seventeenth-century scientists to accept the very idea of forces acting at a distance. He realised, however, that the single force from the Sun could not produce the elliptical orbits, and the changes of speeds of the planets, so that there had to be *an interplay of forces*. Not only the earth, said Kepler, but also the planets and the sun are magnets, and the attractions and repulsions of their various poles determine the paths in which the planets move. Indeed, he rightly suggested that attraction was proportional to mass, and was mutual.

In the *New Astronomy* of 1609 he was now able to formulate the first two of his three great laws of planetary motion: the first says that planets move in elliptical orbits, and the second follows from it, saying planets will appear from earth to move faster when they are closer to the sun than when they are farther away, or more precisely, that an imaginary line drawn from each planet to the sun sweeps an equal area in an equal time. On this basis it was now possible to explain all those irregularities of celestial motion that had, until then, had been insoluble on the assumption of circular orbits. He only discovered his third law much later, while writing *Harmonice Mundi* (1619), in order to prove that the planets obey certain basic laws of musical harmony. [17]

Kepler himself did not rate these laws very highly because, as they stood, there was no obvious reason why they should be true. But it was this substitution of the ellipse for the circle as the basic form of the planetary orbit that was the fundamental breakthrough in allowing the ancient geometrical laws of astronomy to be replaced by mechanical laws, because the ellipse must be produced, as Kepler had surmised, by an interaction between two forces of some kind. In astronomy he was therefore the real watershed between Aristotle and Newton, because 'he was the first astronomer whose analysis reached laws of nature that were physical rather than mathematical or philosophical'. [18]

Galileo had realised that a body in motion will naturally continue moving in a straight line at the same speed, and Robert Hooke, like Kepler, saw that the curved orbits of the planets must therefore be the effect of a second force, which was the attraction between them and the sun. He proved this in 1666, when he released a pendulum, which swung in a straight line, and then gave it a horizontal push, which made it swing in an elliptical orbit like the planets, [19] suggesting that planetary orbits were governed by the same laws of mechanics as cannon balls and pendulums. [20]

All these ideas, however, needed to be proved mathematically, which was beyond Hooke's powers and required Isaac Newton (1642–1727). Newton was not only a rigorous experimentalist in the Galilean tradition, as shown in his reflecting telescope, and his beautiful experiments with prisms to prove that white light is composed of the colours of the spectrum in his *Opticks*, but a mathematical genius. He had absorbed all the latest mathematics, and had also invented the calculus, to handle change and motion. He was also thoroughly familiar with Kepler's work, and with these conceptual tools he was able finally to resolve all the problems of planetary motion in the *Principia* of 1687, which we shall come to in a moment.

Newton has always been regarded as the personification of modern science, but his reputation in this respect, which has been carefully cultivated by scientists since the eighteenth century, is a complete distortion of the facts. He actually believed that rigorous experimental science was a means of rediscovering truths that had been known to the ancients, such as the Pythagoreans, and the many years which he spent on alchemical research, [21] and on Biblical prophecy and chronology were not eccentric hobbies. They were integral parts of his grand enterprise, which was to solve the mystery of the pristine knowledge of how God works in the world that He had given to Adam.

> In his studies of the Old Testament prophecies, Newton was tracing the pristine knowledge of the historical events of future ages; in his alchemical studies, the pristine knowledge of the constitution of things; in his studies of ancient natural philosophy, the pristine knowledge of physical nature and the system of the world. [22]

His mental world, like that of Kepler and Copernicus, was therefore very different from that of modern science, but was nevertheless an essential part of his achievement.

His first law of motion in the *Principia* says that an object moving in a straight line at a certain velocity will continue to do so indefinitely, just as it would remain at rest, unless some external force were applied. This was the principle of inertia, which implies the *resistance* of a body, either at rest or in motion, to having that state changed, and the greater the mass, the greater the inertia. It not only did away with the old idea that some force had to be continually applied to the planets to keep them moving, but showed that both rest and motion could be 'normal' states of matter. [23]

The second law says that force equals mass times acceleration. Not only, then, will it take a greater force to accelerate a large object from rest than a small one, which Aristotle knew, but to change the direction of a moving object will also require a force, because 'acceleration' can involve a change of direction as well as of velocity. Force, then, does not produce motion, as Aristotle thought, but a change of velocity, and a constant force will produce acceleration. So the elliptical orbits of the planets need a second force, *constantly applied,* to keep them in orbit. This force does not make them go faster, but produces a constant change in their direction, acceleration, as they rotate in their orbits, and that force is gravity, which is what Galileo had measured. [24]

By putting precise numbers to gravity and these laws of motion, Newton could now prove that the moon behaved like one of Galileo's cannon balls that, fired with enough velocity, would circle the earth perpetually. It would not fly into space because of the gravitational attraction between the two bodies, and would not fall back to earth because of its velocity, and so would continue orbiting the earth because of its inertia. So the orbits of the moon around the earth, and of the earth and the planets around the sun, are the product of these two forces: if they are perfectly balanced, then the orbit will be circular, but in reality this is highly unlikely, and since they are in fact slightly unbalanced, the result is an ellipse.

He could also now show why Kepler's Three Laws of planetary motion had to be true, as well as solving a whole range of other problems, and completed the Copernican revolution. Most striking of all was the extraordinary precision with which his laws of motion and gravity could predict all planetary motion, and they are still used today in calculating the trajectories of space-craft. Indeed, the thrust of their rockets will only move them through the vacuum of space because of Newton's third law of motion, which is that for every action, there is an equal and opposite reaction.

While Newton did not convince all his fellow scientists for some time, the Newtonian universe marked the final death of the Aristotelian universe. Geometry had been used for two thousand years to explain planetary motion, but this geometry was now revealed instead to be the *product* of interacting forces, which were the real reasons for the ancient difference between linear and circular motion. There were no longer any such things as natural and violent motion, and infinite space was the same everywhere. There was no distinction between the celestial and the earthly realms, and the same law of gravity made Newton's apple fall to the ground and kept the planets in their orbits.

Newton believed that matter was dead and passive with no power of moving itself, and that the true cause of gravity was the direct action of God, but which lay outside the scope of the *Principia*. He still did not believe that one body could act upon another, nor that light could travel, at a distance and though a vacuum, and proposed that gravity and light are transmitted through the aether. He also suggested that the aether could explain how 'muscles are contracted and dilated to cause animal motion', and the idea of the aether will come up again when we come to electricity and magnetism. But the various laws of motion that he had established became the standard of what scientific laws should be like, and the experimental methods by which they were discovered made physics the model that the other branches of science tried to imitate.

2. *Atomic theory and the elements.*

The great physicist, Richard Feynman, once said that if the human race could have just one memorial, it should be atomic theory. While today we think of atoms in relation to nuclear physics, modern atomic theory began with chemistry, and the history of its discovery can be explained fairly concisely.

Hellenistic alchemy had been taken up again by Muslim scholars in Baghdad, and Chinese alchemy also reached Baghdad, notably their idea of the elixir of life. Muslim alchemists made some important advances in chemical knowledge, and when it reached the Latin West, it led to further major discoveries such as the distillation of alcohol [25] and the strong mineral acids, such as nitric and sulphuric acid. These would prove of vital importance in the study of chemical reactions, and alchemy remained a powerful influence on research until the eighteenth century. [26]

Paracelsus (1493–1541) had a major impact by teaching the importance of mineral drugs, such as mercury, but by this time, however, the theory of matter was in complete confusion. [27] The ancient theory of atomism was now reappearing on the scene, however, and while it was initially tainted by suspicion of atheism because of its materialistic and anti-religious background in antiquity, [28] it was too attractive an idea to ignore, and Galileo was one of its principal advocates. Atoms could be treated as indestructible so that matter is conserved throughout all possible chemical transformations, and if they existed they inevitably involved rejection of the traditional four elements. The question was whether they could take their place as a new kind of element, the basic building blocks of nature. [29] We shall see that the major gateway to understanding this was provided by the gases involved in fire, water and air – oxygen, hydrogen and nitrogen – because the chemical combinations involved are very simple, and can easily be discovered by experiment. One of the most important of these involved combustion.

The Hon. Robert Boyle (1627–1691) performed an experiment to test his theory that fire was the result of atoms in rapid motion: he sealed a number of metals in glass flasks and heated them to produce an oxide. He then opened the flasks, and found that the weight of the metals had increased, as metallurgists had known for a long time. He explained this as the result of fire atoms that had penetrated the glass and combined with the metals to increase their weight. But he also realised that air was necessary for this, and his colleague Hooke believed that some part of the air disappeared during combustion. Paracelsus and others also believed that air contained something that was necessary to support life, and the Scottish alchemist Alexander Seton (d.1604) identified this 'vital spirit' with the essence of saltpetre that was necessary for the explosion of gunpowder, as Hooke had also realised. This 'vital spirit' was, of course, oxygen, but no further progress in this direction could be made until much more was known about gases in general.

During the seventeenth century scientists had become convinced that there were other 'airs' or gases besides common air, and this was especially due to the researches of van Helmont (1577–1644), a wealthy physician and alchemist. In the course of various experiments, he collected a number of gases that we now call carbon dioxide, nitrous oxide, carbon monoxide, and methane. Professor Joseph Black of Glasgow University, and James Watt's collaborator in the development of the steam engine, took up this work, and carried out fundamental experiments on gases in 1754. [30] These showed that gases behaved just like liquids and solids in chemical reactions. They were in fact 'chemical individuals', stable entities with predictable behaviour. 'The greatest advance in the eighteenth century was the isolation and identification of gases as chemical individuals. This was the last pillar needed to erect the new chemistry.' [31]

Once Black had demonstrated this, progress depended on refuting the idea of phlogiston, which had begun with Johann Becher (1635–82), physician, alchemist, economist, and general polymath. This was believed to be the essence of combustion, and was the dominant chemical theory of the eighteenth century. [32] The idea of phlogiston meant that metals lost it when they were heated, whereas everyone knew that they got heavier, but the advocates of phlogiston could ignore this completely for many years, because phlogiston was a chemical principle, rather than a definite physical substance. 'It was only later, when the idea of a chemical substance as a physical entity became accepted by the chemists that this point became crucial for the phlogiston theory.' [33]

Having established what the theory of phlogiston involved, we can take up the experimental trail again. In 1772 the great French chemist Antoine Lavoisier (1743–94) performed an experiment that disproved Boyle's claim that oxides absorb atoms of fire, and proved instead that they had to absorb something from the *air*. [34] In 1774, the English chemist and Unitarian minister Joseph Priestley (1733–1804) heated the oxide of mercury, [35] and found that it gave off a gas that supported combustion very well, and also the respiration of small birds and animals. In October of that year he visited Lavoisier in Paris and told him of his results, which Lavoisier soon confirmed in his laboratory. But Lavoisier was one of those chemists who had come to doubt the existence of phlogiston, because it seemed absurd that any kind of definite chemical substance could have negative weight. Careful experiments with the balance, showing that in all cases of combustion oxygen combines with the substance that is burnt, convinced him that this gas was the component of air that supported both combustion and respiration, and in 1779 he named it 'oxygen', 'maker of acid', because he wrongly supposed that it was also a component of all acids.

But it was also clear that there was at least one other gas in common air that did not support combustion, and in 1772 Rutherford, a student of Black, had discovered nitrogen. The work of Henry Cavendish (1731–1810) brought in the closely related problem of water and its composition. In 1766 he had isolated hydrogen, and in 1781 Priestley noted that when hydrogen burned it produced condensation on the walls of the glass vessel, which Cavendish proved was pure water. Lavoisier realised that in fact water is the combination of oxygen and hydrogen, 'the water maker'. The ancient elements of fire, air, and water had now been revealed experimentally as different combinations of three gases – oxygen, hydrogen, and nitrogen.

Although Lavoisier had shown that oxygen, hydrogen, and nitrogen were chemical individuals, he did not think this proved that they each had a distinct type of atom. [36] The final relationship between atoms and elements was established by the Quaker John Dalton (1766–1844). As we have seen, the simple chemistry of air and water, and their gases, had been the easy gateway into a deep understanding of matter, and it was Dalton's interest in meteorology that drew him into this research. The absorption of water by air, of the gases of air by water, and the relationship between nitrogen and oxygen in the atmosphere were especially important problems in meteorology. In particular, since oxygen is heavier than nitrogen, why did they not separate out and form distinct layers?

The favoured answer was that they formed a chemical compound, as oxygen and hydrogen do to form water, but Dalton rejected this because these compounds have very different properties from their components. He believed that oxygen and nitrogen in the air are a simple mechanical mixture, like different sorts of grain in a sack, just as he believed that water vapour was also absorbed mechanically by the air. As the result of further experiments, he concluded that the atoms of oxygen and nitrogen have different weights. [37] He also studied the absorption of different gases by water, and found that at a given pressure, a given volume of water would absorb different amounts of each gas. His explanation was that the different weights and complexity of the different particles might be responsible. [38]

By this time he was thinking of the atom of each element as a tiny, indivisible ball of a specific weight relative to hydrogen. Taking hydrogen as having the basic weight of 1, by use of the balance he established the relative atomic weight of nitrogen as 5, oxygen as 7, phosphorous as 9, sulphur as 13, and so on, and in his *New System of Chemical Philosophy* (1808) he formulated a set of laws about how the atoms of these different elements had to combine. [39]

All the rules of chemical combination follow from these assumptions, the Law of Fixed Proportions, and also the Laws of Multiple and Reciprocal Proportions [40]. While his theory needed revision, notably by Avogadro's work on molecules, this, and Mendeleyev's Table of the Elements were still directly based on Dalton's model. The atomic theory of the elements became the basic principle of chemical analysis in the nineteenth century, and the production of synthetic chemicals by the growing chemical industry also depended on it. [41] Many physicists, however, regarded chemical atoms simply as useful fictions, and not real entities, until J.J. Thomson discovered the electron in 1897, and Rutherford proved the existence of the atomic nucleus in 1911. But these discoveries were far beyond the powers of science at the beginning of the nineteenth century: Dalton's chemical atom had to come first, and was the essential basis for later research into the atomic and sub-atomic worlds.

*3. Electricity and magnetism.*

The discovery of electricity was one of the great transforming events of human history, but had its roots in the natural magic tradition. We can take up the story with the introduction of the magnetic compass to Europe in the late twelfth century. The medical astrologer Pierre de Maricourt, or Peregrinus, carried out a remarkable investigation of the lodestone or magnet, described in his *Letter on the Magnet* of 1269 [42]. He performed a series of experiments showing that a magnet has two poles, that like poles repel and opposite poles attract each other, that a magnet cut in half becomes two complete magnets, and that when floated on wood the magnet will align itself north and south.

Now, the poles of the magnet had great astronomical significance. Astronomers believed that the sphere of the heavens rotated, and since this must be around an axis, a sphere must have two poles. Because he did not imagine that the earth rotated, de Maricourt supposed that the compass needle aligned itself to the poles of the heavens. He therefore carved a lodestone into a sphere, and showed that small needles followed the meridians of this model globe, converging at the poles, where the needles stand upright, so demonstrating the physical reality of the poles around which the heavens rotated. He also believed that magnetism was the cosmic force that caused this rotation of the heavens.

The physician William Gilbert (1540-1603) came out of the same natural magic tradition, but as a Copernican thought that the earth rotates, not the heavens. He believed that the earth was animated by a magnetic soul, and that if he could show by experiment that a magnetic sphere behaved like the earth, this would prove that the earth was a giant magnet and that this would also explain its rotation – something that Copernicus had not been able to do. [43] Why else would God have given the earth poles unless it rotated? He knew de Maricourt's work, published in 1558, and also Norman's on the compass, and used this to carry out a really thorough piece of experimental research in the modern manner, published as *De Magnete* in 1600, proving that the earth indeed works like a giant magnet. He also investigated the magnetic properties of non-metals, 'electrics' such as amber, glass, sulphur, and resin, and measured the different amounts of attraction they produced. This led him to make the fundamental distinction between magnetism and what we now know as static electricity.

Otto von Guericke (1602–1686), [44] took up Gilbert's work on magnetism because, as a Copernican, he wanted to know how the heavenly bodies attracted each other. But he was much more interested in the 'electrics' than magnetite, and therefore made a model globe primarily of sulphur which he mounted in a frame so that it could be rotated. When rubbed he found that the globe not only produced a strong attraction but sparks, a new phenomenon that further distinguished electricity from magnetism. Electrical machines attracted a great deal of attention, and increasingly elaborate ones began to be made in many parts of Europe. During the eighteenth century it was discovered that the charge from electrical machines could be stored in capacitors, such as the Leyden jar. These could deliver powerful shocks and their effects on the body, especially muscular contraction, opened the way to a whole new field of research.

Since Aristotle, the subtle animal spirits responsible for this had been associated with the aether, and the idea of the aether as an elastic, subtle fluid permeating the universe had been promoted by Newton, who had also suggested that the aether could be involved in muscular contractions. The anatomist Galvani (1737–1798) was stimulated by this basic Aristotelian problem, [45] and in the course of his experiments with electricity on frog muscles, concluded that the muscles were a unique source of movement by generating electricity within themselves. In the course of this work he fixed a brass hook through a frog's legs and attached it to an iron rail. This also produced a jerk which Galvani explained as the action of an inherent force within the muscle, animal electricity.

This created a sensation when published in 1791, but while Galvani has ultimately been proved correct in his belief that cells produce an electric charge, Professor Volta's experiments led him to conclude that the legs twitched because of an exterior stimulus coming from the brass hook and the iron rail, 'metallic electricity', generated by the contact of different metals in a conducting fluid. [46] So he constructed a 'pile' of silver and zinc plates between pads of cloth soaked in brine and for the first time produced an electric current. He communicated his results in 1800 in a letter to Sir Joseph Banks, President of the Royal Society, and when William Nicholson and Anthony Carlisle read this letter they immediately constructed their own pile of 36 silver half-crowns and zinc discs. They noted a drop of water bubbling on one of the terminals and realised its cause, because it was already known that water is composed of oxygen and hydrogen. So they were soon able to produce, for the first time, oxygen and hydrogen by passing an electric current through water (electrolysis). Sir Humphrey Davy then installed a huge Voltaic battery in the cellar of the Royal Institution and used this for decomposing a wide range of chemical compounds into their elements, which laid the foundations of a new chemical industry.

Now that a supply of current electricity was available, it at last opened the way for fundamental experiments on the relations between electricity and magnetism, which had been impossible with static electricity. The Danish physicist Oersted discovered in 1820 that the needle of a compass would deviate whenever a current flowed in a wire adjacent to it, so demonstrating that magnetism and electricity were somehow related. Stimulated by this, Faraday in 1821 showed that a wire carrying a current would rotate in a magnetic field, and in 1831 produced an electric current by rotating a copper disc between the poles of a magnet, so discovering the electric motor and the dynamo.

The evolutionary potential of electricity was obviously enormous for energy and lighting, communications by telegraph and radio, the whole electronics industry, medicine, and so on, producing fundamental changes in our society. But to achieve this potential there had to be a new source of power that could generate the vast quantities of electricity needed to support an industrial society, going far beyond the capacities of the water wheel, and this was the steam engine.

*4. The Void, atmospheric pressure, and the steam engine.*

Materialists regard the development of the steam engine as a triumph of applied technology, in which theory was basically irrelevant, but this is a serious distortion of the facts. Early seventeenth century attempts to use steam power were initially in the wonder-working tradition of entertaining gadgets, such as fountains worked by steam pressure, based on Hero's work and the natural magic tradition. [47] These encouraged a series of attempts in England to use steam for pumping water out of tin and coal mines. As these went deeper they were encountering increasing problems from flooding, and existing pumps could only be driven by water or horses.

All these devices were based on the power of steam not only to force water up a pipe, but to suck up water as it condensed. Vessels of water were heated and then cooled, and by the end of the seventeenth century this principle culminated in the sophisticated pumping engine of Thomas Savery (*c.*1650–1715), patented in 1698. [48] But although it was intended for draining mines, the limitations of suction, and the fact that it could only raise water 50–60 feet was a fatal obstacle to its use in mines, many of which were 150–300 feet deep, and it was a failure. [49]

Savery, in *The Miner's Friend* (1702), was very clear that his machine worked by the pressure of the atmosphere forcing the water up into the vacuum created by the condensed steam. But this was fairly recent knowledge, and we should not suppose that earlier experimenters knew anything about the relation between atmospheric pressure and suction. Those who used steam for pumping were applying a purely practical know-how of what steam and water could do. It was only when the nature of atmospheric pressure was realised that it eventually became clear that the cylinder and piston provided a much more effective way of using the vacuum, and we shall now see how this occurred.

By the beginning of the seventeenth century, the vacuum was being discussed again, in relation to the Aristotelian Void, and to problems of suction pumps. Like many other people, Galileo knew that they could not raise water higher than about twenty-eight feet but, although he had verified that air has weight (1/400$^{th}$ that of water), he positively rejected the suggestion that atmospheric pressure could be connected with this. The image of suction as *pulling up* the water is very powerful, and if a scientist of Galileo's calibre could not see the connection between suction and atmospheric pressure, we can be sure that hardly anyone else could have done so, either. He preferred the mechanical explanation that water did not have the tensile strength to support a longer column of water than this, and broke under its own weight.

It was his young colleague Torricelli (1608–47) who correctly understood the significance of Galileo's work on atmospheric weight, and that it *pushed up* the water into the vacuum. He took a glass tube 4 feet long and closed at one end. This he filled with mercury, applied his finger to the open end, and inverted it in a basin or mercury. The mercury in the tube sank at once to 2½ feet above the basin, leaving 1½ feet apparently empty. Torricelli deduced that it was the pressure of the atmosphere that maintained the level of mercury, and the tube of mercury in its trough also worked as a barometer, another fundamental scientific instrument. It was soon shown that atmospheric pressure varies with altitude by taking Torricelli's barometer up a mountain. [50]

Von Guericke was not aware of Torricelli's work, but from his Copernican interests he wanted to test Aristotle's theory of the Void, and that a vacuum is impossible. If this were true, he predicted that if he pumped the air out of a copper sphere, the sphere would collapse. He built a piston-and-cylinder air-pump and, after several attempts, he did create a vacuum in a sphere, so disproving Aristotle. He followed this up at Ratisbon in 1654 where he set up a vertical cylinder 20" in diameter in which a well-fitting piston was suspended by a rope passing over a pulley. Connecting an air pump to the cylinder he created a vacuum below the piston, causing it to descend in spite of the joint efforts of 50 men to hold it up. He made a further demonstration in front of the Emperor Maximilian at Magdeburg in 1657, in which two iron hemispheres were placed together and from which he again pumped out the air. Teams of eight horses attached to each hemisphere were unable to pull them apart. The 'Magdeburg hemispheres' became the talk of Europe, and the reality and power of atmospheric pressure became a well-known and amazing fact. It is, after all, quite extraordinary that this insubstantial air around us actually bears down on a piston of 20 inches diameter with about 2 tons of force.

Von Guericke had used an air pump to create his vacuums, but the pump was slow, and itself needed energy to work it, so could never have been the basis of a practical atmospheric engine. Denis Papin (1647–1712) was a French Huguenot, who fled to London in 1675, and collaborated with Robert Hooke in various experiments. [51] He invented a pressure-cooker, and this may have reminded him that steam creates a vacuum when it condenses, because in 1690 and 1695 he published accounts on the Continent of an experimental model of a steam engine. He placed some water in the bottom of a small cylinder, and then heated it; the steam forced the piston up to the top of the cylinder, where it was held by a catch. The fire was then removed,

## Modern Science and Industrialism

the steam condensed, and the vacuum produced allowed atmospheric pressure to force the piston down to the bottom again. In March 1697 these publications were reviewed in the *Philosophical Transactions* of the Royal Society, with the conclusion that such an engine could be used to pump water out of mines. In this way the essential idea of the atmospheric steam engine, with its practical possibilities, had now entered the public domain of informed discussion in England.

There was, however, an enormous gap between Papin's simple model, and a real-life engine that could be of practical help to miners, and the size of this gap can be judged by the picture of one of Thomas Newcomen's first atmospheric engines. Papin was a pure theorist, with no commercial experience, who depended on government patronage for a living, whereas Newcomen (1663–1729) was a business-man and skilled engineer. He is described as an 'ironmonger' of Dartmouth, now a small fishing port on the south Devon coast, which suggests an uneducated local tradesman dealing in packets of

nails and door-hinges.

He was in fact descended from country gentry in Lincolnshire who had forfeited their lands under Henry VIII, and his father and grandfather had been ship-owners and prominent citizens of Dartmouth, which was then an important trading port. In those days, 'ironmonger' meant an iron merchant, and Newcomen was engaged in the supply of tools and pumps to the tin mines of Devon and Cornwall; in 1698-9, for example, he bought 25 tons of iron from foundries in Worcestershire, [53] which would have come by boat via the Severn, and then round the coast to Dartmouth. He was also a practical engineer with a workshop, and a thorough knowledge of pump technology in particular, and he was also well-travelled, not only in the course of his commercial activities but also because, as a devout Baptist, he had a wide network of fellow Baptists, especially in the Midlands and London. He is recorded by his contemporaries as well-read, and there is nothing at all surprising that a man in his position, and in a society where new inventions were eagerly discussed by artisans as well as the learned, should have quickly learned of Papin's discovery. [54] In view of the ten years or so of experiments that he had to spend overcoming the profound technical problems involved, [55] and the considerable sums he had to borrow to finance them, he would have needed to be certain, from the outset, that what he was attempting had already been proved scientifically to be possible.

He seems initially to have been unaware of Savery when he began his experiments in around 1698, and devised an atmospheric piston engine that only needed low pressure steam, which was easily handled by the relatively simple technology of the day; and since it was only intended to pump water from mines, it did not need to produce rotary motion either. He apparently built his first engine in Cornwall in 1710, and perhaps one more there as well, but coal was expensive in Cornwall and his engines burned a great deal of it, and he used his Baptist connections to gain the contract for an engine to drain a coal mine near Dudley castle, in the Midlands, in 1712, which was commercially successful. Here, coal was plentiful and cheap, and all the necessary craft skills for building his engine were locally available.

The engine at Dudley had a 21-inch diameter cylinder, and produced about 5½ h.p., pumping 120 gallons per minute from a depth of 150 feet, though 300 feet was quite possible for such engines. Since effective atmospheric pressure was only about 8 lbs per square inch, cylinders had to be very large, from around 20 to 40 inches in diameter, to produce sufficient power, and the cost was formidable. The price of an engine alone, apart from its brick house, was around £1000, with a further £250 a year for its operation and maintenance, and half the colliery's profits as well. But these figures also indicate the engine's value to the mining industry, and at Newcomen's death around 100 of them were working in Britain, and many more in Europe.

The development of the steam engine therefore took place in an intensely capitalist atmosphere, in which cutting the costs of fuel and of manufacture, and improving efficiency, were paramount. Like guns, steam-engines did things that could be measured, and right from the start Newcomen had to calculate the power of his engines, and within a few years tables of performance were being compiled, correlating gallons pumped with engine dimensions, [56] while James Watt later performed many experiments to measure the efficiency of engines, and developed the notion of 'horse-power' (1hp=33,000 foot-pounds per minute, or 750 watts). The experience of the steam engine therefore became an integral part of experimental science, and especially the growing awareness of the fundamental importance of energy in nature, like that of force in understanding motion.

James Watt, son of a master carpenter and shipwright, was mathematical instrument maker to the University of Glasgow. This was an institution, as Gresham College had been in the sixteenth and seventeenth centuries, that brought practical skills such as engineering into close contact with scientific theory in the eighteenth century. Watt, who had been given a model of a Newcomen engine to repair, discovered that cooling the steam in the cylinder by injecting cold water at each stroke was immensely inefficient, because it wasted a great amount of heat in cooling and condensing the steam. [57] He performed a number of experiments, in which he consulted Professor Joseph Black at the University, and who had carried out fundamental studies of heat. Watt used this knowledge to make his most important improvement to the steam engine, by cooling the steam in a separate condenser from the main cylinder, which was always kept hot.

Watt also made his new version of the steam engine double-acting (admitting steam at both ends of the cylinder instead of just one), and devised a means of converting the reciprocal motion of the Newcomen beam to rotary motion so that it could now be used to drive machinery. The steam engine could not only develop much greater power than the water wheel, but could be located anywhere, and Watts' improvements made his engine far more economical to run than the Newcomen version, a crucial advantage for manufacturers. Watt went into partnership with Matthew Boulton, a factory owner and entrepreneur of Birmingham, to manufacture and sell his improved engines.

During the eighteenth century small textile mills had already been developed using water wheels to drive new types of machinery for spinning and weaving cotton and the fulling of wool. These machines, unlike mine pumps, needed rotary motion to drive them, and once Watt had solved the problem of converting linear into rotary motion there were even more industrial opportunities for Watt's engines than there had been for Newcomen's. Very large numbers of machines could now be concentrated into the great factories of the industrial revolution, with all that implied for urban life.

Watt's engine still used low-pressure steam, however, but the development of high-pressure steam by Trevithick fundamentally changed the situation. Using a vacuum had limited the power available to that of atmospheric pressure, around 14 lbs per square inch in theory, and 7–8 in practice, so that eliminating the need for the vacuum now allowed the *expansive* power of steam to be fully exploited instead. This was the true 'steam', as distinct from 'atmospheric' engine, and because it was much more powerful it could also be much smaller and therefore become mobile for the first time, notably as the railway locomotive operating at around 120 lbs per square inch. George Stephenson made improvements to the locomotive as important as Watt had made to the Newcomen engine. While the reciprocating steam-engine gave place to the internal combustion engine and the electric motor, it remains, in the form of the steam turbine, the essential generator of electricity.

*5. Statistical thinking and evolution.*

During the sixteenth and seventeenth centuries, commercial calculations of risk, insurance policies, annuities, and the study of mortality tables had led to a great increase in the understanding of probability (the details of which can be found in Chapter XV, note 21). [58] By the eighteenth century, mathematicians such as Laplace, de Moivre, and Bernoulli had made scientists much more familiar with the whole idea of statistical reasoning, and de Maupertuis (1698–1759) applied it to the problem of how the order of Creation had been produced.

He wanted to go beyond the alternatives of direct design and blind chance, and maintained instead that order could be generated by the operation of very simple principles, with sufficiently large populations and very long periods of time. Looking at all the populations of animal species, he suggested that over time, the accumulation of very small, random changes, and the selection of those that were better adapted, could produce both the diversity of species, and their excellent adaptation to their specific environments. He imagined the same explanation being applied not only to plants, but to the increasing complexity of the inorganic world as well, and for him this Law of Large Numbers showed how it is possible to reconcile randomness with order. Indeed, for de Maupertuis, statistics were the basis of Divine Providence, showing how the Creator had allowed order to be generated out of the simplest principles.

In the same way, Adam Smith saw the economy as Providentially designed for the greatest efficiency: 'Business in competition faced the options of survival in various degrees or exclusion through the statistical accumulations of gains or losses, or of transformation to meet new circumstances' [59]. It is not therefore very surprising that the first published account of the biological theory of natural selection actually appeared in 1831, [60] 28 years before Darwin's *Origin of Species*, by Patrick Matthew, a well-educated Scottish landowner who had a very wide practical and theoretical knowledge of trees and their uses.

He was a disciple of Adam Smith, and obsessed by the value of competition, and the need for a vigorous struggle for survival as a means of selecting the best human beings, as well as plants and animals. The Royal Navy was essential for protecting British commerce, and his book *Naval Timber and Arboriculture* was intended to show that inferior trees were being systematically bred for shipbuilding, with the theory of natural selection in an appendix. Marxists may treasure this as the one good case of a member of the bourgeoisie producing a scientific theory while solving a technological problem in the economic interests of his own class. For our purposes, however, it is just another example of how science has advanced by the union of theory with craft knowledge.

His theory of natural selection flowed quite easily from his basic assumptions about competition, combined with his expert knowledge of arboriculture. As he wrote in a letter of 1860, 'To me the conception of this law came intuitively as a self-evident fact, almost without the effort of concentrated thought'. [61] Matthew believed, in short, like Darwin, that just as the breeder of animals and plants selects, from the endless variations between individual specimens, those whose characteristics he wishes to propagate to form a new variety, so Nature, on a vastly larger scale, does essentially the same thing when forming new species, but of course without conscious purpose.

The basic steps in his argument are as follows:

1. All organisms have a power of reproduction far beyond that needed to maintain a stable population, and this increase of numbers will press on the means of subsistence.

2. Since there are only limited natural resources, it follows that those individuals that are best adapted will survive to maturity and leave offspring, while the least well adapted will be prematurely destroyed in the competition for survival, before they can reproduce.

3. There is enormous variability in each individual plant or animal, and selection operates upon all these slight variations. Those variations that encourage survival will be inherited by the offspring.

4. The result will be that, over time, 'the breed will gradually acquire the very best possible adaptation to its condition which it is susceptible of, and when alteration of circumstance occurs, thus changing in character to suit these as far as its nature is susceptible of change'. [62] By 'alteration of circumstance' he had in mind especially the evidence of the fossil record, and the fact that geologists '. . . discover an almost complete difference to exist between the species or stamp of life, of one epoch from that of every other. We are therefore led to admit, either of a repeated miraculous creation; or of a power of change, under a change of circumstances, to belong to living organized matter.' [63]

This is the theory of natural selection, but while, of course, Darwin was not the first to think of it, he did provide its first thorough exposition to the scientific world. Just as Copernican theory had been radically incomplete without the laws of motion and gravity, Darwinian theory was radically incomplete without genetics, but, like Copernican theory, natural selection initially convinced by its simplicity and elegance rather than by the actual evidence for it, which accumulated later.

## 6. Conclusions.

These chapters on the evolution of science have, in one way, been a prolonged commentary on Mr Gradgrind's simple materialist belief that Facts alone are all that the reasoning animal needs. His adulation of the Facts expresses the widely held view that our senses are the basic source of scientific knowledge. As Sir Peter Medawar put it:

> plain factual truth is what scientific reasoning is supposed to begin with. We start . . . with an exact apprehension of the facts of the case, with a reliable transcription of the evidence of the senses which inductive reasoning can then compound into more general truths or natural laws. . . Error is due to an indistinctness of vision, a false reading of that Book of Nature in which truth resides and can be got at if only we can retain or reacquire the innocent, candid, childlike faculty of grasping what is in fact the case.
>
> I share Karl Popper's view that this conception of truth and error is utterly unrealistic. Scientific theories . . . begin as imaginative constructions. They begin, if you like, as stories, and the purpose of the critical or rectifying episode in scientific reasoning is precisely to find out whether these stories are about real life. Literal or empiric truthfulness is not therefore the starting-point of scientific enquiry, but rather the direction in which scientific reasoning moves. If this is a fair statement, it follows that scientific and poetic or imaginative accounts of the world are not distinguishable in their origins. They start in parallel, but diverge from one another at a later stage. [64]

The natural philosophy of the ancient world was just that poetic, imaginative account of the world with which science had to begin, the universe according to the Pythagoreans, Plato, Aristotle and the Atomists, of astrology, alchemy, and natural magic. In the last two chapters we have seen how the 'critical or rectifying' aspect of science in the form of experimental method then gradually developed in the sixteenth and seventeenth centuries. These experiments overthrew Aristotle's physical theory in every detail, but it provides a splendid example of the survival of the mediocre. Wrong about everything, it flourished for so long because intellectual competition had been weak. But it had enormous evolutionary potential because it was so clear and testable, and because the issues Aristotle discussed were central for modern science.

The experiments that disproved it discovered (relatively) easy gateways into some of the most important secrets of nature. Rolling balls down planks allowed Galileo to establish some basic laws of motion; a few experiments with gases overthrew the elements of fire, air, earth, and water, and were the basis of chemical atomic theory; investigating the magnet was the only feasible route to the understanding of electricity; and some simple experiments with air disproved Aristotle's ideas about the vacuum and paved the way to steam power. Technologically, all these experiments could have been performed in the Museum at Alexandria, but no one would have seen the point of doing them because the philosophical assumptions of the Hellenistic world were very different from those of the seventeenth century, and did not include the idea of testing theory by well-designed experiments.

This chapter, however, has provided more good opportunities to test the 'mutation and selection' theory of scientific evolution. In this view, for example, the elliptical model of planetary orbits was a mutation of the circular model, and was then selected in the competition of scientific ideas. The fallacies here are only too obvious, however. In the first place, by no stretch of the imagination could the theory of elliptical orbits be described as a random mutation. Secondly, circular orbits and elliptical orbits are not simple ideas at all. Each expresses a whole scientific theory, one that celestial bodies obey geometrical laws, and the other that they obey physical laws of motion and force. The basic model of the mutation as a change in a single idea which is then selected for is completely unrealistic, because in science we are not dealing with populations of atomistic ideas, but with systems of inter-related ideas.

The evolution of science that we have traced is a prime example of an enormous human endeavour that, until modern times, had no practical pay-off or, as in the cases of alchemy and astrology, the pay-off was imaginary. The unproveable assumption that nature is based on reason was an essential motivator for the whole enterprise, and a number of false ideas, but with great evolutionary potential, were the basis of advance on a number of fronts. Experimental science was the product of a whole culture, not of unaided technology, and despite the belief that science should be useful, the actual motivations of seventeenth and eighteenth century scientists were intellectual curiosity, and the belief that the universe made sense in some deep way because it was governed by the rational laws of its Creator. It would be hard to find more effective refutations of materialism and the Darwinian model of scientific evolution.

# Conclusions

In the nineteenth century, experimental science finally achieved a really effective combination with industrial capitalism, but this new civilisation that was developing was now superbly disqualified from understanding its own origins. The Age of Steam assumed that European history was the model for the history of the whole human race, and that Victorian man, economic man, was the epitome of human nature so that material progress was inevitable, like the victory of individual liberty over the tyranny of priests and kings, and of reason and science over superstition. We still live under this cultural shadow today.

Newtonian physics had set the standard of what natural science could achieve and so, especially when it became clear that man himself had evolved from the animals, it was determined to have a natural science of man as well:

> The empire of science was to be extended to every facet of man's nature; to the workings of men's minds as well as their bodies and to the their social as well as their individual behaviour; law, custom, morality, religious faith and practice, political institutions, economic progress, language, art, indeed every form of human activity and mode of social organization, were to be explained in scientific terms. . . [1]

In particular, a science of man should be able to predict, because that was what all proper science did, and some of the first fruits of this ambition were various theories of social evolution, including historical inevitability.

Natural science had succeeded by going beneath the ordinary appearances of things, removing the inessential details and reducing them to their simplest forms. The secondary qualities of things, for example, were stripped away from their primary properties; the ancient elements of fire, air, earth, and water lost all their familiar everyday associations, and dissolved into their underlying atomic structures; and the flight of birds, arrows, and comets was reduced to the basic laws governing matter in motion. The Darwinian model of social evolution, reducing everything to the simple model of random variation, and selection based on our material needs, is a classic example of this programme of natural science in action.

Darwin's theory was concerned with the evolution of plant and animal populations, and in this context its basic ideas of variation, selection, competition, and adaptation make perfectly good sense. Social systems, however, are a radically new kind of phenomenon because they are not just animal groups, but groups organized by ideas, of culture communicated by language. Culture also includes knowledge and belief systems, ranging from mathematics and logic, through political and religious beliefs, to symbolism and art. These various social, intellectual, and technological environments do not so much select, as provide a set of constraints and opportunities for the choices of individuals.

We have seen that the *origins* of novelty are of supreme importance. People do not innovate randomly, but tend to do or think what is easiest in these particular environments. Exploring the easy gateways into particular innovations has given us a far deeper understanding of why they occurred than simply attributing them to blind variation. Some innovations may have great evolutionary potential, but may, however, be thoroughly mediocre, and they can survive because competition may be weak. Competition itself has evolved in the course of history: in the range of choices, in their importance, and in the efficiency of the competitive process itself, such as the emergence of professional warfare, rational argument and logic, scientific experiment, and capitalism.

The concept of adaptation, so central to Darwinism, becomes confused and incoherent when applied to social evolution. Much of our technology and our social institutions are useful because we have consciously designed them to be so, and in this sense are obviously adaptive. This kind of adaptation, however, is not Darwinian at all, because that is only concerned with how the physical adaptations of plants and animals were produced by the *unconscious* processes of natural selection. But when Darwinian theory tries to explain the unintended benefits of an institution by natural selection, we have seen that these benefits are often speculative, and that if the people are unaware of them, they cannot explain the institution's survival. 'Adaptation' is also ambiguous, and can mean either that an institution survives because it is useful to real people, or because it fits in well in a particular social system, although, like slavery, it may not be beneficial at all.

Instead of puzzling ourselves by looking for adaptive explanations for some practice or institution, we have done much better to ask how it came about, and how it then changed its social environment, and made possible or inhibited further developments. Here we are talking the language of construction rather than selection. Social institutions and systems of knowledge and belief are not atomistic bundles of individual adaptations to the here-and-now, that were each selected; they were *constructed* in a certain order, by the accumulation of necessary conditions; and the whole process of social evolution has been thoroughly un-Darwinian. Ultimately, the test of a good theory is that it can explain the evidence better than its rivals, and Darwinism has simply failed this test.

Meaning and purpose are also at the very heart of culture, but they are precisely what experimental science had to abandon in order to explain nature. We are certainly physical beings, constantly subject in all sorts of ways to the laws of cause and effect, and I have taken full account of the material factors of subsistence and technology. But the meaning and social significance of natural phenomena also depend to a significant degree on how we interpret them. The history of science and technology has shown us that they were not simply a set of responses to 'the facts' of nature, so beloved of Mr Gradgrind, that forced themselves willy-nilly into human awareness, but the result of a cultural and creative process that was only possible for human beings.

We can still, then, give a rational explanation of how human society and technology have evolved, but it must also include all those uniquely human characteristics – language, reason, creative imagination, play, honour and glory, religion, the love of beauty, music and the arts, and so on – some of which are non-rational, or of no material significance. These are not irrelevant details, however, like the colours or smells of falling objects for the laws of motion, but fundamental in explaining social evolution.

The account of social evolution I have given is therefore scientific in the sense that it attempts to give a rational explanation for a wide range of phenomena in terms of a few general principles, supported by evidence. But it cannot be *natural* science, because culture goes beyond the material basis of life and has its own unique properties.

This theory of social evolution cannot be used, either, to predict anything significant about the future of man. Marxism claimed to be a natural science of society, and therefore not only said that every society had to pass through certain stages, but that it could predict the future development of society as well – the Communist utopia when the state would wither away, and all would enjoy a world of equality and plenty. But evolutionary biology is a natural science, and yet is unable to predict the future of any species, because mutations, and the selective circumstances in which they will appear, are inherently unpredictable.

In the same way, it is impossible to predict inventions before they have been made, but, especially in a highly developed technology, there are all sorts of inventions that would radically change society. More generally, we have only been able to explain something like the emergence of experimental science by establishing the necessary conditions for it. But we could only do this because we already knew what experimental science looked like, and we could only know that after it had emerged. The necessary conditions for world government may already exist, but we will only know which ones they are *after* world government has actually appeared, perhaps in a form that we had not imagined.

If we cannot predict, we can at least learn some lessons from the course of social evolution. One of the most important is that politicians, state control, bureaucracy, 'harmonisation', 'unification', and official orthodoxies are the enemies of creativity, which is unpredictable, wasteful, eccentric and sometimes deranged, and flourishes best in fairly chaotic conditions. The Greek city-states, the 'Hundred Flowers' of China, and the Renaissance and the Industrial Revolution are cases in point, and would not have been possible under the European Union, for example. The state has certainly been an important patron of science and technology in history, and modern science is also extremely expensive, so that it requires public funds. But how many of the scientists we have studied would ever have achieved what they did in fundamental research if they had been required to apply for government grants, and justify their research projects by the practical benefits they would produce?

The remarkable progress of experimental science, however, has encouraged many scientists to exaggerate its powers, and claim that it is the *only* genuine form of knowledge, which can answer all the ultimate questions about the meaning of life, why we are here, and why the universe and its laws are as they are. But they forget that this success was achieved at a price. This was the abandonment of anything like Plato's and Aristotle's attempts to answer these ultimate questions, because they are outside the scope of science. [2] The basic reason is not just that we can't perform experiments to answer them but, more generally, that we can't fully explain any system from within itself, even in mathematics, where some theories are believed to be true, but are intrinsically unproveable. We can understand rationally why the game of Monopoly has certain features, such as the players taking turns, and rolling dice to decide the number of squares of each move. But it is impossible, simply by studying the game, to explain why it has 'Community Chest', or to know if it is a celebration of capitalism, or a Marxist satire on it instead. The only way to discover this would be to go outside the system and ask Mr Darrow, who patented it, why he designed it as he did.

In the same way, science can tell us a great deal about how the universe works, but it can't explain why its laws are precisely appropriate for the emergence of life. In this case, however, there is no equivalent of Mr Darrow whom we can go and ask. But people need to base their lives on what they think is the case now, without having to wait for ever until science can give a definite verdict one way or the other. (And in any case, scientific verdicts can never be definite, because they can always be overturned by subsequent research.) Those who believe that life, the order and beauty of nature, and the existence of man, who alone can understand the laws of nature, point to the existence of God are therefore perfectly entitled to do so, just as they are free to believe that spiritual experience is not an illusion. 'Pointing to' is not, obviously, the same as 'hard scientific evidence for', but it is not the same as pure fantasy either, and in practical terms religion provides many millions of people with a meaningful basis for life.

Extreme claims for a natural science of man, like all extremism, have bred an equally extreme reaction – that the study of man should not involve the search for truth at all:

> Social anthropologists should not see themselves as seekers after objective truth; their purpose is to gain insight into other people's behaviour, or, for that matter, into their own. 'Insight' may seem a very vague concept but it is one that we admire in other contexts; it has the quality of deep understanding which, as critics, we attribute to those whom we regard as *great* artists, dramatists, novelists, composers... [3]

This is a rallying call for the woolly minded, with the additional attractions of being intellectually undemanding, while allowing the initiated to speak an esoteric language that elevates them above the vulgar herd. Cultural meaning only exists in the context of social relations and institutions, and of interactions with the physical environment: in a universe of objective constraints, in other words. So between the extremes of bogus natural science and 'artistic insight', lies the moderate path of rational explanation.

# Notes

CHAPTER I HOW SOCIAL EVOLUTION WORKS

1. In theory, an important distinction can be made between culture and society: culture includes systems of ideas and values like religion, language, and mathematics, whereas society comprises groups of actual people, institutions, and networks of relationships. But in practice, since one cannot have either culture or society without the other, it is not a distinction that we need to go into here.
2. See Brown 1991 for some very useful material on human universals.
3. Sahlins 1974:8.
4. See Kitcher 1985 for an excellent analysis of the fallacies of sociobiology, and also Hallpike 1984. Evolutionists have made many attempts to explain how those distinctive features of human nature such as art, self-decoration, dancing, laughter and tears, and of course language and general intelligence, evolved by natural selection over millions of years in Africa. (This was the 'environment of evolutionary adaptation'). But there is little point in evolutionary psychologists always telling us that 'The human mind is adapted to the conditions of the Stone Age', when we know so little about what these conditions actually were. Even in the case of *Homo sapiens sapiens* from around 200,000 years ago we do not know when they first acquired the ability to speak, or what sort of things they might have said to each other, what made them laugh, what they quarrelled about or how they maintained peaceful relations within the group. Nor do we have any idea when they first had personal names, or when they could form the ideas of 'mother', 'father', or 'mother's brother', or when they developed the idea of some sort of official union between adult men and women, or if they exchanged women between bands, or how hunting co-operation was organized, or what sort of leadership existed. Nor do we know when man first had ideas of magic and symbolism, gods, ghosts, and spirits, or when or why he first performed religious rituals and disposed of the dead in a more than merely physical manner. But without this basic evidence, the theory of natural selection is unable to explain how man evolved his unique traits, and evolutionary psychologists have instead created an almost entirely imaginary world for our prehistoric ancestors, which provides an inexhaustible supply of glib answers to any question one can think of about the human condition. Why is only one of the basic human emotions, joy, positive while all the rest, such as fear and sorrow, are negative? 'Ah well, you see, during the Pleistocene epoch when most of human evolution occurred, life was very grim so this made pessimism adaptive.' Why do we

have nightmares? No problem: 'In the ancestral environment, human life was short and full of threats', so that 'A dream-production mechanism that tends to select threatening events, and to simulate them over and over again in various combinations, would have been valuable for the development of threat-avoiding skills'. These 'explanations' are not my own invention, but have appeared in *New Scientist*.

5. Darwinian thinking has had a long history in anthropology, as in E.Durkheim's *Rules of Sociological Method* (1895), A.R.Radcliffe-Brown's *Structure and Function in Primitive Society* (1952), and Gordon Childe's *What Happened in History* (1954). Over the last thirty years, in particular, a great deal has been written, and the idea of using Darwinian theory to explain human society in general has become increasingly popular, especially with the fashionable idea of the meme, popularised by Richard Dawkins in *The Selfish Gene*, and which led to *The Journal of Memetics*, and books like Susan Blackmore's *The Meme Machine* (2000). The following is intended as a useful sample of this Darwinian literature: R.Aunger (ed.) *Darwinizing Culture. The status of memetics as a science*, 2000. G.Basalla *The Evolution of Technology*,1988. R.Boyd & P.J.Richerson *Culture and the Evolutionary Process*, 1985. L.L.Cavalli-Sforza & M.W.Feldman *Cultural Transmission and Evolution. A quantitative approach*, 1981. D.Dennett. *Darwin's Dangerous Idea*, 1996. U.J.Jensen & R.Harré (eds.) *The Philosophy of Evolution*, 1981. P.Munz *Philosophical Darwinism. On the origin of knowledge by means of natural selection*,1993. A.T.Rambo & K.Gillogly (eds.) *Profiles in Cultural Evolution*, 1991. R.Rindos *The Origins of Agriculture. An evolutionary perspective*, 1984. W.G.Runciman *A Treatise on Social Theory* (II), 1989. W.G.Runciman *The Social Animal*, 1998. S.K.Sanderson *The Evolution of Human Sociality. A Darwinian conflict perspective*, 2001. E.A.Smith & B.Winterhalder (eds.) *Evolutionary Ecology and Human Behavior*, 1992. See also Mesoudi *et al.* 2004 for a very concise summary.

6. The Darwinian notion of variation can therefore be applied quite simply to human society and culture in the way that I have described. This avoids the problems of those who claim, following the lead of Richard Dawkins among others, that there are social equivalents of the gene, fundamental particles of culture usually referred to as 'memes', that reproduce with greater or lesser success. Social and cultural evolution can therefore be understood as changes in the relative frequencies of competing memes in the 'meme-pool', like genes in the gene-pool. Far from advancing the cause of Darwinism, however, this merely introduces an entirely unnecessary confusion between biological and social inheritance, which are quite different types of process. Living organisms need genes because they have to reproduce, but obviously societies, institutions, and customs, and ideas in general, do not have babies.

The internal combustion engine may have developed from the steam engine, but it was not its offspring in any physical sense. There is therefore no reason at all to think that memes could or should exist, unlike genes, atoms, and phonemes, which really do exist, and are essential scientific ideas. They are essential because, once we have identified the particular phonemes in a language, for example, linguists can discover the rules by which they combine to form words. In the same way, identifying the different types of atom allows chemists to discover the rules by which they combine to form different substances, and the same is true of genes. But memes are not fundamental particles of culture, since *anything* in human society and culture can count as a meme, from 'science' and 'religion' to a trivial tune or catch-phrase. So no rules about their combination, comparable to those of linguistics or chemistry, could possibly exist, and it is therefore not surprising that enthusiasts for the meme have not been able to produce a real problem in the social sciences, and then show that it can be solved better by meme theory than by any other alternative. On this, see for example Hallpike 1999. I have discussed the general question of memes in Hallpike 1986:37-46.

7. G.K.Zipf proposed what he called The Principle of Least Effort: 'the person will strive to minimize the probable average rate of his work-expenditure over time', and claimed that 'the structure and organization of an individual's entire being will tend always to be such that his entire behavior will be governed by this Principle' (1949:1). Where effort is a means to an end, there is probably a good deal of truth in this, but it does not explain why people choose some *ends*, such as gaining political office, that involve far more effort than peaceful obscurity. When, as in all creative work, effort is valued for itself, the Principle obviously breaks down completely. Again, it is psychologically much easier to vote for a proposal that has overwhelming majority support, than to vote against it, but no difference of *effort* is involved here. Zipf's grandiose Principle was based on a crude extension of physics to human behaviour that was fashionable at the time, and everything of value in it is contained in the simple proposition that people tend to do what is easiest in the circumstances.

8. We shall see that the whole idea of blind or random variation gives us a remarkably inaccurate picture of the major innovations in science, thought, technology, and social organization generally. On the contrary, we find again and again, that some *combination* of factors has normally been involved, a coming together of different elements of technology, ideas, beliefs, values, and institutions, and so on.

9. See Edgerton 1992 for a very useful critique of adaptive myths in anthropology.

10. Rappaport 1968:5.

11. As the famous archaeologist Gordon Childe put it, 'Evidently societies of men "cannot live by bread alone". But if "every word that proceedeth out of the mouth of God" does not directly or indirectly promote the growth and the biological and economic prosperity of the society that sanctifies them, that society and its god will vanish ultimately. It is this natural selection that guarantees that *in the long run* the ideals of a society are "just translations and inversions in men's minds of the material".' (Childe 1954:16)

Cultural materialists, such as Marvin Harris (1980) and S.K. Sanderson (2001), regard human groups as based on an *infrastructure* comprising:

The biostructure: that is, our nature as biological organisms, and its needs, and

The ecostructure: comprising ecological, technological, and economic structures in particular.

The structure: social institutions of all kinds,

And finally, the superstructure: the primary forms of mental life, including religion, philosophy, and science.

'The flow of causation' says Sanderson, 'is primarily from the biostructure to the ecostructure, then from the ecostructure to the structure, and finally from the structure to the superstructure'. (Sanderson 2001: 150.) The biostructure and the ecostructure have this causal priority because they concern vital needs, so that the superstructure of ideas, which includes religion, philosophy, and science, has least causal impact on the social structure, and in any case the objective requirements of the natural world cannot be altered by our mental constructs, which he refers to as 'The priority of nature over mind' (p.143). I am not, of course, simply reversing this model and claiming that everything is caused by ideas instead. On the contrary, the fundamental fallacy of the whole materialist model is to think that ultimately there is only one basic set of causes, those of the infrastructure, from which everything else follows. In reality, there is a constant interaction between all these various 'structures', none of which, by itself, can explain the rest.

12. See Hallpike 1977.

13. First published in 1890, and eventually comprising many volumes, it is an immense collection of information about magic and religion from all over the world. While it is often a useful starting-point when seeking information on specific topics, such as the evil-eye or sacred stones, its evolutionary scheme is entirely obsolete.

14. This should not be confused with the Law of Evolutionary Potential of Sahlins and Service, which said that 'The more specialized and adapted a form in a given evolutionary stage, the smaller is its potential for passing to the next stage'. (Sahlins & Service 1960:97) My concept of evolutionary potential is clearly not concerned at all with specialisation and generalisation,

but emphasises instead that things chosen for one reason may have multiple properties and functions, that may manifest themselves later when conditions are appropriate.

15. The claim that important social, cultural, and technological innovations occur in a regular sequence has been repeatedly demonstrated by anthropologists. See for example Naroll 1956; Freeman & Winch1957; Murdock & Provost 1973; Hallpike 1986:1–3; Peregrine, Ember & Ember 2004.

16. I have written very extensively about primitive thought in *The Foundations of Primitive Thought*, and *The Evolution of Moral Understanding*, and readers who would like to know more about the subject than I have been able to include in this book should consult them.

17. Fisher 1936: v.

18. I do not consider Marx's stages of social evolution in this book. 'Primitive communism' is a misconception (see Chapter V, note 5); 'slavery' was not a universal stage, and it was fairly unimportant in China and India, for example; 'feudalism' was merely an administrative technique and did not necessarily require serfdom and manorial production; and capitalism and socialism can to some extent co-exist. His idea that the state would eventually 'wither away' requires no further comment.

19. Popper 1957:149.

## Chapter II The Simplest Societies

1. 'Seismic shocker: Gilbert-Louis Duprez's history-making high-C' *Opera News*, January 1st, 1983.

2. Sanderson 2001:147.

3. See Sahlins 1974:1-39.

4. Throughout the continent, we find that many dialect groups are divided into a number of small, usually patrilineal clans, each of which has its own sacred place, and that these clans often belong to one of two groups in the group, called 'moieties' (moiety = half); the rule being that members of one moiety can only marry someone from the other moiety. These moieties are also used to classify the plants and animals of the local environment. Marriage itself was strictly limited not only by the moiety to which the men and women belonged but also in many cases to the sub-sections of each moiety. (This reinforces what I was saying earlier about the constraints on freedom of marriage even among hunter-gatherers.) This extremely elaborate set of lineal groups and marriage classes had, however, no connection with the bands in which people actually lived together and which, like all other hunter-gatherer bands, were small and flexible associations of people, which often changed their membership. So despite the extreme simplicity of Australian material

culture they were nevertheless able to construct very elaborate systems of lineal descent groups that had no basis in the economics of property ownership. This is an excellent illustration of how even at the level of hunter-gatherer society there can be very distinctive cultural traditions that could have influenced the adoption of agriculture.

5. Evolutionary biologists who speculate that such preferences were selected because they relate to fertility and reproduction seem quite oblivious to the actual conditions of marriage in primitive society generally. They seem to suppose that they were like those of modern cities, where millions of men and women freely intermingle, and a well-turned ankle may catch a man's eye on the boulevards of Paris or New York.

6. Lee 1979:49.
7. Lee 1984:58.
8. Turnbull 1965:228.
9. Marshall 1976:370
10. The question of how co-operation evolved has been debated endlessly by evolutionary biologists, because Darwinian theory might seem to predict that selfishness must have the adaptive advantage. See Hallpike 1984 for a discussion of this.
11. Marshall 1976:350.
12. Howell 1989:38.
13. Gardner 1966:394.
14. Woodburn 1968:91.
15. It has also been a general finding that children in foraging societies are brought up to be assertive and individualistic, whereas in farming societies there is far more pressure to be obedient and conformist. See Barry, Child, and Bacon 1959.
16. Balikci 1970:158.
17. Lee 1984:131.
18. Denny 1986:133.
19. Morris 1976:544.
20. Gardner 1966:398.
21. Gardner 1966:394.
22. The basic moral orientation of hunter-gatherers is well summed-up as 'Pre-Conventional', in Lawrence Kohlberg's scheme of moral development:

| | |
|---|---|
| What is right: | Following rules only when it is to someone's immediate interests and needs and letting others do the same. Right is also what's fair, what's an equal exchange, a deal, an agreement. |
| Reasons for doing right: | To serve one's own needs or interests in |

Social perspective of Stage: a world where you have to recognize that other people have interests, too. Concrete Individualistic Stage. Aware that everybody has his own interest to pursue and these conflict, so that right is relative (in the concrete individualistic sense). (Kohlberg 1984:174).

23. For a full discussion of this, see Hallpike 1986:116-19.

## Chapter III  The Agricultural Revolution

1. China and south-east Asia 6000 BC (rice and taro), north China 4000 BC (millet), sub-Saharan Africa 3-4000 BC (yams and millet), Meso-America 5000BC (maize, squash, and pumpkins), and South America 5500 BC (beans, manioc, and potatoes around 2000 BC), all of which then diffused from these primary centres to other regions. These dates are simply based on the earliest evidence that has managed to survive in areas actually studied by archaeologists, and so it is very likely that they underestimate the antiquity of agriculture in these various places.

2. See Reed 1977a, Ucko & Dimbleby 1969, Ucko, Tringham & Dimbleby 1972, Bray & Needham 1984:29-39.

3. See, for example, Golson 1977:8-9.

4. See in particular Cohen 1977.

5. Cohen 1977:15.

6. Bray 1984:34.

7. Hunter-gatherer populations have a very low growth-rate because a woman cannot nurse a baby and carry another small child around with her so that there is usually a natural interval between births of about four years. A nursing mother needs about 1000 calories a day for milk, and this means that in hunter-gatherer conditions she cannot accumulate enough body fat to allow conception. 'Stored fat is necessary for another pregnancy to occur, because a minimal relation exists between weight and height; below this ratio menstruation will not be initiated…and ovulation does not occur.' (Reed 1977b:895.) Studies of modern hunter-gatherer populations show that when they begin living in settled villages body-weight increases, menstruation occurs at younger ages, and the interval between births decreases. This simple fact of physiology has profound implications for population growth when the cultivation of crops replaces hunting and gathering as the basic economy.

8. See Simoons 1967.

9. This is particularly true of large herd mammals such as cattle, sheep, goats, for meat and milk, pigs, oxen for pulling the plough, and also horses, camels, onagers, and the ox for transport and, in the case of horses, for military purposes. In the Americas, however, there were very few large mammals

suitable for domestication: the native camels and horses became extinct soon after the end of the Ice Age, and were possibly exterminated by human hunters who had no use for them other than as food. The llama was used as a pack animal, and it and the smaller alpaca also provided meat, hides, and wool. (The wild vicuña, hunted for its wool, will not breed in captivity.) While in North America the bighorn sheep, and perhaps the bison, could have been domesticated, the agricultural base developed too late for this to occur. In Africa few of the indigenous animals can be domesticated, apart from the donkey in Egypt, and the large domesticated mammals that are now found in some parts of Africa – cattle, sheep, and goats – were brought much later from the Middle East and North Africa. The wide range of the tsetse fly in sub-Saharan Africa, however, effectively prevented their use in much of traditional African society, thus depriving them of the animal power for transport and ploughing that has been so significant in technological evolution elsewhere in the world. While the eland, for example, could have been domesticated and is immune to the tsetse fly it would have been useless as a pack animal or for ploughing. The range of domesticable plants in Africa (outside Ethiopia) was also more limited than those of Eurasia.

10. See especially Diamond, *Guns, Germs and Steel* (1997), where he makes this point very well. (But while his book contains some interesting information, it tries to replace a naïve racist view of history by an equally naïve geographical determinism.) Whereas horses and wheat are found all the way from Ireland to Japan, the turkey, in contrast, never spread from Mexico to the Andes, llamas and alpacas did not spread from the Andes to Mexico, and maize which was first domesticated in Mexico took until about 500 AD to be modified for the shorter growing season and changing day length of North America. In Africa cattle, sheep and goats reached the northern edge of the Serengeti plain of East Africa by 3000 BC, but took another 2000 years to cross the Serengeti and reach the Khoisan in the south. In Egypt, the food staples were wheat and barley, but these could spread no further south than Ethiopia. The development of agriculture in sub-Saharan Africa began with yams, sorghum and millet, but these crops could not penetrate the south of the continent whose climate is only suitable for the northern crops of wheat and barley. Many modern tropical African plants came by sea in the last few hundred years – taro, and bananas from Asia, and maize and manioc from South America.

11. There were at least three varieties of horse in Europe and Asia at the time in question: the Przewalski, the tarpan, and the forest wild horse, and it is not certain which of these were the ancestors of modern horses. It is possibly that domestication occurred separately among the Mongols and in the region of the Ukraine 3-4000 BC. Onagers, a related species, were domesticated by

the Persians in the fourth millennium BC, and by 3000 BC the Sumerians were using them to pull chariots in battle, but they were eventually replaced by horses. It is not certain when horses were first ridden, but presumably long before the ninth century BC, when the Assyrians were using cavalry.

12. Sahlins 1974:81-2.

13 For example, when I lived among the Konso of Ethiopia in the 1960's, for a population of 55-60,000 there were only about twenty smiths, and a similar number of potters, who were entirely dependent on the local markets for survival. But hundreds of men had recently taken up weaving because two merchants with lorries could take their cloth to other parts of Ethiopia, and when I returned in 1997 I found that improvements in transport had allowed cloth to become a major export.

14. Although of course the early farmers had no means of knowing this, the much larger and denser populations of agricultural societies would also act as hosts to a range of diseases such as measles, small pox, typhoid, and bubonic plague, that could not sustain themselves in small band populations, while domesticated animals would also become the source of a number of diseases that crossed the species barrier.

### CHAPTER IV THE NEW WORLD OF TRIBAL SOCIETY

1. Firth 1957:414-15.

2. '…matriliny is most likely to develop on a horticultural base, [i.e. with digging-stick and hoe, not the plough] with women doing the agricultural labour. It is most likely to disappear in the face of large-scale coordination of male labour; increased importance of property such as domesticates in the hands of males – property, that is, which is divisible and which can multiply; male control of the major tools of production [e.g. the plough]; and the regulation of economic and political life through non-kinship devices. Most, if not all, of these features are likely to be associated with increased productivity. For this reason, matrilineal systems tend to cluster in a relatively narrow range of levels of productivity. They are not found among the largest political units based on high agricultural productivity, and are not found at all at the highest levels of productivity.' (Aberle 1961:670) But this is only a very rough guide to the occurrence of matrilineal systems, and there are very many horticultural societies where women do almost all the agricultural work, but which are still patrilineal, as in most of New Guinea, for example.

3. See Hallpike 2008.

4. See Bernardi 1985 for a general survey of age-systems and how they work.

5. When anthropologists have looked at all this elaboration in tribal society,

the clan and lineage systems, the complex forms of ceremonial exchange, the rituals, the age-grading systems, and so on, those who dislike the idea of social evolution have concluded that these societies are just as complex as ours, but in different ways. But this misses one of the main points about social evolution, which is that some ways of doing things have much more evolutionary potential than others. However complex an age-grading system or a cycle of sacred drums or a system of ceremonial exchange may be, none of them can really lead us anywhere new because they do not have the potential to produce fundamental social change, unlike those really powerful innovations such as agriculture, urbanisation, trade, money, political centralisation, or conquest warfare. We will look at some of these issues in the next chapter.

6. Huizinga 1949: 10.

7. See in particular R.Needham 1978 for an excellent account of the symbolic principles of order.

8. See Hertz 1960 and R.Needham 1967, 1973.

9. In her well-known book *Purity and Danger*, Professor Mary Douglas has claimed that dirt is 'matter out of place', and that impurity is really a way of thinking about disorder. The obvious and fatal weakness of this theory is that there are many examples of what we would consider dirt, such as glasses and plates from which people have drunk and eaten, or soiled handkerchiefs and undergarments, or a rim of scum round the bath, where the dirt is matter very much *in* place. The pollution here obviously comes from the human body and has nothing at all to do with disorder. See Hallpike 2004:233-36.

10. Hallpike 1978.

11. Gell 1975:125, from whom this account of the Umeda is taken.

12. Gell 1975:133.

13. Another good example of how supernatural beings can only be understood within a wider context of order is the Tauade notion of *ago*. The basic image of *ago* is 'stone', which is an archetype of power not only through its weight and solidity, but through its use for tools and weapons, and when heated for cooking. But it is also a complex of meanings that covers our notions of culture-hero, prototype, wild form, source of power and fertility, independence, and immortality. The *agotevaun* as culture-heroes are regarded as non-human beings of great power who eventually became stones, but the different species of animals and plants are each also said to have their *ago*, immortal ancestors or prototypes which, in the case of plants are found in the forest, while a self-seeded tree is also be said to be an *ago*. The prototype of anything is also an *ago*, and the first string-bag, for example, was an *ago* called Ila [string-bag] Otauruma, and though it no longer exists as such, it survives in the actual string-bags made ever since. Certain types of rain and wind that come every year are *ago*, as are the sun, moon, and stars. The corners of houses are also

considered *ago* because they appear every time someone builds a house. The Tauade are clearly trying to express an impersonal, proto-philosophical idea here that goes well beyond the doings of the culture-heroes. See Hallpike 1977:254-70.
14. Baxter 1978:179.


## CHAPTER V ECONOMICS, WAR, AND POLITICS IN TRIBAL SOCIETY

1. Karl Polanyi 1957 was probably the first economist to realise the fundamental differences between subsistence and market economies, but as Dalton points out, economists (and materialist evolutionists) still tend to base their idea of human nature on modern capitalist society:

'The pursuit of material gain compelled by laissez-faire market rules is still not seen as behavior forced on people as the only way to earn a livelihood in a market system, but as an expression of their inner being;... *man* is seen as a utilitarian atom having an innate propensity to truck, barter, and exchange; material maximization and the primacy of material self-interest are assumed to be constants in all human societies.' Dalton 1968:xxvii.

2. See Sahlins 1974:41-99 for a convenient summary of the facts in support of this.

3. See Hallpike 1977.

4. A feast and dance would be held about every fifteen years or so, when they invited their neighbours to come and stay for two or three months; large villages of a hundred huts or so were specially built for these occasions by all men from the different hamlets of the tribe, together with two vast men's club houses at each end of the village, with a long dance yard in the middle of the huts. Whole trees were cut down and brought from the forest, and erected on each side of the dance yards like telephone poles and hung with yams and other produce, while tall platforms were also built for the big men deliver their speeches. Yams, smoked pandanus nuts, and pork are the most prestigious forms of food and required on ceremonial occasions. The hosts and their guests danced every night with drums wearing their ceremonial feathers, trying to outdo each other in their athletic prowess, and all sorts of sexual assignations also occurred between the women and the dancers. (In pre-colonial times these often led to homicides by enraged husbands.) The visitors were given string bags containing the bones of the deceased relatives of the hosts, with which they danced to do honour to the dead. The hospitality of the hosts was designed to show how rich and prosperous they were and how superior they were to their guests, who on other occasions were their enemies. So at the conclusion of the festivities the Big Men gave boastful speeches, in which they praised their own power, wealth, generosity and past victories

in battle. After these vainglorious speeches there then followed an orgy of slaughter in which the pigs were beaten to death with clubs and cooked in earth ovens. Piles of yams, sugarcane, and smoked pandanus nuts were also distributed, together with the pork to the guests. I was told that there was so much meat that men would vomit or throw the surplus to the dogs. The guests then tried to destroy or damage the men's houses, and were often sent on their way home with taunts and showers of arrows by their hosts. Tauade feasts were obviously extremely wasteful from the sober economic perspective of Mr Gradgrind, and were no more adaptive than their warfare. They were, however, hugely enjoyable, and because the Tauade economy was very simple and their pigs were of no agricultural use for manure or pulling ploughs, it did not matter if most of them were killed off periodically. The men had plenty of leisure and could afford to build the elaborate dance villages from the plentiful timber, vines, and bark of the forests and it was easy to expand their gardens to provide extra food for their guests. The women, however, were greatly burdened by having to bring the extra sweet potatoes for the pigs, which broke into gardens in increasing numbers and caused numerous disputes and violence between their owners and those who gardens they had destroyed.

5. Inheriting the right to use group land should not, therefore, be interpreted as 'primitive communism', because no sharing of the food produced from the land is involved at all – except perhaps for some ceremonial exchange at feasts. The whole idea of primitive communism is a modern fantasy, which misinterprets customary gift-exchange as altruistic sharing. The closest one gets to genuine primitive communism is a Benedictine monastery, or other small utopian communities, which are self-consciously egalitarian, and apply the principle of 'from each according to his ability, to each according to his needs'. Primitive societies are quite unlike this. In the same way, tribes where land is privately owned, like the Konso, are not 'primitive capitalists', because landowners are not producing a surplus for sale in the market, but only for the subsistence of their own families. Increasing one's wealth alone cannot in any case be a route to power and influence in tribal societies, which are dominated by institutions based on kinship and age. It seems that originally Konso land was owned by the lineage, under the control of the *poqalla*, but over time it has become privately owned. The system of farming is very intensive, with stone terraces and manuring, and it seems likely that this high investment of labour has primarily been responsible for the private ownership of land. Since eldest sons inherited much more of their father's land than their younger brothers, the descendants of some of the most junior lineage branches have eventually ended up as a landless class, having to work for others.

6. Bohannan and Dalton 1962:14.

7. Ekholm 1974:119.
8. Schapera 1956:99-100.
9. Sahlins 1974:139-40, quoting Firth (*Economics of the New Zealand Maori, p.133*).
10. For example, we are told that by at least 50,000 BC 'a second element [of human sociality] which developed from an instinct shared with other primates, was a sense of fairness and reciprocity, extended in human societies to a propensity for exchange and trade with other groups'. (Wade 2007:158). If we study real people, however, rather than speculating about the instincts of our remote prehistoric ancestors, we find that reciprocity comes in significantly different forms. There is 'generalised reciprocity', typical of families and close kin-groups generally, where people help one another on the basis of need, without keeping count of credits and debts. This may indeed have an instinctual origin. Then there is the 'balanced reciprocity' typical of those pork exchanges in New Guinea and other ceremonial exchange we discussed earlier, where individual prestige and *competition* are very much at stake, a strict account of debts and credits is kept, and those who cannot repay are shamed, and finally there is trade, 'negative reciprocity', which is openly self-interested and each party attempts to get the best deal at the expense of the other. (See Sahlins 1974:185-275.) Trying to explain these very different forms of reciprocity by a simple instinct is entirely unilluminating.
11. Goldman 1970: 255.
12. See Bohannan and Dalton 1962:1–3.
13. When Columbus first encountered the inhabitants of the New World on the Caribbean island of Guanahani on 12th October 1492, 'To some of them I gave red caps, and glass beads which they put on their chests, and many other things of small value, in which they took so much pleasure and became so much our friends that it was a marvel'.
14. See Bohannan 1959 on some of the effects of money on a primitive economy.
15. Einzig 1966:318.
16. Turney-High 1971:53. 'Lack of specialization is just as much a mark of primitivity as is lack of specialized tools in the material culture of peace, or lack of division of labour in social and economic life'.
17. Ibid., p. 53.
18. In primitive warfare the aim of subjugation is lacking in many cases, partly because 'The end of the enemy would have meant the end of manhood society, the *raison d'etre* for the male population'. Ibid., p. 104.
19. Gluckman claimed, for example, that civil wars of succession to the throne of some southern African kingdoms had the function of preserving the unity of the state: 'a periodic civil war was *necessary* to preserve national

unity: sections fought for the kingship, and not for independence from it'. (Gluckman 1954:78) See also my 'Functionalist interpretations of primitive warfare' (1973).

20. Anthropologists have been just as ready to claim that a particular custom is functional because it *prevents* warfare. For example, Gluckman argued that the common Papuan taboo on eating one's own pigs led to inter-tribal systems of pork exchange, and that these in turn inhibited warfare. But were the people subconsciously motivated to invent the taboo in order to prevent warfare? This seems implausible, to put it mildly, even to Gluckman, who says 'I cannot help puzzling about how a custom of this sort, with all its widespread effects, originated'. (Gluckman 1967:63).

21. See in particular Koch 1974.

22. See Feil 1987.

23. A modern school of anthropologists (e.g. Arens 1979, Obeyesekere 1998) claims that, other than in situations of extreme starvation, cannibalism is a myth inspired by the racial prejudice of white men. Political correctness is not a substitute for facts, and there is ample eye-witness evidence for cannibalism in tribal societies, especially in Fiji, New Guinea, and New Zealand, for example, for those who take the trouble to look for it. For a more balanced assessment of cannibalism see Brown and Tuzin 1983.

24. Marvin Harris 1980, and see also Diener et al. 1980.

25. Sahlins 2003:4.

26. In many societies there is the belief that ritual leaders must have nothing to do with killing and the shedding of human blood, on the grounds that they are the bringers of life, not death. This idea is found in Indo-European societies, and very widely in East Africa, including the Konso, so that as a result many chiefdoms were never able to convert what was primarily a sacred office into effective political rule. On the other hand, there are many societies where making war is seen as part of the sacred function of a ruler, that Life-giving enterprise that we discussed earlier, so that priestly and warrior functions are regarded as quite compatible. In Bantu-speaking societies in Africa, for example, the chief is not only legitimated by his descent and his religious status and is legislator and judge, but also organizes warfare and may lead his men into battle personally. The same was true of the West Cushitic kings in Ethiopia who combined warrior and priestly functions and were frequently involved in the conquest of their neighbours.

27. I made these points in *The Principles of Social Evolution*, but the Darwinist Azar Gat, in his *War in Human Civilization* (p.112), finds it all very confusing. Why, he complains, do I not ask why young men have aggressive propensities, or why a lack of social control and mediation leads to warfare, or what triggers revenge, and so on? Let me try to explain. If we are investigating why a stretch

of road has many serious traffic accidents, especially at certain times, we look at such factors as road layout and visibility, volume of traffic, the influence of alcohol, and so on, and typically find that some combination of these factors is responsible for the high level of accidents. We don't have to explain why alcohol impairs judgement, why a hidden junction is a more dangerous situation than an open road, or why head-on collisions are dangerous to health. The explanation of the accidents is that a certain set of conditions will affect the probability that people will behave in a particular way – collide with other vehicles. In exactly the same way, another set of conditions can explain why there is a high level of warfare in certain kinds of society.

28. Goldman 1970:xvii-xviii.
29. I discuss core principles at length in Hallpike 1986: 288-371.
30. On the reconstruction of original Polynesian society see Kirch and Green 2001, and Goldman 1970.
31. Howard and Kirkpatrick 1989:59.
32. Until the introduction of the sweet-potato in the last two to three hundred years, in much of New Guinea hunting and gathering were the most important sources of food, so that hereditary leadership did not have the opportunity to develop, and such leaders as existed tended to be fight leaders or shamans. See Feil 1987.
33. Prof. Bruce Trigger, personal communication.
34. Trigger 1990:132.
35. See Hallpike 2007.
36. While they had cattle, their economies were primarily agricultural and based on ensete and cereals, and none of their economies could be described as pastoral, whereas in East Cushitic societies there were some very important pastoral societies. It is therefore very likely that the initial impetus for age-systems came from the division of labour between the older men who cultivated their land, and the young men who tended the herds in the lowlands, but this did not prevent such age-systems being adopted by agricultural societies like the Konso.

## Chapter VI Primitive Thought

1. Many will point out that there is a clear association in our society between a high level of superstition and dangerous/unpredictable jobs where people such as actors, racing-drivers and trawler-men don't feel they are in control. This is quite true, but if we look at the superstitions involved, we typically find that they are about warding off bad luck. The idea of good and bad luck, however, is impossible to disprove scientifically and so can survive without difficulty. So while racing drivers, for example, may carry lucky charms we

do not find them rubbing magic potions on their engines, or trying to cast spells on other drivers because, even if they wished to believe in such things, they know they are impossible.

2. One of the favourite explanations for belief in supernatural beings is that our cavemen ancestors needed to evolve some sort of intellectual 'security blanket' to protect them from the fear of predators and other threats, and that modern religion is a survival of this evolutionary defence mechanism. But if the function of supernatural beliefs has been to give us intellectual comfort, then the human race has generally gone about this in a very odd way, since the terrors of the supernatural have been at least as obvious as its consolations. The deities of hunter-gatherers and early farmers, for example, far from being comforting are generally punitive, arbitrary, and thoroughly nasty, much more concerned with punishing offences against taboos than with rewarding or helping people, while fears of evil spirits, witchcraft, the evil eye, and a great deal of magic have blighted the lives of millions. Witchcraft is a belief that is particularly difficult to explain by any psychological or social advantages that it could produce. The conviction that one's misfortunes are the result of someone else's malice makes them even harder to bear, and the problems of detecting those thought to be responsible for all this imaginary malevolence permeate society with an odour of fear and suspicion that is neither comforting nor functional.

Again, the belief that the soul survives death is pretty well universal, but if it is just the product of wishful thinking, why was the after-life represented in most societies (until the rise of the world religions) as generally a gloomy place, where the dead endured a sad existence and envied all the joys of the living? And, once the idea of some kinds of rewards and punishments after death developed in the world religions, why have the punishments been so important? If, for example, the Christian idea of Heaven was the product of wishful thinking, why (at least until the modern era of Santa Claus and the Easter Bunny) did most Christians traditionally think they were in serious danger of Hell fire? It was not entirely without reason that the Epicureans claimed that by denying the existence of the gods (or at least their power to do anything) they were relieving mankind of a burden that made them miserable. One could argue that, for most of human history, supernatural beliefs have been as much a source of terror as of comfort.

Belief in an afterlife, too, is often explained by the emotional need to be reunited with one's loved ones after death, or to go to a place of compensation for the sufferings of this life. Such hopes may appeal to modern Westerners, but they do not figure very prominently in other cultures and earlier epochs. A happy reunion with one's loved ones in the next world is not, strangely enough, an aspiration that anthropologists come across when they study primitive societies. What really interests members of primitive societies is not if their departed relatives and ancestors are happy in the next world but *if they are angry with the living* – have we honoured our ancestors' bones with a sufficiently lavish dance, or sacrificed an ox of suitable size, or broken a taboo such as marrying someone from the wrong clan? The general impression given by the anthropological evidence is that primitive peoples do not believe in the afterlife because it fills an emotional need (either for personal survival or reunion with loved ones), but simply because it seems an obvious fact, based on the experience of dreams in which they think they meet the dead, on ghostly apparitions, and on out-of-body experiences.

3. Not all beliefs in supernatural agencies, however, are related to levels of cognitive development. It is obvious, for example, that many people in modern industrial societies, as well as in primitive societies, believe in ghosts, and they do so because they think they have either seen them themselves, or have read accounts of them that seem convincing evidence of their reality. So here we have beliefs that have nothing to do with the believers' levels of intellectual development but simply with how things appear to be. We also have to distinguish between a belief and the reasons why it is held. In early seventeenth century England the peasantry and the educated classes believed in witches, but learned men, such as King James I, did so primarily because the Bible said that witches existed, and they believed on theological grounds that every word in the Bible was literally true, whereas the reasons why the country folk believed in witches had much more resemblance intellectually to those of tribal peoples. This distinction between the outward form of a belief and its intellectual foundations is of particular importance when we come to religious ideas that can be developed to considerable levels of conceptual sophistication. A Jesuit theologian and an illiterate peasant can both be said to believe in God, for example, but they will justify these beliefs at very different intellectual levels, which is one of the reasons why religion has not disappeared despite the domination of Western culture by science.

4. Otherwise known as 'pre-operational' thought.

5. Building a house can be spread over a month or more when a man feels like it – and with materials that are often available in unlimited abundance. He does not have to make any explicit calculations to decide whether, within his budget, he would be better off obtaining more materials for walls and less for roofing materials, or vice versa; or whether to employ more men to finish the work sooner, or to use fewer men and take longer. His materials come to him in ready-made sizes and qualities – usually the only modification he will need to make is to the length of timbers – while the use of each type of material – thatching grass or leaves for the roof, canes for the rafters, aerial roots of pandanus trees for binding ties, logs for walls, and bark for floors

– is decided for him by the natural properties of the materials themselves. He builds according to well-tried and conventional principles which he does not need to adapt or experiment with. He needs no plan to be drawn before work can begin, nor does he need to calculate breaking strains of lintels or posts, or the necessary thickness of beams to support possible loads, as it might be in a granary. The size of openings, the height of verandas from the ground, and the floor area, are decided by the size of the human body and its customary activities, while the limitations of his technology ensure that he will be unable to set himself architectural problems that he cannot cope with. Nor should it need emphasizing that, in non-monetary economies, there is no opportunity or need for the quantitative analysis that precise costing imposes on the house-builder in our society.

6. Prince 1969:134.

7. While pieces of bone with rows of notches dating to 20-30,000 BC have been found in Europe we have no idea what the notches signified. When verbal numbers are restricted to 'one' and 'two', as often occurs in New Guinea and Australia, they are not real number words at all but refer instead to configurations that would be more accurately translated as 'single' and 'pair'. Configurations exist in their own right, rather like shapes: we can think of a 'brace' of pheasants without thinking of it as equal to one less than three, or one more than one, for example

8. From Gay and Cole 1967.

9. Prince 1969:31.

10. Littlejohn 1963:9

11. It is a very popular fallacy that people needed calendars when they became farmers so that they could plan their agricultural activities properly. In fact, nature provides many more accurate and useful predictors of the seasons than the moon, such as the flowering of different plants, the migrations of birds, and the rising positions throughout the year of the sun and a variety of stars and constellations. (See, for example, Best 1922.) The Pleiades, for example, are used generally in the tropics for predicting the onset of the main rainy season. So the practical significance of calendars can easily be exaggerated. Even where people like the Konso do have named months, unlike us they are not continually thinking about which one it is, and when asked the name of the current month, they may not know and will have to reflect about it or ask someone else. The primary use of named months is actually for co-ordinating social events and as a guide to when religious ceremonies should be performed, but since religious ceremonies are fundamentally concerned with good harvests and the health and fertility of men and beasts, there is a need for these calendars to be linked in a permanent way to the seasons and to agricultural activities. (See Bohannan 1967 and Hallpike 2008.)

Calendars were therefore motivated more by that general human need for cosmic order which, as we have seen, permeated tribal society, so that we often find that some months are auspicious and others inauspicious, and that particular ceremonial activities are linked with certain months. (It is in fact quite easy to keep a lunar calendar in a fairly stable relation with the seasons by repeating or adding a month every three years or so.) Just as the directions of East, West, North and South had profound symbolic meaning, so too did the divisions of time. Calendars were therefore like mental maps charting the territory of time for each tribe, rather than practical necessities of economic planning.

12. The time concepts of the Iraqw of Tanzania, for example, are basically topological. While they have words for years, months, and days, they only use these sequentially, not chronologically: 'They do not use these standards of measurement comparatively to produce a general concept of uniform time, a chronology, that is, against which all events may be compared. Years, months, days, and hours are partitions of the flow of time, and in the way we use the term here, are topological. The creation of a uniform (e.g.Euclidean) space or a uniform time standard (e.g. a chronology) requires a logical act that the Iraqw do not perform.' (Thornton 1980:171).

13. Evans-Pritchard 1940:103.

14. It is not therefore uniform, continuous, or homogeneous. Homogeneous time is common to the whole universe, and not localized or subjective; in continuous time the present is only one moment in a single process in which there are no gaps, while uniform time flows at the same rate.

15. Forde 1954:219. For an excellent study of classification in a tribal society see Cole *et al.* 1971.

16. 'The one unambiguous finding in the study to date is that schooling (and only schooling) contributes to the way in which people describe and explain their own mental operations.' (Cole and Scribner 1974:122). From his research with Uzbek and Khirgiz peasants, Luria concluded that 'There is every reason to think that self-awareness is a product of socio-historical development and that reflection on external and social reality arises first; only later, through its mediating influence, do we find self-awareness in its most complex forms.' (Luria 1976:145).

17. Gay and Cole 1967:16. See also Hallpike1979:105-10.

18. Lienhardt 1961:149.

19. Onians 1954:13, 67-68.

20. Malinowski 1922:408-9.

21. Read 1955:265.

22. Read 1955:265.

23. Piaget 1952:429-430.

24. Piaget 1932:250.
25. Piaget 1930:118-19.
26. Piaget (Piaget and Inhelder 1975) conducted experiments with children to examine their understanding of probability that used, for example, a spinning metal bar that came to rest opposite different numbers. He found that children at the intuitive Stage I of development assumed that each spin of the bar was in some way related to previous ones so that it had a kind of memory and purpose that made it stop at different numbers, and compensate for omitting some numbers by picking them more often later. They could not understand that when the bar produced the same number a number of times in a row, a longer sequence of results could nevertheless be regular in the sense that each number could eventually be expected to come up a more or less equal number of times.

By Stage II the children understood that each trial is distinct from its predecessors, and in an experiment that involved rolling steel balls down a board through a regular pattern of nails into a set of slots at the bottom, the children could see that the balls had an equal chance of going in either direction when they hit a nail, and so of going into any of the slots. So, the greater the number of collisions the more regular the final distribution of the balls would be, but only up to around twenty or thirty trials. They still believed, however, that if there were many more trials, then the irregularity of the results might increase. They also began to develop a sense of the general probabilities involved, such as of a ball falling into a particular slot, but still did not understand that there are different degrees of probability. Finally, in Stage III, that of formal operations, there was a full understanding of the law of large numbers, and of degrees of probability that could be quantified in terms of permutations and combinations.

27. Marshall 1976:368; Howell 1989:38; Holmberg 1969:260; Woodburn 1968:91 Barry, Child, and Bacon 1959.
28. von Fürer-Haimendorf 1967:208.
29. The significance of this increase in IQ scores in the industrialised nations over the last hundred years is discussed in detail in Flynn 2007.

## Chapter VII The State and Civilisation

1. Materialists such as Johnson and Earle (2000:29-37) claim that growing population pressure inexorably produces higher and higher levels of political integration to manage the increasingly complex systems of production and resource distribution that are needed to sustain these larger populations. There are many reasons not to believe this, but one of the most compelling is that there is no simple correlation between population density and complexity

of political organization, once we leave foraging societies out of the picture. Many early states, especially where there is little urbanization, had population densities considerably less than those recorded for many tribal societies. I discuss this in more detail in Hallpike 1986:237-38. States can, of course, co-ordinate much larger *total* populations than tribal societies, but that is another matter altogether.

2. Some have argued that, since a large quantity of water can only be controlled by mass labour, this would have to be organized by some sort of central authority. So farmers who wanted to expand their settlements into arid areas would thus have had no choice but to submit themselves to a directing authority, and this necessarily produced state organization. (Wittfogel 1957:18).

The theory is basically flawed, however, because even irrigation works that control large quantities of water do not have to be constructed as a single massive project. Studies of many ancient and modern irrigation systems around the world have shown that they can begin and be maintained simply as local forms of water conservation with construction of dykes and channels for flood-water control. The huge irrigation systems of Sri Lanka, for example, took over a thousand years to construct in piece-meal fashion, and were organized at village level before and after the rise of the state there.

On the Nile no large-scale irrigation system at all was needed, since the annual flood was the main agency of irrigation. On the Euphrates, dykes were built under local control, while in the Indus Valley the basic crops of wheat and barley were sown, as in Egypt, after the flood-water had receded. And in China large-scale irrigation projects were only begun around 400-300 BC, long after the development of centralized state organization in that region. While, as in China, such projects certainly intensify the power of the state, by increasing administrative control and making more land available for settlement (and so producing more potential soldiers for the state army), irrigation nevertheless can only intensify state control, not create it. See Hallpike 1986:260-66 for further discussion of the irrigation theory of the origins of the state.

3. States appeared in a continuous band across Europe, North Africa and the Nile Valley, the Far East and South Asia including India China and Japan; (2) Central America; (3) South America in the high Andes; (4) a West African zone; (5) an East African zone in the lake regions; (6) Southern Arabia and the Ethiopian Highlands (where the literate Axumite civilisation was of south Arabian origin); (7) a Southern African cluster of states; and (8) in Polynesia the incipient states of Tonga and Hawaii. (Cohen 1978; Hallpike 1986:286). The state emerged late in sub-Saharan Africa because this part of the continent was very under-populated, with infertile soil and lack of

draught animals, so that the conditions for trade and urbanism were very unfavourable. Conquest warfare, often involving the capture of slaves, was of great importance in Africa, but the state was, in material terms, usually very simple. The 'capital', apart from the city states of West Africa, was often just the royal residence, together with dwellings for the large number of servants and artisans necessary to perform the rituals and maintain the royal household, and was only seldom a place with a large number of the king's ordinary subjects. These were subsistence farmers, who paid their tribute in the form of manpower when needed for fighting, or for repairing the palace, or perhaps working on the roads. Some cattle would be collected as tribute in states where sufficient existed, but usually communications were too poor to allow the transportation of large amounts of tribute. So it was not necessary to have large numbers of civil servants to keep records of taxation, and literacy never developed south of the Sahara in Africa, except under late European and Islamic influence.

4. Cohen 1978:61.
5. '. . . trade played a definite role in all the [21] states. In thirteen cases it appeared to be of great importance, and in eight cases was found to be minor importance' (Claessen and Skalnik 1978:542).
6. Trigger 2003:671.
7. Because of the king's basic responsibility for maintaining cosmic as well as social order, in some early states the royal residence is indistinguishable from a temple. When the first Pharaoh, Menes, united Upper and Lower Egypt he established his palace-temple at Memphis, dedicated to the god Ptah; the focus of the Sumerian city was the palace-temple of the ruler, *en*, and in Meso-America the palace-temple was the centre of the city and government. The same also seems to have been true of the Minoans.
8. Analysis of the great mass of Shang dynasty oracle bone inscriptions over the period 1200 – 1050 BC 'suggests an evolution from an outlook which stressed a heavy reliance on divination, an enormous emphasis on sacrifices to nature deities and the mythic or semimythic predynastic ancestors, irregularity and "ad hoc arrangements" in the performance of ancestral sacrifices, and the reliance on a vast group of diviners to an outlook in which the volume of divinations is markedly diminished, sacrifices to nature deities and remote ancestors cease, the king becomes the main diviner, and the sacrificial rites to ancestors are regularized in an annual five-phase cycle of sacrifices.' (Schwartz 1985:37)
9. Edwards 1904:23.
10. Trigger 2003:233.
11. A good example of the drastic difference between tribal warfare and the type of true warfare that was the basis of the state is provided by the Zulu

at the beginning of the nineteenth century. Up to this time tribes had been small and scattered under their different chiefs, and conflict was formalized and produced few casualties. Warriors met for battle by appointment, and fought with throwing spears and tall shields. Selected warriors would advance ahead of their comrades, shout abuse at the enemy, and then hurl their spears. More and more warriors were drawn into the fighting until one side fled; if the defeated dropped their spears it was a sign of surrender and no more blood was shed.

But by 1800 population pressure meant that a dynastic dispute could no longer be resolved by splitting up the tribes involved and by resettling them on vacant land. Chiefs began to impose their will on dissident tribes by force, and Dingiswayo of the Mtetwa tribe made the first organizational improvement in warfare by forming the young men's age grades into military regiments. Shaka, who was chief of the Zulu, and a friend of Dingiswayo, then invented a new technique of fighting. This involved the use of a short, broad-bladed thrusting spear to replace the long throwing javelins (the only actual change in weaponry), the abolition of sandals to improve mobility, and the technique of using the shield to hook away the opponent's shield and expose his chest to the stabbing spear. Shaka also organized his soldiers into companies of 100 men in a close formation, the main body in the centre with two 'horns' on the flanks to encircle the enemy, while the reserves remained at the rear to exploit the opportunities of battle.

As soon as boys were recruited into their age-regiments they were subjected to rigorous training and discipline: they had to live in temporary campaign huts, built under the supervision of the regimental commander, were sent on difficult missions to gather kraal building material, being allowed to take no food and having to live off the country. They were also drilled and instructed in the martial arts. Discipline in the army as a whole was very harsh, and all defeated troops or those who retreated were massacred en masse, and a Zulu who returned without his spear was executed. They were taught to bear pain unflinchingly from youth, and unthinking obedience and loyalty were exacted, such that a man might be tested by being ordered to kill his own brother.

A proper hierarchy of command was established. A colonel or *induna* was placed at the head of each regiment, and below him were the commanders of two wings or battalions, each of whom commanded a number of company commanders, these in turn being assisted by 1 - 3 junior officers. The commander not only devised special plans for particular campaigns, but there were standard battle plans for general fighting, and the commander gave much thought to planning the entire campaign beforehand. The Zulu were naturally able, on this basis, to mount substantial campaigns. The supply function was well organized: troops of girls carried food and beer for the expeditionary force until it was consumed when they returned home. Male porters were provided for the principal men, otherwise troops carried their own supplies, but a herd of cattle was marched two miles on the flank, both for food and in the hope that they would succeed in finding enemy cattle.

In safe country, regiments marched in open order, with scouts to the front and rear, but close order was maintained in enemy country to prevent stragglers being picked off. About ten companies preceded the main body, for security and to mislead the enemy into thinking that they were the main body. Pass-words and countersigns were used for security, and camps were well guarded by sentinels at night. The army was said to be able to withdraw from battle in as good order as the best European Guards regiments.

Shaka was excellent at exploiting victory, and spared no effectives of the enemy side: the men were slaughtered, the women and cattle taken away prisoner. The defeated region was systematically devastated and terrorized, as security against surprise. He knew that the object of war was not only the defeat of the enemy army, but the destruction of his materials and morale, and those of the civilian population as well.

The development of these methods of true warfare reaped immediate and spectacular political rewards. From 100 square miles in 1816, his territory had increased to 200,000 square miles by 1820. With one tribe in 1810 he eventually subjugated 300, and the size of his army increased from 500 to 50,000. So did the number of casualties: under the old system of formalized, duelling battle they had been about 3%, but under the new system of warfare they rose to about 80%. (This account is based on Ritter 1955.)

12. Competition between neighbouring societies would have intensified the level of military organization and made people more willing to cede power to their leader in the first place. Two of the best known examples in modern times where the threat of external conquest made people more aware of the advantages of centralized leadership are the Cherokee chiefdoms of North America which developed centralised political leadership in the eighteenth century in response to the English settlers, and the Maori chiefs of the North Island of New Zealand who established a king from among their number in 1857 for very similar reasons (Gorst 2001).

13. Claessen's survey of twenty-one early states shows clearly that the sovereign has a very close association with warfare. In eighteen cases he was supreme commander (three cases of insufficient data), and in ten he personally participated in combat as well (Claessen and Skalnik 1978:563).

14. Trigger 2003:241.
15. Trigger 2003:679.
16. Trigger 1972:578.

17. 'Although cities in ancient China were prominent not only as cult centres but also as military strongholds and political centres extending their control over agrarian hinterlands, there is little clear evidence in the ancient period of an inner political life of the urban community as such.' (Schwartz 1985: 28).

18. The Sumerians, for example, lived in the delta of the Euphrates in southern Mesopotamia, probably as fishermen initially, and then cultivating the wheat and barley that, much earlier, had been domesticated in the uplands of the Zagros mountains. With cattle, sheep, and goats they had an abundance of agricultural products, and also had a supply of obsidian, all of which they traded for the metal, stone, and wood that they lacked. They developed urban centres such as Uruk very early around 3500 BC, and these were based on temples with the king as a key figure as well. The temples were also economic centres to which the people brought offerings and which formed the basis on which the state developed. Early Sumerian cities did not have walls, so it appears that warfare was a later development as competition between city-states intensified.

19. Curtin 1984:6.

20. Trigger 2003:349.

21. Curtin 1984:6

22. I discuss the core principles of Indo-European and Chinese society in more detail in Hallpike 1986:288-371.

23. '. . . among the Hittites the king was in a degree limited by a "general assembly" [*pankuš*] of nobles and warriors which maintained against him in particular the right of trial by peers and of restraining him from jealous violence in his own family. They could exact even the capital penalty in requital of such murder committed wilfully.' (Gadd 1948:41). 'Although the kings of the Hittites may have been less absolute than those of other realms, and in their choice of successors partly dependent upon the assent of their subjects...' (ibid, p. 47).

24. Sharma 1968.

25. Schwartz 1985:31.

26. Tuden and Marshall 1972:454-56; Hallpike 1986:270-272.

27. Many archaeologists believe, for example, that prolonged droughts between 800 – 900 AD could have led to the collapse of Mayan civilisation.

## Chapter VIII Technology and Invention

1. Corlett 1990:17. Unfortunately for Humphreys, however, Brunel then decided that the newly invented screw-propeller would be more efficient than paddle-wheels. Since the propeller would revolve much faster than paddle-wheels, this also involved replacing the engine that Humphreys had designed

with one designed by Brunel and his father. Humphreys was so mortified by the loss of this commission that he developed brain-fever, and died soon afterwards.

2. Indeed, the idea that technology is a response to human needs, taken literally, implies that people with the least developed technology for some strange reason had fewer needs than everyone else, which is obviously nonsense. For example, the Aborigines never invented the bow and arrow during the 40,000 or more years that they inhabited Australia, and the even more remote Tasmanians never even developed hafted tools or discovered how to make fire (as distinct from preserving fire that occurred naturally). Are we to suppose their needs were somehow different from those of other peoples, or that something special in the Australian environment created a need for the boomerang, but not for the bow and arrow? The underlying assumption here is the inherently unproveable belief: if a society invents something they must have needed it, and if they didn't invent something then they could not have needed it.

3. Needham 1965:258-75.

4. Even though many forms of clay can easily be moulded into different shapes, and are found all over the world, many groups never discovered how to make clay into pots. In addition, most of those who did make pots were content to fire them by piling brushwood over them and setting it alight. This produces enough heat to make serviceable cooking pots (as I saw for myself among the Konso), but is a typical example of the survival of the mediocre, because much harder and more durable pots can be made by firing clay in a kiln at around 1000° C, instead of the 500° - 600° C of an ordinary fire. (The basic kiln was a chimney of brick or stone with an air inlet at the bottom and was a special invention restricted to the Middle East around 4000 BC. Bellows were invented much later) The Romans, the finest builders of bridges and aqueducts in antiquity, thought that arches had to be semi-circular so that the approaches to their bridges typically rose up very steeply from the road on either side. Unlike the Chinese engineers, they never discovered that it is possible to build a much flatter arch based on the segment of a circle rather than a semi-circle. Again, the throat-and-girth horse harness used all over the ancient world for millennia was remarkably inefficient because it constricted the horse's air-passage, and in all these and many other cases the mediocre could survive despite what we would consider the obvious need for improvement.

5. For example, Basalla in his book *The Evolution of Technology*, evades the issue by saying, in his Darwinian account of technological evolution, that 'novelty is present wherever, whenever, and by whatever means humans choose to make things', but that 'the lack of a satisfactory theoretical approach to novelty does

not affect the evolutionary theory presented in this book. That theory calls for an adequate supply of novel artefacts, or ideas of novel artefacts, from which a selection can be made.' Basalla 1988:134.

6. Finley 1973:10.

7. See Forbes 1966, 1972.

8. Dr Bruce Bradley, *New Scientist* 24 June 2006, p.55.

9. Forbes 1971:10.

10. Here we need to note that pottery played a very important part in metallurgy, once this had got going, not only because the kiln was the basis for smelting ores, but because fired clays can withstand much higher temperatures than metals. Clay crucibles, in particular, became essential tools in the melting, refining, and casting of metals. When the draught of the kilns was improved by bellows, clay nozzles directed the blast of air into the heart of the fire; clay provided a fireproof lining for the inside surface of the kiln; and the molten metal was poured into moulds of clay, as well as sand.

11. Trigger 2003: 280-1.

12. Forbes 1972:228.

13 Ritter 1955:27. See Eliade 1978 for a very useful survey of the magical and religious significance of metal-working in pre-modern societies.

14. Forbes 1966:115.

15. Crystal and glass lenses were used by craftsmen in the ancient world as magnifying glasses. See Chapter XIII, pp. 250-51, and note 19.

16. This section is based on Needham 1980 and 1986. Although the Chinese priority in the discovery of gunpowder and cannon was definitively established by Joseph Needham and his colleagues more than 20 years ago, one still comes across many modern statements in various media that its origins are unknown, or were European or Islamic, or that even if the Chinese did discover it, they only used it for fireworks.

17. Needham 1980:40.

18. See Landels 2000 and White 1984.

19. The vehicle wheel makes one unique and strange appearance in the New World that may be relevant to spindle-whorls and the use of clay. As is well known, in an Aztec tomb in 1949, there were discovered clay models of animals whose legs were mounted on clay discs that functioned as wheels. Since the Aztecs had neither vehicles nor animals to pull them, and transported large quantities of goods by boats while small loads were carried by porters along simple tracks how, then, did they ever come to think of the vehicular wheel at all? They were skilful weavers and potters, with fine spindle-whorls of clay, and it therefore seems possible that here, too, the spindle-whorl was the simple model for the vehicle wheel which, in the technological circumstances of the Aztecs could never become anything more than a child's toy.

20. The water-wheel came in two forms, the vertical and the horizontal. The ancestor of the vertical wheel that predominated in the West may have been the *noria*, a water-wheel with a set of jars fixed to its rim that is rotated by the force of a river or stream to lift the water to a higher level for irrigation and may have come from India. The horizontal wheel was particularly favoured by the Chinese and was driven by the force of the water on its spokes in the manner of a turbine.

21. The south-pointing carriage was an example of court gadgetry, 'a two-wheeled cart with a train of gears so arranged as to keep a figure pointing due south, no matter what excursions the horse-drawn vehicle made from this direction'. See Needham 1965:286-303.

22. In a corbelled arch, each course of stones, on both sides of the arch, is made to project farther than the course below, so that they eventually meet in the middle and support each other. It seems that while this was the standard Mayan form, archaeologists may have found what appears to be one case of the true arch (see Befu and Ekholm 1963).

23. Even simple discoveries, however, were never made by everyone but were borrowed from other groups: the Konso told me, for example, that not only had iron-working, weaving, and pottery been brought by immigrants, but even the knowledge of curing skins and brewing beer. The early centres of innovation naturally had a head start over other areas, and this is especially true of agriculture. The main routes of diffusion were across Eurasia, initially between the Middle East and later the Mediterranean and Western Europe, and Asia so that those regions which were remote from these routes such as Africa, and Australia and the Pacific were inevitably at a major disadvantage. With trade went new crafts, techniques and ideas as well, and this meant that those areas of the world which lay off these new trade routes after the agricultural revolution became increasingly isolated and stagnant. It is surely no accident that Australia, for example, thinly populated and with an essentially homogeneous lifestyle and range of products, and lying off the main trade routes, should have been so intensely conservative in technology.

The Americas were not, however, as completely cut off from this process of diffusion as some scholars have liked to imagine. For example, similar tools and techniques in preparing bark paper are found in South East Asia and in Central America, as are the pitches and scales of pan-pipes, while Chinese and Mayan astronomers used exactly the same calculations for lunar and solar eclipses, and there are also similar sequences of animals in lunar zodiacs, and similar correlations between colours, animals, and compass points. Since Chinese cultural developments were prior to those of the Maya, and their boat building techniques were much superior, it is reasonable to infer that there were a number of Chinese and South East Asian influences on the Americas.

For a convenient summary of this evidence see Needham 1971:540–553, and Coe 1999:57-8.

Some scholars, *e.g.* Elliot Smith 1933, have taken the extreme view that every important invention, including agriculture and the state, was so difficult it could only ever have happened once in each case (usually in Egypt), and had then spread throughout the rest of the world by diffusion. Those who took the opposite view, and believed that societies evolve by themselves, like different plants in their own pots, in a necessary sequence of fixed stages, were naturally unwilling to admit the importance of diffusion. But when we realise that what matters is simply the accumulation of necessary conditions, no matter whether those conditions are home-grown or imported, then it can be seen why diffusion has been of fundamental importance to the evolution of culture.

24. Skinner 1999:27.

## Chapter IX Writing, Mathematics, and High Culture

1. Rudgley 1999:73.
2. Trigger 2004:44.
3. Mesopotamia provides the only example where we can trace the earliest forms of a script, and in this case it was clearly primarily economic in origin, to keep temple records of tribute, and clay tablets from Uruk in Sumer of 3400 BC record the details of goods in the temple accounts. The earliest examples of writing in Egypt occur in the tomb of a king/chief at Abydos from about 3250 BC. Since there was extensive trade with Mesopotamia at this period, and examples of Mesopotamian cylinder-seals with writing occur in Egypt, at least the idea of writing must have been common to both societies, but the degree of mutual influence between them remains unclear. Although, unlike Mesopotamia, there is no earlier record of 'pre-writing' in Egypt, the use of phonograms appeared much earlier than in Sumeria, suggesting that writing had been developing well before the Abydos tomb, and the two scripts developed in very different ways as well. For example, unlike Mesopotamia, in Egypt a simpler 'hieratic' script developed together with the hieroglyphic system. It eventually came to be used for a wide range of documents: tax accounts, records of workers on jobs, court records, leases, wills, and also literary works: stories, poems, spells, medical texts, literary teachings and etiquette, temple inventories, records of rituals, guide books for the afterlife, and private letters. The use of the earliest Egyptian writing also seems different from that of the Sumerians. Nearly 200 labels of wood and ivory were found in the tomb at Abydos, some merely with numbers on them, but others with references to names of places or activities such as bird

hunting, fish catching, or cloth production. About fifty signs were used, some with sound values and others with meaning values. Egyptian tombs were intended as replicas of the home so that in the afterlife the dead would be properly provided for, and it seems that these were labels on different sorts of goods that the dead king would need. Since this was only the tomb of a local ruler in the Pre-Dynastic period, it seems unlikely that he would have had a large administrative bureaucracy which needed to keep extensive written records, so that here, unlike Mesopotamia, the use of writing seems to have been more religious in nature.

4. The Aztecs certainly used writing for recording tribute, but also for genealogies, historical events, calendrical matter, and divination. But Aztec writing seems always to have retained a mnemonic element so that one had to know what the pictures were about before one could understand them.

5. In Assyria, too, inscriptions on temples and palaces were addressed to the gods, not to men. Gadd 1948:60-61.

6. Wilson 2003:22.

7. Sampson 1985:163. The fact that we recognise words primarily by shape and not by looking at the letters individually can easily be demonstrated. Although in the next sentence the words are written from right to left, the reader will have no difficulty in understanding it: letters the on not words the of shape the on concentrate we because is this ←
But if the letters as well as the words are written from right to left,
elbarehpicedni emoceb yeht neht ←

8. Gelb 1963:167.

9. The Maya and the Mesopotamians used signs to represent syllables, usually consonant-vowel, while the Egyptians only had signs for consonants. The Chinese had separate logograms for each word, but to write more abstract words they had to add signs to indicate pronunciation, and so developed complex signs combining one sign for meaning (the radical) on the left, and the other for pronunciation on the right. This system also allowed them to distinguish between many words with the same pronunciation but different meanings which are very common in a monosyllabic language like Chinese.

10. Machinist1986:202.

11. Trigger 2004:66.

12. Gelb 1963:196.

13. Gelb 1963:196.

14. Here a great deal of light is thrown on the borrowing of writing by pre-state societies by the well known case of the invention of a script by the Vai of Liberia, in West Africa, in the first half of the nineteenth century. (My account is taken from Scribner and Cole 1981.) The earliest and generally accepted account of the origin of this script was written by a German by the

name of Koelle in 1854 about 15 years after the event. He was told by a man named Dualu Bukele that he had dreamed of a white man who brought him a book of signs; he could not remember them on waking, so with five friends he set about inventing a new set of signs. Bukele clearly understood the basic principle of using signs to represent sounds, and with this he and his friends produced a syllabary with approximately 210 characters of 30 consonants in 7 different vowel forms. The set of signs made use of a long West African tradition of symbols used in ritual and magic, and personal communication, and Bukele would also have been broadly familiar with the idea of writing as a system of reproducing sounds by signs because the region had had Portuguese traders and the Roman alphabet since fifteenth century, Islam and Arabic was established by the beginning of the nineteenth century, the colony of Liberia had been founded 1822 fifty miles to the south-east, using English as its official language, and Christian missionaries had come from Sierra Leone. The new script was initially taught in schools, but these were destroyed in tribal warfare, so since that time it has been taught informally on an individual basis.

It is important to realise that the Vai were not merely subsistence farmers living in traditional villages in the bush. Trade and commerce were very important in Vai society, which was a stratified society with various forms of domestic servitude. 'One is tempted to speculate that the existence of an agricultural labour force supporting a number of free-born families played a role in constituting a social group with the time available to invest in systematisation and teaching of the script…Smith convincingly argues that the script's enthusiastic adoption by the Vai was due to the complexity of the economic and social transactions the people engaged in, and the dispersal of kin groups, friends, and trading partners across Liberia and Sierra Leone' [for which oral communications were insufficient]. Scribner and Cole 1981:269.

Only about 20% of the adult male population knew the script, which took them from about two to three months to a year to learn, depending on the man's abilities and circumstances. It was used for writing letters to friends, relatives, and business partners, sending news of funerals, so that people could bring gifts, and was also used by farmers and craftsmen for business ledgers and technical plans. The few men who might qualify as Vai scholars wrote family and clan histories, kept diaries, and recorded maxims and traditional tales in copybooks.

15. 'Once the idea of writing was understood, once its social potential was recognised, an island elite evidently usurped the idea as soon as possible to reaffirm their disintegrating role. Doubtless *rongorongo* began as a superficial appendage to priestly paraphernalia. [It was mainly used for recording fertility

chants.] It was the social symbolism of *rongorongo* to support those who were still in authority that was *rongorongo*'s principal *raison d'être*.' Fischer 1997:375–6. Special schools for teaching *rongorongo* to the sons of elite families were established, and since there were only about 120 basic signs, considerable progress could be made in only a few months. The 120 basic signs or 'glyphs' could be combined to produce several hundred compound glyphs, standing for sounds as well as words, but some signs could also be used as ideographs. The very limited uses of the script meant that it needed only to be capable of expressing a portion of the Rapanui language.

16. For the degree of literacy in ancient society see Harris 1989.

17. I am most grateful to Dr Stephen Chrisomalis for making his Ph.D Thesis *The Comparative History of Mathematical Notation* (2003a) available to me, and I have also benefited from his articles (Chrisomalis 2003b, 2004). The classification of number systems in this section is based on his work.

18. In the Middle East the cubit was the distance from the point of the elbow to the tip of the middle finger, and half this is the span (from the tip of the outstretched thumb to that of the outstretched little finger). The palm of the hand (across the width) goes 6 or 7 times into the cubit (the Egyptians said 7) and the width of 4 fingers or digits equals the palm, so for the Egyptians one cubit equalled 7 palms or 28 digits. The height of a man also equals the distance between the finger-tips of the extended left and right hands, or about 4 cubits – the fathom – while for measuring distance the pace was universal.

19 'In the Ganges valley and probably throughout most of northern India during classical times the standard unit was the weight of the Gunja creeper seed, *Abrus precatorius*'. (Mitchiner 1973:10). In mediaeval England, 'An English penny. . .round and without clipping, shall weight thirty-two wheat corns in the midst of the ear', and 'It is ordained that 3 barleycorns, round and dry, make an inch, 12 inches make a foot, etc.' Statutes of 51 Henry III (1266), and 17 Edward II (1324). There was nevertheless a good deal of variation from one state to another, and one of the functions of rulers was to establish standards for weights and measures. Measures of length were fundamental for commerce and land surveying, but weighing by scales was first used for precious metals – gold in Egypt, and silver in Mesopotamia and China – while ordinary goods were not weighed in the market-place for a very long time but were sold by measures of capacity.

20. Archaeologists have found significant numbers of small baked clay tokens all over the Fertile Crescent, dating from about 8000 BC onwards, which have a number of standard geometric shapes, such as cylinders, cones, spheres, and so on. Schmandt-Besserat (1992) has claimed that, because they appeared at a time and place where agriculture also originated, these tokens must have been used by early farmers for keeping records of the distribution of stored grain. She

believes this because she imagines, wrongly, that the introduction of farming would have immediately produced a ranked society, with chiefs responsible for redistribution. This is an utterly implausible development at such an early date in relatively simple societies, and an even more decisive objection is that, to my knowledge, no use of tokens for such a purpose has ever been found by anthropologists among farmers in modern tribal societies, even those with chiefdoms. If modern pre-literate cultivators are quite capable of organizing their farming without the use of tokens, why should we assume that farmers 8000 years ago could not have done so too? A much more likely explanation for these tokens is that they were used for divination and gambling, and there are many examples from all over the world of standard objects, both natural such as coffee beans and yarrow stalks, and manufactured, that are used for precisely these purposes.

21. So to express 843, for example, they would have written, from right to left, the hieroglyphic equivalent of:

UUU T T H H H H
T T H H H H

where U = units, T = tens, and H = hundreds.

1
2
3
4
5
6
7
8
9
1s
10s
100s
1000s
N/A
N/A

From Chrisomalis 2003b:486

So 843 would be written as 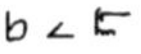 , (reading from right to left).

23. Chrisomalis 2003b.

24. Neugebauer 1969:19

25. Indian mathematicians were also concerned with commercial problems such as compound interest.

26. The Chinese also had developed a positional system, which illustrates very well the importance of the counting board in the development of positional notations. They used rods for their counting boards, and had developed a system of 'rod numerals' to write down the totals of their counting boards.

27.

Table 4. *Two-dimensional typology of numerical notation systems.*

| | Additive<br>The sum of the values of each power is taken to obtain the total value of the numeral-phrase. | Positional<br>The value of each power must be multiplied by a value dependent on its position before taking the sum of the numeral-phrase |
|---|---|---|
| **Cumulative**<br>Many signs per power of the base, which are added to obtain the total value of that power. | Egyptian hieroglyphic<br>1434 = 𓆼𓆼𓆼𓆼𓍢𓍢𓍢𓎊𓎊𓎊𓎊𓏺𓏺𓏺𓏺<br>$(4 \times 1 + 3 \times 10 + 4 \times 100 + 1 \times 1000)$ | Babylonian sexagesimal cuneiform<br>1434 = «𐤕𐤕𐤕  ‹:"𐤉'<br>$(2 \times 10 + 3) \times 60 + (5 \times 10 + 4) \times 1$ |
| **Ciphered**<br>Only one sign per power of the base, which alone represents the total value of that power. | Greek alphabetic<br>1434 = ͵αυλδ<br>$(1000 + 400 + 30 + 4)$ | Bengali<br>1434 = ১৪৩৪<br>$(1 \times 1000 + 4 \times 100 + 3 \times 10 + 4 \times 1)$ |
| **Multiplicative**<br>Two components per power, unit-sign(s) and a power-sign, multiplied together, give that power's total value. | Chinese (traditional)<br>1434 = 一千四百三十四<br>1 1000 4 100 3 10 4 | LOGICALLY EXCLUDED |

[from Chrisomalis 2004:42]

28. For example, 'Observations on the stars proved so successful in foretelling when to begin agricultural operations that the Sumerians were induced by the same means to predict the unpredictable. In other words astronomy led to astrology. . .' (Childe 1954:110).

29. Sen 1979:59.

30. Based on Coe 1999:59-63.

31. Neugebauer 1969: 102.

32. The Indians were generally obsessed by gigantic numbers in their astronomical and cosmic theories. The Great Cycle of 4,320,000 years had long been used for astronomical calculations, but this was small by comparison with the population of the world which was calculated as $2^{96}$, while Jain mathematicians postulated a number of the order of 8,400,000[28]. (In very much the same spirit, Archimedes developed a notation for immense numbers, using a base of 100,000,000, for calculating the number of grains of sand in the universe.)

33. Neugebauer 1969: 100.

34. Vinette 1986.

35. Chattopadhyaya 1986:152-3.

36. The following are examples of the sorts of geometrical problems involved:

*1.* How to construct (or draw) a square, the length of its side being given. The text gives two methods for the purpose, adding subsequently also a third one.

*2*. How to construct an oblong or a rectangle *(dirgha-catu-rasra)*, its length and breadth being given.

*3*. The proposition that the square on the diagonal *(aksnaya)* of a given square is twice as large as that of the given square.

*4*. To construct a square whose area is three times the area of a given square. The square on the diagonal of an oblong (rectangle) is equal to the sum of the two squares on the two sides. This is shown in the cases of the oblongs the two sides of which are (a) 3 and *4* ; (b) 12 and 5;(c) 15 and 8; (d) 7 and *24*; (e) 12 and 35; (f) 15 and 36. This, it may be noted, is in essence the proposition usually associated with the name of Pythagoras, though in our text we have a hint also of the formulation of the proposition in general terms.

*5*. The way of making a square equal in area to the combined areas of two other squares of different sizes.

*6*. The way of making a square having an area equal to the difference of two given squares.

*7*. To construct a rectangle (or an oblong) whose area is equal to the area of a given square.

*8*. To construct a square whose area is equal to that of a given rectangle (or an oblong).

*9*. The way of transforming a square into an isosceles trapezium, whose shorter side is given as lesser than the side of the square.

*10*. To construct a triangle equal in area of a given square.

*11*. To construct a rhombus equal in area of a given square.

*12*. The way of turning a square into a circle.

*13*. The way of turning a circle into a square. (Chattopadhyaya 1986:159-60) The mathematics involved in working out these problems is given by S.Ray in Appendix I (457-80), and are very far from elementary.

37. By Chattopadhyaya 1986, in particular, much influenced by Gordon Childe's version of materialism. 38. Chattopadhyaya1986:170.

39. Probably under Babylonian influence. He introduced some abbreviated forms of words for the unknown (our *x*), the exponents ( $a^2$, $b^3$ ) and a symbol for the minus sign, although he considered actual negative numbers absurd.

40. The *Epic of Gilgamesh* is quite unphilosophical, and deals with such basic themes as the wild and the tame, friendship, and the search for immortality, which are typical of the myths of tribal societies.

## Chapter X  Social Crisis and the Need to Think

1. In this connection it is interesting to note the conventional dates of the major religious and ethical thinkers of this period: Buddha c. 563–483 BC, Socrates 469–399, Confucius 551–479, Second Isaiah and other major Hebrew prophets c. 625-580. The conventional dates of the Iranian prophet Zoroaster, 630–553 BC, would place him squarely within the Axial Age, and his teachings were in many respects typical of the period. Unfortunately, the scanty evidence about his life is conflicting, and some of it points to a much earlier date, nor has very much about his life or the society in which he lived been preserved, so his position among the prophets and thinkers of the age must remain obscure. 'One line of inquiry, based on the linguistic antiquity and the socio-cultural allusions within the Gathas [earliest hymns] puts the date of Zoroaster anywhere between 1500 and 1000 BCE. Another line of inquiry, based on literary, historical, and theological evidence, puts the date of Zoroaster anywhere between 900 and 400 BCE.' (Nigosian 1993:16, and see Boyce 1975:3, 190-1.)

2. A term first used by Karl Jaspers 1953. See especially pp.1-21. See also Eisenstadt 1986.

3. Momigliano 1975:8-9.

4. But not elsewhere among early states, which retained the various forms of primitive money such as cowrie shells, cacao beans, bars of salt, and so on. These were quite adequate for local markets but for large payments and long-distance trade gold and silver were more appropriate.

Since metal, unlike shells or beans, does not come in fixed sizes the easiest way of establishing a standard amount is to weigh it. By around 2000 BC the Babylonians had established the gold shekel at 8.18 gms. One gold shekel was equal to ten heavy silver shekels of 10.91 gms., or fifteen light shekels of 7.27 gms. These were not actual coins, however, but weights that merchants would check on their scales, and ingots of standard weight could then become accepted units of trade. Coinage began when a symbol of guaranteed weight was stamped into an ingot, and the Lydians of Asia Minor, a trading kingdom with skilled metallurgists and control of large deposits of gold and silver, seem to have been the first, around 630 BC, to stamp ingots of bullion with the lion's head emblem of the royal dynasty. These rapidly evolved into round coins that were soon copied by many Greek city-states and by Persia, and subsequently spread more slowly to Italy. Gold and silver coins were of high value and it supposed that they may have been used by the state for paying mercenaries and for taxation, and by merchants dealing in bulk, not for everyday transactions in the market-place, but bronze coinage with much lower values developed later.

India had long been engaged in trade from Mesopotamia to the trading kingdom of Taxila across the Iranian-Afghan plateau, and across the Indian Ocean to Avanti, and by the beginning of the sixth century BC stamped silver ingots of the same weight as the Babylonian heavy shekel had begun appearing in Taxila, and the light shekel further south. (Mitchiner 1973) So it seems that the two basic weights of Indian coins were of Mesopotamian origin, while the Lydian idea of stamping the ingots came from Asia Minor via Mesopotamia.

In China coinage developed in the opposite manner, not from gold and silver but from bronze imitations of cowries as far back as the Shang period in the twelfth century BC. At this stage the coinage was not stamped or inscribed. By the seventh century BC miniature bronze knives, spades, and hoes were being minted with inscriptions sometimes giving the place of origin and the denomination, and this seems to have been an initiative of merchants rather than the state. Round coins with holes in them developed in the second century BC and then spread though East and South-East Asia. But gold and silver were never used for coins until the late nineteenth century AD and before then these metals, especially silver, were weighed and cut up when used for payments as had been done in Mesopotamia.

5. As we shall see, the widespread use of money in daily life also acclimatises people to the use of number and measurement and applying these to the physical world. Another revolutionary aspect of coins was that for the first time they made it possible to detach the nominal value of the medium of exchange from the actual value of the 'stuff' from which they were made, so that the state could now to some degree control the supply of money by other means than simply mining more of it. The very early use of banknotes by the Chinese (c.6$^{th}$ century AD) was a further extension of this principle. They were originally receipts issued by merchants for large deposits of coins by their customers, and were only later taken over by the state. The debasement of the coinage and other artificial increases of the money supply by use of bank notes were further contributions to the inherent instability of states.

6. Bellah 1970:35.

7. Garelli 1975:50, and Oppenheim 1975:38. 'The scribes. . .far from establishing an independent cultural and political base, simply continued to serve the ruling groups who had brought them into existence in the first place, and who remained in control of public access to them thereafter.' (Machinist 1986:202). While there was certainly some discussion of the meaning of life, it does not seem to have been philosophically sophisticated. For example, in a dialogue between a master and his slave from eighth century BC Assyria, the slave says, 'Nothing is of any use, including sacrifice.' 'Well, what is good, then?' asks his master. 'To break my neck, thy neck, to throw oneself in the river, that is good.'

8. There had been debate about the fundamental issues of life, man, and the gods among the Aztecs in the century preceding the Spanish conquest. Evidence for this comes, in particular, from 'the hundreds of folios written by some of the natives in the Nahuatl language using the Latin alphabet they had learned, by which means they preserved the poems, songs, chronicles, and traditions they had been made to learn by rote in their schools prior to the Conquest'. (León-Portilla 1964:36). The Aztecs had imposed their form of Nahuatl culture on the Toltecs by conquest in 1438, and there seems to have been much debate among the aristocratic wise men about the value of warfare and blood sacrifice, and a more general pessimism and scepticism. There are important parallels here with some of the thinking in the Axial Age, but it seems to have been expressed in an essentially poetic idiom, with none of the rationalistic analysis that was so important in the thought of the Old World.
9. Humphreys 1975:112.
10. They 'originated from various parts of the Greek world: Protagoras from Abdera, Gorgias from Leontini, Prodicus from Ceos, and Hippias from Elis, to name the four most important. They travelled from city to city, imparting their instruction for very substantial fees to the young men of the wealthier citizen families, and they naturally tended to gravitate to Athens. The central subject of their courses was rhetoric, the art of persuasion by eloquent speech, and they claimed to be able to teach their pupils to speak persuasively to any "brief", to argue both sides, of any case. With rhetoric went the subordinate verbal arts, grammar and the study of the meanings of words . . . and a general emphasis on all the elegant arts and accomplishments of life. An all-round virtuosity and a supreme skill in the art of persuasion were their ideals for themselves and their pupils. . . Their attitude to life might be described as a humanist agnosticism. Protagoras . . . expressed it very well in a couple of sentences which have been preserved: "Man is the measure of all things, of the reality of those which are, and the unreality of those which are not"; "As to the gods, I cannot know whether they exist or not; too many obstacles are in the way, the obscurity of the subject and the shortness of life".' (Armstrong 1968:23).
11. Rhys Davids 1955:71-2.
12. Kroll 1985-87:118.
13. See also Lloyd 2002:131-2.
14. Graham 1978:15.
15. Dasgupta 1932:64.
16. Hourani 1985:95.
17. Humphreys 1975:91.
18. See, for example, Scharf 1996.
19. Cornford 1950:43.

20. For example, 'Because he believed that one could not have a science of politics without clarity of language, over and over Kautilya tried to define matters rigorously.' (Boesche 2002:31). It is in this sense that when Confucius was asked to take office he said 'Would it not be necessary to correct names? . . .If names are not correct then one's words will not be in accord [with one's actions]. If words are not in accord, then what is to be done cannot be [correctly] implemented. . .' and ultimately the social order fails. Schwartz 1985:93.

21. *Republic* 444e.

22. . . . whoever is devoid of the heart of compassion is not human, whoever is devoid of the heart of shame is not human, whoever is devoid of the heart of courtesy and modesty is not human, and whoever is devoid of the heart of right and wrong is not human. The heart of compassion is the germ of benevolence *[jen]*; the heart of shame, of dutifulness *[i]*; the heart of courtesy and modesty, of observance of the rites *[li]*; the heart of right and wrong, of wisdom *[chih]*. Man has these four germs just as he has four limbs. For a man possessing these four germs to deny his own potentialities is for him to cripple himself. (*Mencius* 2.A.6).

23 The Buddhist virtues were developed in the context of the fundamental teaching that suffering was rooted in desire and in egotistical selfishness: faith (*saddha*), watchfulness (*sati*), energy (*viriya*), contemplation (*samadhi*), vision (or wisdom, *panna*).

24. 'He has not only to put a curb on ethically wrong actions but also, through conscious effort, to constantly train his mind to deter it from harbouring ethically wrong notions and desires. There should be perfect harmony between his actions and his thoughts, ethically pure actions springing forth from an ethically pure mind.' Misra 1984: 92.

25. In Confucianism, for example, the idea of introspection was basic to the attainment of virtue. It had three main aims: it taught a person to recognise his own failings and not to delude himself that he had no faults; it taught him to understand his own character, and 'Thirdly, people basically resemble one another so in order to know how to treat them, a person must look within and try to understand how he would feel in a similar situation'. Munro 1969:95.

26. One is not saying that this was the first time that the idea of conscience had appeared, but rather that it now became a central feature of moral awareness. In Mesopotamia and Egypt there had certainly been some idea of inner-self-condemnation among the educated elite; this, for example is a copy-book maxim from Babylonia in the eighteenth century BC: 'the thievish boy said to his mother, "come on, nobody's looking". But his mother answered, "he that sees thee is carried (in thy heart)". If the last phrase is correctly translated, this is as near a description of 'conscience' as we have from antiquity.' (Gadd

1948:83)
27. Finley 1973:45.
28. Tambiah 1976:20.
29. Ibid, 21.
30. Rhys Davids 1955:159.
31. Boesche 2002.
32. Thapar 1975:123.
33. The Brahmins were teachers of the sacred scriptures and advisers as well as sacrificers to the king: the Kshatriyas were the warrior nobility led by the king whose characteristic virtue was courage but who could acquire wisdom from their teachers; and the Vaishya, the farmers, craftsmen, and merchants who were concerned only with bodily needs, though we saw that Kautilya disapproved of merchants, and of accumulations of private wealth in general. The Shudra, the servants of the higher castes and forbidden even to hear the Vedas, of course were outside the moral scheme entirely. Since it was believed that the caste into which one was born reflected one's moral and spiritual development in previous lives it represented a scale of virtue in this respect as well.
34. Ames 1993:87.
35. Munro 1969:2.
36. The Muslim philosopher al-Farabi (A.D.870 - 950) later wrote in a similar vein:
'It is through this diversity [of natural aptitudes among individuals] that various classes arise within the state, which is a necessary form of association answering man's basic needs, which he cannot gratify without the assistance of his fellow men. Being analogous to the human body, the state requires a ruler, together with a series of subordinates, corresponding to the heart and the subordinate organs of the body respectively.' (Fakhry 1970:143).
37. Bodde 1991: 373.
38. *Republic* 369.
39 This obvious similarity has of course been noted before by many scholars. Plato's rulers, the Guardians, are the epitome of reason and Brahminic virtue, owning no property because wealth is a temptation to vice, and controlling and keeping in balance the other two classes who represent the warlike spirit and the physical appetites, neither of which are able to govern themselves through lack of reason. The Auxiliaries, who carry out the military, police, and executive functions of the State are the equivalent of the Kshatriyas; the third class of farmers, tradesmen, and artisans are the equivalent of the Vaishyas; while the slaves are essentially the Shudras. The only difference between this and India is that Plato's rulers are selected, not born to their position. Plato's notion of justice, which he defines as 'minding your own business', that is,

fulfilling your proper function in society, is also essentially like that of the Indian *dharma*, in which each caste should perform the duties appropriate to its place in the social hierarchy. (The *Laws*, written much later, takes a more pragmatic view of how the state should be organized, such as the need for the rule of law and a system of checks and balances to prevent the misuse of political authority.)
40. *Politics*, 1295b35-40.
41. Randall 1962:266.
42. Rostovzteff 1957:46.
43. Aristotle's idea of a 'ladder of souls', whereby plants have a vegetative soul, animals a sensitive soul as well, while man has a rational soul in addition to the vegetative and sensitive souls, appeared in China and India.
44. Baldry 1965: 53.
45. ibid., 56.
46. Baldry 1965:19.
47 Ibid., 33.
48. 'Equity bids us to be merciful to the weakness of human nature; to think less about the laws than about the man who framed them, and less about what he said then about what he meant; not to consider the actions of the accused so much as his intentions.' 1374b9-23.
49. Needham 1956:520.


## CHAPTER XI THE NEW RELIGIONS
1. Schwartz 1985:83.
2. Bellah 1970:35.
3. In the previous chapter we noted the great importance of debate in religious controversy, which can be summed up by Aquinas, who said of debate between Christians, pagans and Muslims 'We must, therefore, have recourse to natural reason, to which all men are forced to give their assent'. While Muhammad did not resort to argument or disputation, this is reflected in the word for 'theology', *kalam*, which meant 'speech' or 'conversation': 'it is based upon the idea that truth is found via a question and answer process. Someone proposes a thesis, and somebody else questions it'. (Leaman 1985:8). When the Muslims became acquainted with Greek philosophy as the result of their conquests of hitherto Christian regions they greatly admired Aristotle's work on logic, which they adopted despite the great cultural differences between Greek and Islamic culture. At the Abbasid Court of Baghdad in the eighth and ninth centuries debate was fostered by the Caliphs: "The court were not so much interested in theology itself as in listening to disputation: they liked to have representatives of different religions and confessions [Zoroastrianism,

Buddhism, Christianity, Judaism, and Islam] argue against each other.' (van Ess 1987:221-222) The Mughal Emperor Akbar (1556-1605) also established a similar institution in which representatives of all the religions of India were invited to take part in debate.

4. Dodds 1963:192-3.

5. Armstrong 1968:6. The obvious resemblance between Orphic and Indian ideas has been noted many times, and there are good reasons to believe that Orphism was strongly influenced by Indian thought: its main beliefs were both very different from traditional Greek ideas and very similar to Indian ones, and Orphism seems to have come from Thrace, between the Greek world and the Persians, whose empire in turn abutted on northern India. Indeed, the domain of the Emperor Darius (521–486 BC) included both Thrace and part of north-west India.

6. Bellah 1970:34.

7. The history of Islam is dominated by conquest and warfare from the very beginning until the nineteenth century: the conquest of Syria 636 AD; Egypt 642; Persia 637-51; North Africa (the Maghreb) 639; Spain 710-97 (but driven out of France at the battle of Tours 732); the prolonged invasion of India beginning in 710; the capture of Constantinople 1453; the invasion of Abyssinia (Ethiopia) 1517; and a permanent state of *jihad* with Western Europe, including the unsuccessful siege of Vienna in 1683, until 1816 when the fleet of Sir Edward Pellew destroyed Algiers. On the other hand, it should be pointed out that Islamic rule tolerated other religions, albeit in a subordinate status, whereas Christianity, until modern times, did not.

8. 'Heraclitus' assumption that it is one and the same *logos* which determines patterns of thought and the structure of reality is perhaps the most important single influence upon Stoic philosophy.' (Long 1986:131). 'The Stoic universe is a world determined by law, by immanent *logos*. This is a fundamental concept in Stoicism, and it runs through all three aspects of their philosophy [physics, logic, and ethics]. After all, these are only aspects, ways of presenting something which in the last resort is a unity – nature, the universe, or God.' (Ibid.,144).

9. Hussey 1972:47.

10. Randall 1962:105.

11. In early Greek thought the word that came to be used for the soul in this new sense, *psyche*, merely meant the life-principle, located in the head which was also believed to be the source of semen. While it survived bodily death and went to Hades, it was not associated with consciousness: this was *thymos*, and was associated with the breath (and so with thought), but this disappeared at death. (For a full account see Onians 1954.) The first school of thought in Greece to break with this tradition entirely was the Orphic

cult in the fifth century BC which, apart from denouncing the traditional Greek gods, and the limitations of the city-state, preached that man's soul is of divine origin, imprisoned within the impure and polluting body. The Orphics were influential in some parts of Greece, especially Athens, and in southern Italy, where they had important links with the Pythagoreans, and who in turn had a powerful influence on Plato and Socrates. The Pythagorean physician Alcmaeon concluded from his dissection of the head, and the connection between the optic nerves and the brain, that the head is the seat of consciousness and reason, and in the teachings of Socrates and Plato the soul, *psyche*, is the divine and immortal element of reason that resides in the head, and should rule the body. Plato says in the *Phaedo*, 'The body fills us with desires, passions, and fears, all kinds of imaginings and nonsense so that we can never understand by means of it anything in truth and reality. . . if we are ever to have any pure knowledge we must escape from the body and consider things in themselves with our soul alone'.

12. Benveniste 1973:232.
13. Dasgupta 1932:27.
14. Ibid.,26-7.
15. Ibid.,42 In India there was an idea in Vedic times that the non-material part of the person survives death and flies like a bird either to the realm of the ancestors or the gods, depending on the quality of the rituals they had performed on earth. In the later period that concerns us (600 – 300 BC), 'behind or beyond all of the whirling flux of one's personal existence, deep within the living and dying body, exists an eternal, unchanging, intelligent, incorporeal, and joyful "self" [*atman*], and they saw that this essential reality is identical to the very ground of being of the universe itself [*bráhman*]'. (Mahony 1987:440).
16. In the words of the Taoist philosopher Chuang Tzu: 'The Tao has reality and evidence, but no action and no form. It may be transmitted but cannot be received. It may be attained but cannot be seen. It exists by and through itself. It existed before Heaven and Earth, and indeed for all eternity. It causes the gods to be divine and the world to be produced. It is above the zenith but it is not high. It is beneath the nadir but it is not low. Though prior to heaven an earth, it is not ancient. Though older than the most ancient, it is not old.' (Cited in Needham 1956:38).
17. Needham 1956:582. Bodde points out, however, that while this is certainly correct, there are a few passages in early Chinese texts 'that in varying ways suggest the existence in early China of ideas not too far removed from those that in Europe led to the developed concept of "laws of nature"' (Bodde 1991:334). The recently discovered silk manuscripts of Huang Lao (lost for 2000 years) show that this influential philosopher of the Former

Han also believed in a version of natural law comparable to that of the Stoics. (Peerenboom 1993).
18. Needham 1956: 581.
19. Munro 1969:155. In China at this period there was also an idea of the soul that was somewhat similar to that which developed in India and Greece. The earthly soul *p'o* comes into existence with life itself, while the heavenly soul, *hun*, emerges more gradually. *Hun* is the spirit of a person's vital force that is expressed in consciousness and intelligence, and *p'o* is the spirit of a person's physical nature that is expressed in bodily strength and movements. Both *hun* and *p'o* require the nourishment of the vital forces of the cosmos to stay healthy. When a person dies a natural death, his or her *hun* gradually disperses in heaven, and the *p'o*, perhaps in a similar way, returns to earth.
20. Whybray 1973:8.
21. Ibid.,10.
22. Crombie 1996:470.
23. Dodds 1963:189-9.
24. Needham 1956:417.
25. Eamon 1994:60. 'In order to protect the secrets of nature from man's prying eyes, Lactantius pointed out, God made Adam the last of his creations so that he should not acquire any knowledge of the process of creation.' On the other hand, we shall see there was another tradition that God had revealed all the secrets of nature to Adam, and that this knowledge could be recovered by human investigation.
26. White 1978:27.
27. Needham 1956:582.

## Chapter XII Natural Philosophy

1. Materialist historians of science, such as J.D.Bernal in his *Science in History*, think of science as just an offshoot of technology: 'Because the *essential character* [my emphasis] of natural science is its concern with the effective manipulations and transformations of matter, the main stream of science flows from the techniques of primitive man. . . ' (1969:61, and see also pp.30-31.) This is a serious confusion. It is quite normal, for example, to be familiar with levers and boats at a craft level, and yet be quite unable to explain how they work, or have any interest in doing so. But without making this basic distinction between knowing what and knowing why, science simply merges into technology, and early man making flint axes also becomes a kind of scientist. The kind of theoretical explanations of nature that we shall be looking at in this chapter, however, were radical innovations in human history, and the significance of this will completely escape us if

we do not draw a clear distinction between science and technology, just as we must distinguish between naturalistic and mythical explanations of the physical world.

2. In China, 'The Taoists, though profoundly interested in Nature, distrusted reason and logic. The Mohists and the Logicians fully believed in reason and logic, but if they were interested in Nature it was only for practical purposes. The Legalists were not interested in Nature at all. [It was in fact the Mohists, reputedly recruited originally from the artisan class, who of all the ancient Chinese schools of philosophers did most in the scientific sphere, and their influence did not last very long.] The Confucians were only interested in the reform of society and never discussed problems of natural science, and their lack of interest in nature was to exercise a permanent influence on pre-modern Chinese culture. (Needham 1956:580). A story dating from somewhere between the fourth and first centuries BC illustrates what Needham calls Confucianism's 'chilling lack of interest in scientific problems': one day, Confucius came across two boys who were arguing about the sun. One said that it was closest to the earth at dawn because it was obviously much larger than it was at noon (In fact this is an optical illusion). The other maintained that it must be closest to the earth at noon, because it was much hotter then than at dawn. They asked the Sage which was the right answer, but he was unable to tell them, so the boys laughed at him, saying 'Why do people pretend that you are so learned?' (Needham 1959:225-6).

Socrates said that 'speculators on the Universe and on the laws of the heavenly bodies were no better than madmen'. He was a major influence in changing the focus of philosophers' attention from physics to ethics. Xenophon, in the *Anabasis,* says : 'With regard to astronomy Socrates considered a knowledge of it desirable to the extent of determining the day of the year or of the month and the hour of the night; but as for learning the courses of the stars, occupying oneself with the planets or enquiring about their distance from the earth or about their orbits or the causes of their movements, to all these he strongly objected as a waste of time'. See also Plato's *Phaedo,* where just before his death Socrates is described as lamenting the time he wasted on science in his youth. From the time of Zeno of Citium, the founder of Stoicism, and Epicurus, ethics dominated logic and physics.

3. 'We know that there was in ancient India a large amount of literature dealing with the practical affairs of life, with technical arts and crafts, and with specific sciences. Much of this has been lost; a large part of what has been preserved is still unedited; and most of the edited texts have not yet been studied critically…Critical study is difficult because of the unhistorical character of Sanskrit literature. . .This lack of precise dating makes it difficult to compare Indian achievements with the achievements of the Greeks and

the Arabs, or to determine priority of invention or the direction of borrowing if borrowing can be proved.' (Clark 1937:335). While considerable progress has been made since Clark wrote this (see for example Bose 1971, Sen 1979, Chattopadhyaya 1986), we still know very much less about the development of science in India than in China. It should be noted that, in the case of the history of Chinese science, the early development of printing (which never occurred in India) was responsible for the preservation of almost all the ancient texts that survive, and there was also a very strong historical bent in traditional Chinese culture, which was not the case in India. Perrett 1998 considers that a major reason for this is that traditional Indian philosophy did not regard even written history as a source of real knowledge, because it relied only on hearsay, not on actual evidence that could be produced.

4. We are given a useful clue to this agricultural origin of the elements by an early Chinese text which attributes to the Sage-King, the Emperor Yü, the saying that 'Virtue is seen in good government. Good government is proven by its capacity to maintain the people. There are water, fire, metal, wood (vegetation), earth, and grain.' Grain was later omitted by Chinese thinkers – as a product of agriculture rather than a foundation – leaving the five elements of water, fire, metal, wood, and earth (five was a very important number in China). At about the same time as the Greeks and Chinese, Jain, Hindu, and Buddhist philosophers in India were also formulating theories of the basic elements. From an early date the four elements were earth, water, fire and wind (rather than air) and it seems very likely that the 'air' of the Western elements was originally the wind, something crucial for agriculture, just as water would have been immediately experienced as rain, and fire through cooking. Some proposed that the Absolute Being created fire, fire created water, and water created earth; wind was the fourth element, and space was later added because it was considered that there should be five elements to correspond with the five senses.

The elements also became dynamic in Chinese thought: the word we translate as 'element', *hsing*, really means something like action or movement, and the five *hsing* were early associated with the 5 directions (from which different winds come) and the five planets. In China Tsou Yen (350 – 270 BC), a contemporary of Aristotle, first formulated a five-element model which, like that of the Greeks was a very simple and unified theory attempting to account for the appearance of things and the forces that accounted for their changes:

| | |
|---|---|
| Water | soaking, dripping, descending |
| Fire | heating, burning, ascending |
| Wood | accepting form by cutting and carving |
| Metal | accepting form by being melted and moulded |
| Earth | producing edible vegetation |

The elements were converted into a dynamic system of principles: so metal overcomes wood, as when a sharp axe fells a tree; fire melts metal; water extinguishes fire; earth conquers water by absorption; and wood/vegetation overcomes earth by abundant growth, thus producing a constant cycle of change. In the scheme of Tsou Yen and the Naturalists each dynasty had its special element and colour, and because the elements dominated one another in a cyclical order this proto-science of correlations was applied to the world of human affairs. The School of Naturalists, rather like that of the Pythagoreans, became for a time an influential political force until they were obliterated by the victory of the Legalists in the First Unification.

5. Empedocles, (c.495-435 BC) however, refused to reject the evidence of the senses – they might mislead us, but so, too, could the mind – but he took very seriously the argument that matter could not be created or destroyed, and only appeared to change. He argued that matter had always existed, but in the distinct forms of the four elements which he called the basic roots, *rhizomata*. They had always existed, and could not be changed or decomposed into one another or into anything else, as the earlier theories had maintained. Instead, like the primary colours of a painter, they could be *combined* in a limitless variety of ways, thus getting round Parmenides' main argument against the reality of change. He argued, most significantly, that it was the different proportions in which the elements combined that produced the characteristic features of all the compounds: so bone, for example, was the combination of fire, water, and earth in the proportions of 4 : 2 : 2. This anticipates the modern chemical Law of Fixed Proportions, which states that chemical compounds contain their constituent elements in fixed and invariable proportions by weight. (Lloyd 1970:42). The whole of existence consists of the ceaseless combination and dissolution of things, of many into one, and one into many. This process is brought about by the opposed forces of change, Love and Strife, or attraction and repulsion, and he developed an elaborate account of how the cosmos had been generated in this way.

6. 'He saw that the cyclical changes of nature bring into being and destroy certain opposed qualities. As the seasons of the year change, the fundamental opposites – cold and hot, wet and dry – give place to each other. These are not abstract opposites, but are attached to certain physical states. The ascendancy of the one gives rise to "injustice", but the course of time puts this right by

giving the opposite its turn to be in the ascendancy. None of the opposites can achieve absolute domination and annihilate the others. In this respect they are all limited, i.e. finite in place and time. But they are all products of the unlimited primordial matter which is an infinite reservoir of those never-ending mutations.' (Sambursky 1963:8).

7. Lloyd 1970:22.

8. Infinitesimals gave enormous problems to Greek philosophers, which are too complex to discuss here. Zeno of Elea (see Lee 1936) defended Parmenides' theory that motion is impossible with his famous four paradoxes of motion (one of which, Achilles and the Tortoise, is still familiar to us) and which exploited the same confusion between geometrical lines and real space and time. He was trying to understand a basic problem of moving objects: can space and time be sliced up into infinitely small distances and instants, or must these slices be thought of as having a minimal size, rather like a set of frames in a film-strip of a galloping horse? Zeno seemed to show that *both* these alternatives led to absurdities, and while Aristotle tried to dismiss them in commonsensical fashion, Zeno's paradoxes could only be resolved by modern mathematics using the calculus. Parmenides' and Zeno's technique of proving that a theory is true by showing that its opposite leads to absurdity also became a standard logical tool in philosophy and mathematics.

9. The major Atomists were Leucippus and Democritus (c.450-400 BC), Epicurus (341-270 BC) and much later Lucretius in the first century BC.

10. Atomism also developed independently in India at about the same time as among the Greeks. They do not seem (at least initially) to have made use of the argument of the impossibility of the infinite division of matter; instead, they argued that just as the idea of the absolutely or infinitely large (space) had to be accepted as a logical necessity, so too the idea of the infinitely small had to be accepted on similar logical grounds, and the infinitely large and the infinitely small also had to be eternal, and therefore indestructible. The notion of these irreducibly small and indestructible entities that could combine to produce the different substances of the physical world then tied in with the theory of elements.

11. Sambursky 1963:111.

12. When Lucretius talks about a law of nature he does not refer to a *lex naturae*, but a *foedus naturae*: *foedus* basically means a contract or covenant, as opposed to a law, *lex*, laid down by the authority of the people in the assembly. (Lex was therefore entirely in keeping with the Stoic view of Nature as 'City of men and gods'.) *Foedus* seems very close to our image of atomic and chemical *bonding*, which the atoms sort out among themselves with no reference to a master plan.

Lucretius also says (*De Rerum Natura*, I, 1026–1028) 'thus by trying every kind of motion and combination, at length they fall into such arrangements as this sum of things consists of.' (*omne genus motus et coetus experiundo tandem deveniunt in talis disposituras, qualibus haec rerum consistit summa creata*). But we should not read into this some precocious anticipation of statistical, probabilistic thinking, which was unknown in the ancient world, as we shall see. The basic idea seems to be simple trial and error, rather like finding a key in one's house, and going round all the doors to find which lock it fits. Simple elimination of this sort does not need any understanding of the law of large numbers.

13. A distinction that would be revived by Galileo in the seventeenth century. The atomists' actual attempts to relate specific tastes, colours, and smells to particular configurations of atom were not very convincing, however: angular, small, thin atoms tasted acid; round, moderate sized one tasted sweet; and atoms with sharp angles tasted sharp, for example – naïve associations for which there was not the slightest evidence.

14. Needham 1962:13 notes the lack of atomic theory in the history of Chinese thought, and suggests that the lack of an alphabetic script may have been a factor in this.

15. As Plato said of the Atomists, 'For on a shortsighted view, the whole moving contents of the heavens seemed to them a parcel of stones, earth, and other soulless bodies, though they furnish the source of the world order! It was this that involved the thinkers of those days in so many charges of infidelity and so much unpopularity... but today the position is reversed.' It had been reversed by what Plato had already described as 'The doctrine of the orderliness in the movements of the planets and other bodies surveyed by the mind that has set this whole frame of things in comely array.' (*Laws*, Bk.8, 967c, 966e).

16. Chomsky 1980:250.

17. Fr. 164. 'When we do find [the formula 'like influences like'] stated in the earlier philosophers, we must not mistake it for a principle derived by any rational process from the observation of nature...The philosophers themselves appeal to current proverbs and the gnomic wisdom of the poets... the moral and human meaning of the maxims is older than their application to physics...' (Cornford 1931:46-7).

18. This table from China illustrates exactly the same sort of associations:

| | | | | | |
|---|---|---|---|---|---|
| **Elements** | earth | wood | fire | metal | water |
| **Colours** | yellow | green | red | white | black |
| **Directions** | centre | east | south | west | north |
| **Metals** | gold | lead | copper | silver | iron |
| **Planets** | Saturn | Jupiter | Mars | Venus | Mercury |
| **Tastes** | sweet | sour | bitter | acrid | salty |

19. Anaximenes, for example, 'explained' lightning by comparing it to the flash of an oar in water. Just as the sea flashes when cleft by oars, so clouds flash when cleft by the wind rushing out, and the thunder is the noise made by the clouds when torn by the wind. But as Jowett says, while analogy was a great source of error, it was also the beginning of truth. '…without this crude use of analogy, the ancient physical philosopher would have stood still; he could not have made even "one guess among many" without comparison. . .They were bringing order out of disorder, having a small grain of experience mingled in a confused heap of *a priori* notions. And yet, probably, their first impressions, the illusions and mirages of their fancy, created a greater intellectual activity and made a nearer approach to the truth than any patient investigation of isolated facts, for which the time had not yet come, could have accomplished.' (Benjamin Jowett, commentary on Plato's *Theaetetus*, pp.383-5.)

20. Euclid's system was constructed on the basis of some simple *definitions* of lines, squares, circles, etc; some elementary assumptions ('*postulates*') such as that two straight lines can't enclose a space; and some self-evident truths ('*axioms*') such as the whole is greater than the part, and if two things are equal to the same thing, then they are equal to one another.

21. On Pythagoras and his school, see in particular Kirk, Raven & Schofield 1983:214-238; 322-50.

22. E.g. Philolaus of Croton (born c.470 BC), Hippasus of Metapontum (c.450 BC), and Archytas of Tarentum, a contemporary of Plato (c.429-347BC). Aristotle (384-322 BC) also gives an important account of their ideas, which strongly influenced Plato.

23. They were familiar with prime numbers, which are of exceptional importance in mathematics; and perfect numbers, which are the sum of their factors, such as six (1 + 2 + 3), 28, and 496. It is also possible that they first discovered that $\sqrt{2}$, the ratio of the diagonal to the side of a square, is an irrational number. It is not a normal fraction or ratio, with which, of course, ancient mathematicians had long been familiar, but continues for ever without resolution: 1.4142 etc., (and $\pi$, the ratio of the circumference of a circle to its diameter, was another example: 3.1428 etc.) Irrational numbers were extremely disturbing because they revealed that there are limits to the possibility of explaining the structure of the cosmos in terms of the whole numbers and the ratios between them.

For the Pythagoreans there was a fundamental connection between number and shape that led beyond geometry to types of number: [29] 'triangular numbers', for example, are the result of adding one to the bottom row of a triangle each time:

this produces a particular *series* of numbers: 1, 3, 6, 10, 15, 21, and so on, and series are of fundamental importance in mathematics. But if instead of adding 1, 2, 3, etc. one adds only the series of odd numbers (1, 3, 5, 7 etc) to one, the result is a square, and if one adds the series of even numbers to two, the result is an oblong:

The relative proportions of the square obviously remain the same all the time, whereas with oblong numbers the ratio of length to height changes each time so that in this respect such numbers are infinitely variable. For the Pythagoreans odd numbers were therefore bounded, stable, male, and good, while even numbers are unbounded, unstable, female and bad. (There are also further shapes of number based on different series.)

24. Heath 1981:144-49.

25. In the Renaissance they inspired the astronomer Kepler in his work on planetary orbits, and the artist Jamnitzer to believe that all living shapes were based on the Five Solids.

26. An example of proportion is '12 is to 6 as 6 is to 3', or 12 : 6 :: 6 : 3, because 12 is twice 6, as 6 is twice 3. Here 6 is referred to as the mean, and there are many different forms of proportion, all expressed in terms of the mean. For example, the arithmetic mean (e.g. 3 : 2 : 1), the geometric mean (e.g. 4 : 2 : 1), and the harmonic mean (e.g. 12 : 8 : 6), of which the cube is an important example since it has 12 edges, 8 corners, and 6 faces.

One of the most important means known to Plato and Euclid was that of 'extreme and mean proportion', which today we call the Golden Section. If we imagine a line AB 13 inches long, and which is cut at C, so that AC is 8 inches long, and CB is 5 inches long, as in A_____C_____B, then the ratios of the lengths will be 13 : 8 :: 8 : 5. Putting it more generally, the whole line AB is to the larger segment AC as this is to the smaller segment CB. This turns out to be, in modern form, 1.6180339887. . . or conversely as 0.618 . . . another irrational number, like π or √2. This ratio can also be expressed in the series 1, 1, 2, 3, 5, 8, 13, 21 . . . in which each number is the sum of the two previous numbers. This became known in the Renaissance as a Fibonacci series, and expresses some of the profoundest laws of growth of plants, animals, populations and so on.

These ratios were the principles of musical harmony and proportion, of reason and beauty, that were applied not only throughout music and acoustics, painting, sculpture, and architecture in the Greek world, but to the cosmos as a whole. The distances of the planets from earth, for example, were believed to be in harmonic ratios, so that as they revolved they produced 'the music of the spheres'.

27. What I refer to as Plato's 'Forms' are often, but confusingly, called 'Ideas'. 'It is well known, but cannot be too often repeated, that the word Idea in this connexion is a very misleading transliteration, and in no way a translation, of the Greek word *idea* which, with its synonym *eidos*, Plato frequently applies to these supreme realities. The nearest translation is "form" or "appearance", that is, the "look" of a person or thing. . . "theory of forms" is much nearer the Greek. . . ' (Grube 1958:1).

28. 'This philosophy led Plato, and all the subsequent generations educated on his doctrine, to treat quantitative experimentation with contempt and to give up the aim of expressing physical facts in terms of number.' Sambursky 1963:44.

29. It has been said of the Lyceum under Aristotle, and afterwards Theophrastus and Strato, that 'Here for the first time in the ancient world we can talk of a corporate research effort, planned and implemented as a whole.' There were histories of philosophy, medicine, mathematics, and astronomy, a vast collection of the constitutions of the different city-states, and major contributions to botany, geology, and zoology. Some members of the Lyceum also contributed to early studies of mechanics. 'Indeed, apart from in the Alexandrian Museum, the scope of natural scientific investigations undertaken in the Lyceum was to remain unsurpassed in the whole of antiquity.' (Lloyd 1979:201-2).

30. Lloyd 1966:52. This naturally seems laughable to us, but the ability to deny the obvious if it challenges some fundamental principle is always with us. For example, the great French sociologist Émile Durkheim maintained, as a fundamental principle, that every regularly recurring feature of society must be adaptive, or else it would not have survived. How, then, can this principle explain crime, which is also a universal phenomenon? Because the abhorrence of crime among the law-abiding majority is reinforced whenever a crime is committed, so that crime actually produces greater social cohesion. While this may sometimes be the *effect* of crime, as an explanation of *why* crime exists it must rank with Aristotle's upside-down plants.

31. The Chinese were not interested in dynamics at all. They assumed that motion was natural to matter, but did not think, as did the Greeks, of the components of the cosmos physically acting on one another, but as behaving rather like the members of a dance, performing in a pre-ordained harmony: 'Things behaved in particular ways not necessarily because of prior actions or

impulsions of other things, but because their position in the every-moving cyclical universe was such that they were endowed with intrinsic natures which made that behaviour inevitable for them.' (Needham 1956:281).

32. In his discussion of motion in the Void, Aristotle said something that looks as if he could understand Newton's First Law of Motion. Newton said that an object at rest will stay at rest, and a moving object will continue moving at the same speed and in a straight line unless acted upon by some force. And Aristotle said 'In a void, if something is in motion no one can say why it should stop anywhere; for why here rather than there? *Hence it will either remain at rest or it must move* ad infinitum *unless something stronger prevents it* [my emphasis]' *(Physics,* 215a 19-21). But Aristotle, like all ancient thinkers, thought that either rest or motion had to be the normal state of matter, but not both. Newton, however, was saying the opposite, that *inertia* will ensure both that a stationary object remains at rest, and that a moving object will continue at the same speed in a straight line (or a rotating body will continue to rotate at the same speed). The idea of inertia, however, *resistance to change,* was only really understood in the seventeenth century.

33. Cornford 1931:26-7.

34. Cornford 1950:88.

35. I am grateful to Dr Ulrich Wenzel for sending me his paper '*Dynamism and finalism in the Aristotelian philosophy of nature*', which first drew this point to my attention.

36. Strato, one of Aristotle's successors, noted however that 'If water is poured from a roof and falls from a considerable height, the flow at the top is continuous, but water at the bottom falls in discontinuous parts. This would never happen to it unless it traverses each successive space more swiftly'. Lloyd 1979:222-23.

37. Lloyd 1979:209-10.

38. In the case of weight, it is a *quantity* when we are measuring the absolute weight of things on the scales, but weight is a *quality* when it is thought of as density, so that lead is qualitatively heavy, while feathers are qualitatively light. It was only about a hundred years later that Archimedes cleared up this confusion by establishing a clear notion of *density*, that is, weight in relation to volume, 'specific weight', but this had no impact on Aristotelian theory. But what about Big/Small, and Many/Few? Aristotle said that these are relative terms, not quantities. So B may Big or Many in relation to C, but Small or Few in relation to A.

39. All across the ancient world from prehistoric times the ankle-bones of sheep and goats, 'astragals', had been used as dice. They were irregular in shape and had four sides with the values 1, 3, 4, and 6: 1 was on a narrow flat side; 3 on a broad concave side; 4 on a broad convex side; and 6 on a narrow concave

side. Not surprisingly, the odds of each side turning up were unequal, but the values allotted to each side did not even reflect these different probabilities either. Tests have shown that the odds of 1 and 6, on the narrow sides, were about 10% each, while those of 3 and 4, on the broad sides, were about 40% each. (See David 1955.) Since astragals were usually thrown in groups of four at a time, in single throws, it would have harder to detect this than if a single astragal had been thrown again and again, but it seems quite clear that players was simply thinking about short-term unpredictability.

Even when cubical dice of the type we use were introduced into Greek and Roman society, regularity of shape was only a matter of aesthetics and good craftsmanship, not of any imaginary search for 'the truest die': the most regular dice were simply the most expensive to make, while the cheaper ones could be notably irregular. Tests of three specimens in the British Museum (David 1955:9) showed enormous variations in bias, from a beautifully symmetrical rock-crystal specimen to obviously irregular iron and marble specimens. Each die was thrown 204 times, and the results were as follows:

| Number of pips ... | 1 | 2 | 3 | 4 | 5 | 6 |
| --- | --- | --- | --- | --- | --- | --- |
| Rock crystal | 30 | 38 | 31 | 34 | 34 | 37 |
| Iron | 35 | 39 | 30 | 21 | 37 | 42 |
| Marble | 27 | 28 | 23 | 47 | 25 | 54 |

It might be supposed that the marble die, in particular, was deliberately loaded, but why then would someone have spent much more money on an expensive crystal die that would be *less* successful? It seems obvious that irregularity was simply the result of shoddy workmanship and was an accepted aspect of unpredictability. In any case, it is unlikely that the results of our table would have been noticed in real life, since four or five dice were thrown together, and only in single throws.

40. The highest score with the astragals, the Venus Throw, was achieved when all four were of different values (1,3,4,6), not when they were all the same, even 6, 6, 6, 6. This is a good illustration that they had no clear idea of the relative probabilities involved because the chances of throwing four astragals with *different* values is actually considerably higher (3/32, or about 1/11) than throwing them with all values the same, which is only 1/256. The Greek mathematicians seem to have had no interest in combinations and permutations (although some work was done in China and India), while Aristotle's only comment on probabilities with dice suggests that he gave the subject virtually no thought: 'To succeed in many things, or many times, is difficult; for instance, to repeat the same throw ten thousand times with the

dice would be impossible, whereas to make it once or twice is easy' (292a).
41. Crombie 1996:360-61.
42. Sambursky 1956:3.
43. For Aristotle, 'To occur "by chance" means, not that there is no reason for the accident, but that factors, themselves determined by their own specific causes, do impinge on other processes, and alter and perhaps even destroy them, without being an essential part of those processes, without belonging to their distinctive nature.' (Randall 1962:183).
44. 1027a15-18.
45. I am grateful to Professor Mark Elvin for allowing me to read his unpublished paper *"Personal luck. Why premodern China – probably – did not develop probabilistic thinking."* I alone am responsible for any conclusions drawn from his paper here.
46. Jewish rabbis, who were familiar with some elementary problems involving probability, were one exception. For example, the Talmud considers a case in which some meat has been discovered in a town where there are nine *kosher* butchers, and one that is not *kosher*: is it lawful to eat the meat? The reply is that it is permitted, since it probably came from the majority. (Rabinovitch 1969:15, and see also Rabinovitch 1970). Strangely enough, it was not gambling but written texts which seem first to have led people to think about different frequencies, and their probabilities – in this case of letters and words – especially in India and the Muslim world.

In Vedic India, the study of poetic metre, of which there were many kinds with 6, 8, 9, 11, 12 syllables and so on, had led mathematicians to the study of permutations and combinations. They 'were concerned with the problem of producing different possible types of metres from those of varying syllables by changing the long and short sounds within each syllable group. . . Emerging from the intricacies of the Vedic metres, these rules found immediate applications, e.g. finding possible combinations of the six tastes taking one, two, three, etc., at a time, . . .or the total number of perfumes that can be made from sixteen substances taken one, two, three, or four substances at a time'. (Sen 1979:156). According to Plutarch, 'Xenocrates determined the number of syllables which are produced through mixing the letters of the alphabet up to 1,002,000 millions', but since few syllables in Greek have more than four letters, the estimate seems wildly exaggerated. (Sambursky 1956:10). The rabbinical author of the Book of Creation (*Sefer Yetsirah*) in about the second century AD believed that the twenty-two letters of the Hebrew alphabet represented the building-blocks of creation (an idea that we have already met in connection with the elements and atomism). This led him to consider their permutations: for example, the 3 letters of the Divine Name, JHV, have six permutations, and these correspond to the six directions of north, south, east, west, up, and down. But the author also states the factorial rule for permutations: that $n$ things can be arranged in $n!$ different orders. So if $n = 4$, $n! = 4 \times 3 \times 2 \times 1 = 24$.

Muslim theologians in the ninth century, trying to establish the chronological order of Muhammad's revelations in the Koran, made very significant discoveries about distribution. They believed that certain words were more recent than others, so that revelations containing higher proportions of newer words would have come later than those with fewer of them. This counting of frequencies was extended to letters, and it was discovered that in Arabic (as in other scripts) some letters occur more frequently than others. In English the order is **e** (12.7%), **t** (9.1%), **a** (8.2%), **o** (7.5%), **i** (7.0%), **n** (6.7%), **s** (6.3%), and so on. The philosopher and mathematician al-Kindi applied this principle to the first recorded code-breaking, in which by counting the relative frequency of letters in the coded message he was able to equate these with their true equivalents in the original message. (I am obliged to Simon Singh, *The Code Book,* London 1999:16-21 for this information.)
47. Lloyd 1966:439.
48. Crombie 1961:3.

## CHAPTER XIII ANCIENT SCIENCES
1. The Stoics, founded by Zeno of Citium around 300 BC, were also very important; while in many ways they agreed with Aristotle, especially against the atomists, they made the Logos their central principle, and for them Nature and God were essentially the same. They also took up the idea of the *pneuma*, 'spirit' or 'breath', composed of fire and air, which permeated the whole cosmos as a dynamic principle of order.
2. The Museum was not, as we might suppose, filled with glass cases of fossils, but was dedicated to the Muses, the Greek deities of the arts, philosophy, and the sciences. It was founded by King Ptolemy I in 300 BC, and was located in the royal quarter of Alexandria. 'We hear of dinners and symposia, often attended by the king, at which scientific, philosophical, and literary problems were discussed. Although we know little of the Museum's activities, there is no reason to doubt that it contributed much to the scientific and literary research for which Alexandria was to become famous.' (Barnes 2004:62)
3. 3. It was already known, for example, that solar eclipses only occur at the new moon, while lunar eclipses occur at the full moon in the middle of the month. (Neugebauer 1969:101.)
4. He started with three facts: that at Syene (modern Aswan) on the Nile, at noon on the summer solstice, the sun is directly overhead and shines straight down a well; that Syene is due south of Alexandria on the same longitude, so that noon will be simultaneous in both places; and that the distance from

Syene to Alexandria is about 5000 stades (1 Greek stade = approx. 607 feet, 1 Egyptian stade = 515 feet). He found that at Alexandria a vertical rod cast a shadow at an angle of 7.2°: now Eratosthenes assumed that the sun is sufficiently far away for its rays to be regarded as parallel to one another, and it was also known that where a straight line crosses two parallel lines the alternate angles are equal to one another. Since the angle of the shadow cast by the rod in Alexandria was 7.2°, then the alternate angle produced at the centre of the earth by projecting imaginary lines from the rod and the well to the earth's centre, must be the same. 7.2° is 1/50$^{th}$ of a circle, so the 5000 stades between Syene and Alexandria must be 1/50th$^{th}$ of the total circumference of the earth, or 250,000 stades. If these were Egyptian stades, then this would work out at 24,384 miles, (very close to the actual figure of 24,917 miles), and therefore the earth's diameter would be 7890 miles.

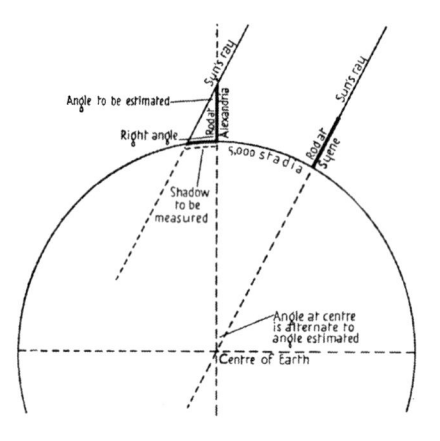

(From Singer 1943:72, Fig. 27)

5. So, from the time that the moon takes to pass through the earth's shadow in a lunar eclipse, Eratosthenes calculated that its diameter is a quarter of the earth's, about 1970 miles, and by simple geometry he calculated its distance as 100 times its diameter, or 197,000 miles.

6. 'Most handbooks of navigation or surveying open with some sentence like this: "For present purposes we shall assume that the earth is a small stationary sphere whose centre coincides with that of a much larger rotating stellar sphere".' Kuhn 1957: 38.

7. The sun and the moon vary somewhat both in the speed of their apparent motion, and in brightness during the course of the year, but much less than the planets. Unlike the planets, moreover, they do not retrogress. While the anomalies of the motions of sun and moon could have been fudged, those of

the planets, however, were a real intellectual challenge.

8. Many readers will wonder how all these geometrical devices could be combined with Aristotle's crystalline spheres. Up to a point, the spheres were seen as a *philosophical* explanation of how the celestial motions were ultimately caused, whereas astronomers were only thought to be concerned with giving a mathematical account of these motions that 'saved the appearances'. But in a book on the planets, Ptolemy did try to reconcile the spheres and the epicycles. He accepted that there could be no Void, and that matter could not overlap matter, and in his model each planet had its maximum and minimum distance from the earth. In other words, each planet's motion was within its own crystalline sphere. 'Ptolemy's idea was simply to make the outermost point reached by a planet in its epicycle equal to the minimum distance reached by the planet next above it. One can therefore imagine that Mercury is always within a certain spherical shell, a sphere with an appreciable thickness, and that Venus is in the shell next outside it'. (North 2005:309, and see 308-12 for a thorough discussion of this difficult point, also Kuhn 1957:81.) Ptolemy does not discuss the spheres in the *Almagest*, but in his *Planetary Hypotheses*, of which the relevant section has only become known since the 1960s. Nevertheless, the relation between spheres and epicycles always remained problematic.

9. Toomer 1978:209.

10. His new geometrical device of the equant, to explain changes in a body's speed, was important because it was this aspect in particular of the Ptolemaic model that, as we shall see in Chapter XVI, Copernicus singled out as going beyond the reasonable limits of what a circular orbit implied.

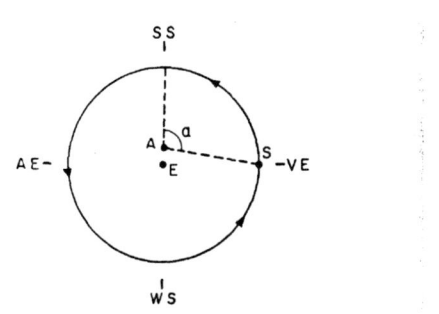

Figure 25. The equant. The sun, S, moves on the earth-centered circle but at an irregular rate determined by the condition that the angle a vary uniformly with time.

(From Kuhn 1957:72)

For example, the sun goes 6 days faster between the equinoxes of autumn and spring than between spring and autumn; in the diagram the centre of the sun's orbit or deferent is still the earth E, but 'the deferent's rate of rotation is now required to be uniform not with respect to its geometric centre E[arth], but with respect to an equant point A, displaced in this case toward the summer solstice.' (Kuhn 1957:71). The crucial point is that angle *a* has to *change at a constant rate*, slightly less than 1° per day. We can see from the diagram that for the sun S to go from VE, the vernal equinox, to AE, the autumnal equinox, is more than 180°, while to go from AE to VE during the winter, is less than 180°. Since the assumption is that each degree of movement must take the same amount of time, it follows that the sun will take more time to go from VE to AE than from AE to VE, because it is travelling farther.

11. Kuhn 1957:73.

12. The eccentric was yet another geometrical device for explaining planetary motion, in which the centre of the planet's orbit is not the centre of the earth, but at some distance from it.

13. Neugebauer 1969:170. In 134 BC Hipparchus had made perhaps his greatest discovery, of a new form of motion in the heavens whereby the equinoxes (and solstices) do not occur permanently in relation to the same constellations, but move slowly eastwards around the sky. This movement, 'the precession of the equinoxes', as we now call it, is extremely slow, and a complete cycle takes about 26,000 years. Hipparchus was only able to discover it because observations of the star Spica from 150 years earlier were available to him. This discovery was of special astrological significance, particularly in relation to the Mithraic cult. See Ulansey 1991.

14. The hot, cold, moist, and dry are the basic causal factors in astrology, and each planet has different combinations of these qualities. The twelve signs of the zodiac are in four groups of three, corresponding to the four elements, and each cluster of stars in every sign also has different attributes which affect the different planets, as do the major stars outside the zodiac.

15. Neugebauer 1969:171.

16. See Forbes 1964.

17. Mercury was certainly known by around the middle of the first millennium BC, but for many centuries was not regarded as a metal at all, and only became associated with the planet Mercury around 500 AD. Not only is it molten above −37° C and so would never have been experienced as a solid, but it has the remarkable property of being able to combine on contact with gold (with which it has a special affinity), silver, tin, and in fact with any common metal except iron, to form a special compound called an amalgam. Mercury, sulphur, and arsenic will also vaporize at much lower temperatures than metals, which makes them very easy to work with. Aristotle believed that metals are formed out of two 'vapours' from the earth: one is dry and smoky, which the alchemists identified with sulphur, and the other is moist, which became identified with mercury, so that all metals were thought to be

composed of sulphur and mercury. (When cinnabar is heated it produces mercury and sulphur.) Sulphur and mercury were therefore very important in alchemy.

18. Although this story about Archimedes was well known in the ancient world, his use of specific weight as a test for gold never seems to have been adopted by assayers, presumably because, as craftsmen, they never had the opportunity to read about it.

19. The still, in particular, underwent fundamental improvements, and 'The Greek alchemists name about eighty pieces of apparatus. Furnaces, lamps, water baths, ash baths, dung beds, reverberatory furnaces, scorifying pans, crucibles, dishes, beakers, jars, flasks, phials, pestles and mortars, filters, strainers, ladles, stirring rods, stills, sublimators, all make their first appearance as laboratory apparatus in their works and have persisted in somewhat modified form to the present day.' (Taylor 1951:46 ).

20. The equation of the incorruptibility of gold with the immortality of man, particularly for the Chinese, was not a mere metaphor or piece of symbolism, but a genuine relationship. In Chinese culture yellow was of great significance: it was the colour of the centre of the five regions of space, corresponding to the Earth, and for most of Chinese history it represented 'the might, mana, and dignity of the Emperor', and it was the colour of the world of the ancestors, so not surprisingly the colour of the incorruptible metal was also the colour of human immortality.

21. Knight 1992:15.

22. For a useful survey of ancient knowledge of magnetism and amber see Roller 1959:13–25.

23. See Trowbridge 1928 for all the references in classical texts to the optical uses of glass. Establishing when glass of sufficient quality for optical lenses was first made is not of great importance, since rock-crystal was always available. Indeed, its magnification is superior to that of glass. Plano-convex lenses dating to at least 1400 BC, and with a magnification of between roughly 3x – 10x, have been found in Crete, (where fine gem engraving for seals is of early date) Greece, and Italy, as well as in Egypt. While some, backed by silver or gold foil, were used as ornaments on furniture, and others no doubt as burning glasses, it seems clear that gem-engravers must have used magnifying lenses because of the extreme fineness of their work. Using a sample of nine Archaic Greek and Etruscan engraved gems and an Etruscan ring, Sines (1992) found that the distances between the hatch marks on the borders ranged from 0.093 mm to 0.033 mm, a thirtieth of a millimetre, which is astonishingly small, and a median value of 0.048 mm. By comparison, 'A skilled machinist working without a lens can achieve a precision of about 0.2 mm, and with a simple magnifying glass about 0.08 mm' (Sines 1992:67) Clearly, then, if the

use of lenses is denied, some considerable abnormality of vision was required. It has been suggested that gem-engravers were young men, whose eyes would have been capable of such short focus. But the opinion of a professional gem-engraver from modern times is that 'The art of engraving in gems is too difficult for a young man to be able to produce a perfect piece, and when he arrives at a proper age to excel in it, his sight begins to fail. It is therefore highly probable that the ancients made use of glasses, or microscopes, to supply this deficit.' (Cited in Sines and Sakellarakis 1987:194).

Failing an adequate supply of suitable young men, it has been suggested, as a last (and rather desperate) resort, that gem-engravers were constitutionally myopic, which might have enabled them to resolve distances as small as 0.033 mm. Even if this were true (which has not been demonstrated), there would have been the basic problem of ensuring that a sufficient number of myopic young men were willing to take up the craft of gem-engraver, and even if they were, that they would have been any good at it. Faced with these alternatives, which also assume that – of course – natural selection would have maintained myopia in families of gem-engravers, it is perhaps more reasonable simply to accept that some of the lenses that have been discovered were in fact used as magnifying glasses by gem-engravers in the ancient world. See also Beck 1928, and Plantzos 1997.

24. Seneca had all the physical means at his disposal to carry out the experiments in optics that were later conducted by the medieval philosophers with great effect, had he thought it worthwhile to do so.

25. Diocles wrote a treatise on burning mirrors not to explain how to make them (that, he says, has already been done by someone else, (Toomer 1976:34), or to discuss their powers of magnification, but to give a geometrical proof that only a parabolic curve will focus the sun's rays to a point, as part of his more general interest in conic sections.

26. Empedocles had compared the eye with a lantern whose inner fire shines out through the pupil (to which Aristotle's retort was that why, then, can't we see in the dark!) Aristotle's view was that the eye allows the soul to see because it is made of water, and therefore transparent. Others, notably the Atomists, said that every object emits an image composed of atoms that detach themselves from the surface of objects, and that these images enter the eye directly – which raises insuperable difficulties in explaining how an image the size of a mountain could get inside an eye. (The Chinese, though not atomists, also supposed that the images of things were transmitted directly to the eye, though they do not seem to have held the visual ray theory.)

27. Piaget 1929:48.

28. Aristotle argued that light travels in straight lines because nature always acts in the most economical way – the *lex parsimoniae* – and a straight line is

the shortest distance between two points.

29. White 1957.

30. The first is by Lucretius, discussing optical illusions, when he says that when we look down a straight colonnade of pillars of the same height, the top and bottom, and left and right seem gradually to merge together into the indistinct point of a cone. (Lucretius, *De Rerum Natura*, 4.426-32) The other ancient reference is a couple of sentences by the architect Vitruvius, when discussing the technique of scene-painting that had begun in the Greek theatre, and had then been adopted by architects for their drawings. He says that scene-painting, 'scenography', is the sketching of the front and receding sides of a building as they converge on a point, and in another place that 'the extension of rays from a certain established centre point ought to correspond in natural ratio to the eye's line of sight, so that they could represent the appearance of buildings in scene paintings, . . . both the surfaces that were depicted frontally, and those that seemed either to be receding or projecting'. (Vitruvius. Bk.7, Preface, §11. [1999, p. 86] .)

31. 'In Athens, when Aeschylus was producing tragedies, Agatharcus was the first to work for the theatre and wrote a treatise about it. Learning from this, Democritus and Anaxagoras wrote on the same subject, namely how the extension of rays from a certain established centre point ought to correspond in a natural ratio to the eyes' line of sight, so that they could represent the appearance of buildings in scene paintings, no longer by some uncertain method, but precisely, both the surfaces that were depicted frontally, and those that seemed either to be receding or projecting.' (Bk.7, para. 11) The editors comment that Agatharcus 'has been credited with the invention of something very like linear perspective. The nature of his invention is still debatable, but his generation was contemporary with developments in pictorial rendering of space, such as shadows and modelling, and a sort of deep landscape space in vertical perspective that has been attributed primarily to Pythagoras of Samos.' (Rowland & Howe 1999:266).

32. *Geography*, Bk. I, Chaps. 20-24. Ptolemy, in his *Optics*, developed the theory of the visual cone further by claiming that not all the visual rays were of equal importance, and that the central ray was the most significant because it travelled the shortest distance to the object. Whereas Euclid said that the visual rays extended in a straight line to infinity, others including Ptolemy thought that they became weaker with distance, which is why very distant objects could not be seen. (Euclid believed that small/distant objects fell between the gaps in the visual rays.) In Ptolemy's view, the central ray, being the strongest, could therefore give the clearest impression of the object, and the notion of the central ray was to make an important contribution to the Renaissance discovery of perspective.

33. For example, he investigated the illusion of double images by means of a board in which he placed a black and a white peg at different intervals, and he devised an instrument to show that the angles of incidence and reflection in a mirror are equal. He performed even more elaborate experiments to investigate the angle of refraction when a light ray passes from air to water, air to glass, and water to glass, and produced some tables with correlations between angles of incidence and those of refraction.

34. It was well known that a spray of water can produce these colours, so it could have been seen that clouds (black or otherwise) were not necessary, and since it was also known that sunlight passing through a glass prism will have a similar effect, it could have been deduced that water droplets were not necessary either. Seneca, in a long discussion of the whole question, considers the case of the prism but maintains that the colours are an illusion produced by the jumbled reflections of the sun within the glass, in the same way as the colours on the neck of a dove or a peacock change when it turns in the light. Since, like all his contemporaries, he could not imagine that light could be composed of different colours, it was impossible for him to regard the prismatic colours as real. Another vital fact about this question was also known but dismissed: that a disc, painted in the primary colours, when spun appears white. Ptolemy held that this is an illusion caused by the rapid motion of the colours that blurred them so that they could no longer be accurately perceived.

35. See Drachmann 1963.

36. He was a general and sometime Governor of Britain to whom the Emperor Nerva entrusted the supervision of the aqueducts of Rome. Frontinus believed that one could not effectively command men unless one thoroughly understood the business they were about, and he wrote a treatise on water administration. Like Vitruvius, he is one of the very few well educated men of antiquity who also wrote books on practical subjects.

37. If it was to fire a bolt, then the diameter D of the spring was given by the formula $D = L/9$, when D is the diameter in *dactyls* (1 *dactyl* = approx. ¾") and L is the length in *dactyls* of the bolt. But in the case of stone-throwing catapults the formula was rather more difficult, and was $D = 1.1(\sqrt[3]{100M})$, (in modern notation): M represents *minas* (1 *mina* = approx 1lb), so for a stone weighing 80 *minas* the diameter of the spring hole would be 22 *dactyls*, 16½ inches. (See Marsden 1969:25 for more details.) Ordinary engineers would have used tables to work this out. Vitruvius gives one, for example, and, if necessary, calculations of cube roots were possible by geometry (the theorem of two mean proportionals), or by a simple mechanical device, the mesolabe, invented by Eratosthenes. (Marsden 1969:39-41.)

38. The normal angle of elevation was 30° or more, rather than the 45° that

attains maximum range, and 'It is possible that stone throwers were roughly aligned on the target, and that adjustments were made after observation of the fall of shot'. (Marsden 1969:92.)

39. For example, his research into the geometry of the spiral was clearly the basis of his invention of the endless screw or worm-drive, and the Archimedean Screw for raising water, both of which use the principle of the rotating spiral. (Drachmann 1958a.) Archimedes seems to have invented the water-snail when he was in Egypt. There is no evidence that the Egyptians or Mesopotamians had any knowledge of the spiral and its mechanical uses before Archimedes. While the Chinese were familiar with gears, they did not know the worm-drive, and it seems very likely that this was because they had no Archimedes.

40. The ruler of Syracuse, Hiero, wanted to discover if he had been cheated by a goldsmith to whom he had given some gold to make him a new crown. Someone told Hiero that the goldsmith had replaced some of the gold with silver and so Hiero asked Archimedes to find out if this was true:
'While it was on his mind, Archimedes happened to go to the bath. On getting in, he observed that the more of his body was immersed, the more water ran over the top. This suggested, the solution. Transported with joy, he ran home shouting Eureka! Eureka! ('I have found It, I have found it!') What he had found, in effect, was the conception of specific gravity… He made two masses of the same weight as the crown, one of gold, the other of silver. Next he filled a vessel to the brim and dropped in the mass of silver. Water ran out in bulk equal to the silver. The measure of this outflow gave the bulk of the silver. The same was done with the gold. The smaller outflow corresponding to the gold was, of course, as much less as the gold was less in bulk than the silver, for gold is heavier than silver. The same operation was now done with the crown. More water ran over for the crown than for the bulk of gold for like weight, less than for the bulk of the silver. Thus was revealed the admixture of silver with the gold.' (Singer 1943:65.)

41. Despite these outstanding practical achievements, Plutarch says that he 'regarded as ignoble and sordid the business of mechanics and every sort of art which is directed to use and profit'. This probably tells us more about Plutarch's priorities than those of Archimedes, but there is no doubt that he wished to be remembered as a mathematician rather than as an engineer. He directed his heirs to erect a tombstone for him with a diagram of a cylinder circumscribing a sphere, and giving the ratio $3/2$ of the volume of the cylinder to the sphere which he regarded as his supreme achievement. Cicero records that he found this tomb and restored it when he was Quaestor of Sicily.

42. It was on this occasion that he is said to have claimed to Hiero that if only he had a place to stand he could move the earth, meaning that there

was no limit to the power of the endless screw and the windlass. Drachmann 1958b.
43. See Hodge 1960.
44. Price 1964:15.
45. See Drachmann 1948. The Chinese also knew the piston and cylinder pump, which they seem to have devised from their familiarity with bamboo.
46. As Seneca puts it, 'The atmosphere borrows its qualities from its surroundings. The highest part of it is extremely dry and hot, and so, very rare also, from the proximity of the eternal fires, the endless motion of the stars, and the constant revolution of the heavens. But the lowest portion next the earth is dense and dark, because it forms a receptacle for the exhalations of the earth'. (Seneca, *Natural Questions*, Bk II, 10.) But later he uses a quite different image, that of solids settling in a liquid: 'All air is the denser the nearer it is to the earth. In water and other liquids the dregs are always at the bottom; in like manner in the atmosphere the thickest parts settle down to the lowest part nearest the earth.' ( Ibid., Bk IV, 10.)
47. Landels 2000:29.
48. Drachmann 1961.
49. There were precision instruments for measurement in the ancient world, but these were invariably either for measuring *angles*, or were complex water-clocks. For example, Ptolemy describes how to make an armillary sphere for astronomical sighting, and an instrument for measuring stellar angles, and Hero describes an elaborate theodolite. But these were all extremely rare and costly pieces, and Hero's theodolite may never even have been made. There was nothing comparable to the enormous variety of measuring instruments that began to develop in seventeenth century Europe, and certainly no trade of precision instrument makers in the ancient world of the kind that had developed by the time of the Renaissance.
50. It is worth noting that most of the experimental apparatus used by scientists in the seventeenth, and even the eighteenth century, was simple, and either existed, or could have been constructed, in the workshops of the Museum at Alexandria. It was not, then, technology that hindered the development of experimental apparatus, but lack of theoretical reasons for performing the experiments in the first place. The Voltaic battery, for example, was simplicity itself to construct, but with no conception of electricity, no one in the ancient world would have seen the point of making one.

## CHAPTER XIV THE UNIQUENESS OF WESTERN SOCIETY
1. Bernal 1969:47.
2. Lloyd 1973:112.

3. Ibid.,112.
4. Ibid.,111.
5. Finley 1973:148.
6. Crombie 1990:28.
7. Seneca, for example, when discussing how snow is formed, says that his readers will probably object to such a frivolous waste of time, when he should really be explaining why it is morally degenerate to buy snow to cool one's drinks. Seneca, *Natural Questions* Bk. IV, 13.
8. Stahl 1962:8.
9. Lloyd 1973:165.
10. cited by Lloyd 1973:167.
11. Lloyd 1973:167.
12. For the following dates of Islamic scientists I follow Nasr 1987. Most had encyclopaedic interests: Jabir ibn Hayyam (c. 721-815) wrote on logic, medicine, and physics, but is most famous as the founder of Islamic alchemy, and was known as Geber in the West. Al-Kindi (c.801-873), among his many interests, is supposed to have written 36 books on chemistry and technology, but was most renowned as a philosopher. (He was one of the very few Arab scientists. Most of them were Persian.) Hunain ibn Ishaq (810-877) was a very important translator of Greek scientific texts; Thabit ibn Qurrah (826-901) was a distinguished mathematician and became court astronomer; al-Khwarazmi (died c. 863) was probably the most important Islamic mathematician, who was responsible for the adoption of the Indian numerical system, and made very significant contributions to algebra and also to trigonometry; al-Razi (865-925) was a notable alchemist and physician, becoming director of the hospital at Baghdad, and was known from his books on medicine as Rhazes in the West; al-Farabi (870-950) is considered as a philosopher to rank second only to al-Kindi, and was the first great commentator on Aristotle, while al-Masudi (died 956) was a major natural historian and geologist.

After this period the greatest scientists mostly worked in Persia: ibn Sina (980-1037) was a philosopher and scientist, whose encyclopaedic works were later very influential in the West where he was known as Avicenna; al-Haitham (965-1039), or Alhazen, was the greatest Muslim physicist, an exact observer and experimenter, as well as a theorist, who made particularly important contributions to optics; al-Biruni (973-1051) was an astronomer and mathematician who considered the heliocentric theory as an alternative to the Ptolemaic model, and calculated the specific weights of various metals, but also wrote a notable work on India; al-Khazini (died c.1125) studied mechanics and hydrostatics, especially the centre of gravity as applied to the balance, and he was followed by al-Jazari (1136–1206), who wrote the definitive Muslim work on mechanics, especially in his study of the balance. In Cordoba, ibn

Rushd (1126-1198), known in the West as Averroes, was a notable expert on Aristotle, who tried to refute al-Ghazzali's attack on philosophy, and, much later, we have ibn Khaldun (1332-1406) in Egypt, historian and political philosopher, speculating on the rise and fall of civilisations.

13. 'This final and deliberate rejection of all causality, once generally accepted, marked the end of free speculation and research, both in philosophy and in the natural sciences. . . ' (Lewis 1993:155) . '[al-Ghazzali] helped more than any other single individual to bring about the intellectual transformation that took place during the . . . twelfth century'. (Nasr 1987:52).

14. Nasr 1987:21.
15. Lewis 1993:145.
16. Lewis 1982:229-30.
17. Nasr 1987:174.
18. Lewis 1982:224.
19. Findley 2005:93-4.
20. Lewis 1982:301-2.
21. Makdisi 1980:26.
22. Eamon 1994:40.
23. See Grimm 1977.
24. A convenient list of Chinese inventions and discoveries can be found in Needham 2004:217-224.
25. The imperial astronomer Su Sung, in about 1088, had built an extremely elaborate astronomical clock driven by water, in which the buckets were fixed around a wheel so that, when one bucket had been filled, this caused a lever to be depressed which then allowed the wheel to rotate and bring the next bucket under the water-nozzle, at which point the wheel was stopped again. (Needham 1965:446-63.)
26. Needham 1959:212-23.
27. Needham 1959:399.
28. Needham 1962:59.
29. Needham 1956:281.
30. Bodde 1991:367.
31. Elvin 2004:xxxii.
32. These were huge ships of 1500 tons (by comparison with the 200 – 300 ton ships of the Spaniards and Portuguese), multi-masted with slotted sails and rigs that allowed tacking into the wind, with water-tight compartments, stern-post rudders and the compass, and with facilities for growing food on board that prevented scurvy. The whole purpose of these voyages was to explore, to collect flora and fauna, and to promote respect and knowledge of China at the places they visited. The Chinese had no special need to find new sources of trade, and the use of the imperial fleet was not open to merchants

to make private profits. The Chinese had no proselytising religion to which they wished to convert the rest of the world, and no use for foreign conquests, although perfectly capable of making them.

Cheng Ho was a eunuch, a class of imperial servants who had been rivals, since the Han dynasty, of the Confucian bureaucracy which strongly deplored the whole policy of maritime exploration as a waste of money. China, they held, with its silk, its porcelain, its lacquer-work, its tea, and the best craftsmen in the world, had no need of foreign trade, while Confucian ethics in any case frowned on the pursuit of luxury for its own sake. They considered that the money consumed by the fleet would be much better spent on ensuring regular grain supplies for the peasants, and on the internal canal system of China and its shipping. The champion of retrenchment was Hsia Yuan-Chi, who when released from prison on the accession of Emperor Jen Tsung in 1424 immediately stood the fleet down. Although Jen Tsung soon died, which allowed Cheng Ho one last voyage, the next emperor soon died, too, and under his successor all further voyages were prohibited. In 1479 all the records of Cheng Ho's voyages were burnt by a senior official at the Ministry of War as 'deceitful exaggerations of bizarre things'. The navy fell to pieces, the shipyards closed, and by 1500 it was a capital offence to build a sea-going junk with more than two masts. By 1551 anyone who went to sea in a multiple-masted ship, even to trade, was committing a crime analogous to espionage by communicating with foreigners. (This account of Cheng Ho is based on Needham 1971:487-528.)

33. Bodde 1991:194.
34. Bodde 1991:251.
35. See Heather 1998, 2005.
36. Brett 1983:159.
37. Some slaves were criminals, or those who could not afford to pay compensation for homicide, but many would have been, like the native Britons, Romano-Celts captured in battle.
38. Dubuy 1980:150.
39. Whitelock 1952:99.
40. Brett 1980:158.
41. Brett 1980:158.
42 The first written evidence for the doctrine that the state needs three types of men is provided by King Alfred at the end of the ninth century, in his translation of Boethius' *Consolations of Philosophy*, where he said that the king must have three kinds of tools: 'He must have praying men, fighting men, and working men', a theme developed by later Saxon writers in the tenth and eleventh centuries, and also by two Frankish bishops, and the Three Estates became a basic political idea of medieval Europe. It is possible,

of course, that Alfred and others were simply making a practical comment on how kingdoms have to be run, but warriors, priests and peasants were standard components of all agrarian states. Why, then, do we not find the idea of the three orders in Japan, for example, the Buddhist kingdoms of south-east Asia, or Russia, or anywhere else outside India? The medieval idea could not have come from classical or Christian sources, so it is clearly possible that it was a commonplace of traditional Germanic culture that these thinkers were applying to contemporary social conditions, although there is no documentary evidence that this was so. The problems of explaining the origins of the medieval doctrine of the Three Estates are discussed at length in Constable 1995:251-341.

43 Holmes 1992:x.

44. Crombie 1996:476.

45 As we have noted before, Western Europe was chronically deficient in goods that were valued in India and the Far East. When Vasco da Gama reached Calcutta, for example, his woollens, honey, and olive oil were laughed at in the markets, and he was told to pay in gold or silver.

46. Eamon 1994:89.

## CHAPTER XV How We Learned to Experiment

1. Williams 1997:391.

2. Eamon 1994:213.

3. See Williams 2003. It is conceivable that the section on how the wise prince should rule was based on the lost treatise on kingship that Aristotle is supposed to have written for Alexander. It agrees with Aristotle's known views on kingship quoted in Chapter X, and Alexander's rule over the Persians was also consistent with Aristotelian teaching. But the rest of the material was foisted on to Aristotle to give it greater authority.

4. A complete Latin translation had become available in the 1240s. It has been described as the most popular book of the Middle Ages, and about five hundred manuscripts survive, an extraordinary number

5. The claim that reading it led him to give up his Chair of Philosophy at Paris has been disputed. See Crombie 1996:57; Weisheipl 1984:454-5; Williams 1997. For Bacon's ideas on alchemy see Newman 1997.

6. Eamon 1994:53.

7. Rienstra 1974:97.

8 Those in the artist-engineer tradition of Galilean science, the 'mechanical philosophers', like the natural magicians, were also opposed to the Aristotelian philosophers: '. . .the goals of the natural magicians, the alchemists, and the astrologers [were] not far removed from those of the mechanical philosophers.

They, too, sought a universal science, and they, too, placed an emphasis on new observations and new experiments as a replacement for the sterile teaching of nature at the universities.' Debus 1967:133.

9. From Macaulay's essay on Bacon in the *Edinburgh Review*.

10. Yates 1972:119.

11. See in particular Rossi 1978.

12. See Inwood 2003. The Royal Society not only conducted a vigorous experimental research programme, but collected histories of trades, in which the whole process used in many crafts was recorded. Leading figures such as Robert Hooke and the Hon. Robert Boyle believed that science could profit from the experience of artisans, just as it had long been accepted that artisans needed to know some science. Galileo said that mechanics could profitably be studied in the Arsenal in Venice, while the chemists, to profit from the experience of metal-workers, studied Agricola's *De Re Metallica* and Biringuccio's *De la Pirotechnia*.

13. See for example Murray 1978:162-87.

14. *The Times*, 5th December 2003.

15. Leonardo Fibonacci (1170–1240), for example, had studied Arab mathematics in North Africa, and published the *Liber Abaci*, Book of the Abacus, in 1202. Many of his examples are drawn from trade, but since only three MSS of the book survive this suggests that it was not widely read. Again, Fra Luca Pacioli wrote text-books on pure and applied commercial arithmetic, notably on double-entry book-keeping. But no known medieval specialist mathematicians were merchants (except Giovanni Villani), and nothing suggests that any other major mathematics writers learnt from merchants. Crosby 1997:215.

16. Kaplan1960:42.

17. Kaye 1988:259.

18. Kaye 1988:262. But Kaye admits that that 'nowhere have I found a statement to show that [the philosophers] were specifically conscious of money as a species of continuum, except in its broadest sense as a scale, or measure, or medium.' (ibid, 264) It seems, then, that money acted as an unconscious cultural influence on thought, rather than as a conscious model such as clockwork or artillery, but we shall look at this in more detail later.

19. His first book, *The Grounde of Artes* (1540) was on arithmetic, and focused on practical problems like how to estimate the number of loads of bricks needed to construct walls of given dimensions. It was followed by *The Pathway to Knowledge, or the first Principles of Geometry* (1551), and *The Whetstone of Witte* (1557) on algebra, the first of its kind in England, as was his book on geometry, and both illustrated theory with practical problems. Recorde clearly distinguished the sober and intelligent unlearned from the vulgar and

ignorant masses, whom he detested. He refers to them as 'the ignorant sort, which hate all things that they know not'; "The ignorant multitude doth, but as it were ever wont, envy that knowledge which they cannot attain, and wish all men ignorant like unto themselves, but all gentle nature contemneth such malice; and despiseth them as blind worms, whom nature doth plague to stay the poison of their venomous sting.' Cited by Kaplan 1960:88.

20. North 2005:173.

21. The economic expansion of the period led moral philosophers to argue that profits and interest were justified by the risk to the capital invested, and risk involved probability. Calculations of risk 'became accepted practice in fourteenth century Italian marine insurance, with graded premiums, estimated from accumulated experience, for distance and season and dangers from storms and pirates. The rational pursuit of profit from any of its sources thus required both personal enterprise and the habit of quantitative order, assisted technically by the new commercial arithmetic and the new financial methods of double-entry bookkeeping and the bill of exchange.' Crombie 1996: 374-75. The development of life assurance, and later on the invention of annuities in the seventeenth century, which required the use of the tables of mortality that were then being produced, were further incentives to the rational calculation of risk and probabilities.

In this new intellectual atmosphere, where people had begun to think of applying mathematical reasoning to accident, there was, for writers on commercial arithmetic, an obvious parallel between the merchant and the gambler: 'At any moment of time, they argued, a partner who had invested a certain amount in a company was in the same position as a player who had gained a certain number of points in a game of chance. What was the value of their investment or stake at that moment?' (No doubt Chinese, Indian and Arab merchants, too, reflected on the mathematics of risk, but in those civilisations this seems to have had no broader intellectual impact.) The sixteenth century gambler, mathematician and astrologer Cardano established the principle that for partners in businesses, or gamblers at the table, it was rational to assume that there would be 'equal possible outcomes under equal conditions.' Crombie 1996:377.

But thinking of risk being calculated under equal conditions, in which the dice ran true, immediately made the mathematics of gambling very relevant to calculations of probability, which is what Cardano did, but the foundations of this had been laid in the thirteenth century. The earliest known attempt in medieval Europe to calculate the possible falls of dice is the *De Vetula*, probably by Richard de Fournival (1200–1250), Chancellor of the cathedral of Amiens, and so written somewhere between 1220 and 1250. It contains a long section dealing with sports and games, and dicing is given special attention. One crucial problem considered is the number of possible ways in which three dice can fall, and the manuscript has a table in which all the possible permutations are written out, just as in the simple table we saw in Chapter XII. From the table, which requires no mathematical understanding of combinations and permutations, the correct frequencies of the different possible throws are easily worked out. See Kendall 1956:22-5.

There are various references to similar calculations by other writers, until we come to Cardano's *De Ludo Aleae*, written about 1526, which takes the mathematics considerably further. Tartaglia, Galileo, and other Italian mathematicians also wrote on the mathematical problems of dice in the sixteenth and early seventeenth centuries, so there was a substantial literature on the subject by the time of Pascal (1623–62), who is popularly, but wrongly, credited with inventing the mathematics of probability.

Unlike the ancient world, then, we find in medieval and Renaissance Europe that the learned took an early interest in the mathematics of dice, which, if it existed in the other civilisations at all, was confined to croupiers, shopkeepers, and others at the artisan level of society. But gambling also had a special resonance with capitalist enterprise and the rational estimation of risk that was unique to Western Europe, so that here again, we see the importance of that union of craft and theoretical knowledge in a capitalist environment.

22. Kuhn 1957:111-12.
23. Crombie 1979: (II) 50.
24. Kuhn 1957:123.
25. See Crombie 1953.
26. And by a number of others, notably the Pole Witelo (c.1230–1280), who experimented with the refraction of light, and Theodoric of Freiburg (d. 1311) who used glass globes as experimental models of raindrops to produce an advanced theory of the rainbow.
27. Crombie 1990:68–9.
28. Sylla 1973:256.
29. Wallace 1971:18, n.13.
30. How, asked Philoponus, could the air pushed forward by the stone or the arrow possibly then reverse its motion and come round behind to push it along? It would obviously go on moving in the same direction. Nor could the air itself receive the force of the hand or the bowstring, and not the projectile, since in that case it would not be necessary to fit the arrow to the bowstring at all. Clearly, then, 'It is necessary to assume that some incorporeal force is imparted by the projector to the projectile, and that the air set in motion contributes either nothing at all, or else very little, to this motion of the projectile'. (Cohen and Drabkin 1958: 223). He also discussed Aristotle's theory that heavy objects fall faster than lighter ones. While he agreed that

this would happen, he qualified this by saying ' if you let fall from the same height two weights, of which one is many times as heavy as the other, you will see that the ratio of the times required for the motion does not depend on the ratio of the weights, but that the difference in time is a very small one'. Ibid, 220.

31. Oresme, in fact, used this and all the other arguments later deployed by Copernicus, to demonstrate that there was no way of actually proving that the earth did *not* rotate on its axis, although he drew back from claiming that it actually did rotate.

32. Buridan 'came very close to saying that the gravity (or weight) of a freely falling body impresses equal increments of impetus (and therefore of velocity) upon a body in equal intervals of time'. (Kuhn 1957:122). Impetus is very similar to Galileo's and Newton's idea of momentum, which is mass times velocity, although strictly speaking, speed is the distance travelled in a given time, whereas velocity is distance in a given direction in a given time. But it still retains Aristotle's old idea that all motion needs a force to keep it going. In fact, force produces *acceleration*, either from rest, or from one speed or direction to another. A continuously applied force will not, then, simply produce continuous motion at the same speed, but an *increase* in speed. What was still missing here was Newton's idea of *inertia*, which means the resistance of a body not only to being moved from rest, but, crucially, resistance to having its *existing* motion either slowed down or speeded up, or its direction changed. It is this resistance that explains why a moving body (in space, of course) will go on in a straight line for ever, unless prevented, and why no extra force is needed to keep it going.

33. Kaye 1988:266.

34. One of the rare exceptions was Cardinal Nicholas of Cusa (1401–64), who experimented with growing a plant to prove, by use of the balance, that it absorbed weight from the air. He recommended the general use of the balance in experiment.

35. For example, 'Ores were assayed for economic value, and coins or jewellery to determine their quality and to detect fraud. The product was put through various processes, and weighing was carried out at appropriate stages with balances of various degrees of sensitivity. The most sensitive showed about 0.1 milligram. There were beam-lifting devices to protect the knife-edge from shock, and in the course of time these became more accurate with further refinements.' (Crombie:1990:88).

36. See Crombie 1979 (II):122-25 for a convenient summary.

37. 'Through experiment [experience] he gains knowledge of natural things, medical, chemical, and indeed in everything in the heavens or earth. He is ashamed that things should be known to laymen, old women, soldiers,

ploughmen, of which he is ignorant. Therefore he has looked closely into the doings of those who work in metals and minerals of all kinds. He knows everything relating to the art of war, the making of weapons, and the chase; he has looked closely into agriculture, mensuration, and farming work; he has even taken note of the remedies, lot-casting, and charms used by old women and by wizards and magicians…has been working for last years on mirror to produce combustion at fixed distance.' From the *Opus Tertium* of 1267 cited by Crombie 1990:53.

38. Eamon 1994:94.

39 Four beliefs, in particular, sum up the claim that the experience of the craftsman could be the basis of certain knowledge:

a. Matter is active, and one must struggle bodily with it to extract this knowledge.

b. Nature is primary, and certain knowledge resides within it.

c. This process of struggle is called experience, and is learned through replication.

d. This imitation of nature produces an effect – a work of art that displays the artisan's knowledge of nature and itself constitutes a kind of knowledge. (Smith 2004:149).

40. Price 1990:78.

41. Another example is the scholar William Gilbert while writing his book on the magnet, the *De Magnete* in 1600. He was also quite at home in the world of craftsmen, metallurgists and miners. For instance, he wished to see if the polarity of magnetite already existed while it was in the ground, and went down a mine to find out. "We had a twenty pounds' heavy loadstone dug and hauled out after having first observed and marked its ends in the vein. Then we put the stone in a wooden tub on water, so that it could turn freely. Immediately the surface which had looked to the North in the mine turned itself to the North on the water." Quoted in Henry 2001:101.

42. White 1978 311-12. For example: Vigevano (1335) wrote on prefabricated pontoon bridges, the assault tower with platforms hoisted by pulleys, and the siege-tower propelled by internal machinery; Conrad Keyeser (d.1405), wrote on giant trebuchets, rockets, multiple firing guns, a device for drawing the cross-bow, and the fusee; Fusoris (1365) wrote on clocks, the astrolabe, celestial spheres, and the orrery; Fontina built and wrote on organs, fountains, rockets, sand clocks, surveying methods, parabolic burning mirrors, alchemical furnaces, hydraulic engineering, and the combination lock; Arnault of Zwolle, physician to Philip the Good of Burgundy, wrote on clocks and planetaria, musical instruments, especially on the keyboard mechanisms that were being developed at this time, gem-polishing machines, compound pulleys, and wall-scaling apparatus for sieges. (White 1978:306-

11).

43. See, for example, Drover 1954.

44. In the bell-striking mechanism there was a wheel with a row of pins set alternately around each rim. As the wheel rotated, driven by a weight, these pins would strike first one side, and then the other, of a semicircular piece of metal on the end of a shaft. (See North 2005:180,181 for an illustration.) The shaft, and the cross-piece on top, would therefore oscillate, and the ends of the cross-piece would have struck a bell, or bells, but if weights are attached to each arm of the cross-piece this slows down its action, so that it can be used instead as an 'escapement' to control the speed at which the wheel rotates. 'It is easy to imagine that such a crude bell-ringing device, with a falling-weight drive left to run unchecked, suggested itself as the first mechanical escapement.' (North 2005:183). This controlled rotation of what was now the 'escape' wheel was the basis by which the rest of the gear-train of the mechanical clock could then be regulated, powered of course by a weight on a cord around the axle of the driving wheel. A full account of the invention of the mechanical escapement can be found at pp. 175-85, and Robertson 1931:14–19 also gives an earlier explanation of the escapement as based on an alarm mechanism. In the Whipple Museum of Science in Cambridge there is a working replica of Richard of Wallingford's clock, in which the striking mechanism and the escapement both operate in exactly the same way, in the manner described by North.

45. The only exception was Su Sung's clock in China. Although this was the first case of intermittent motion in time-keeping devices, the speed of the apparatus was still controlled by the amount of the water flow, whereas the medieval European clock was entirely self-regulating. Given the official secrecy that surrounded Chinese astronomy, it is scarcely conceivable that news of Su Sung's clock could ever have reached the West, and in any case we have seen that the Western clock-escapement derived from an alarm mechanism that played no part whatever in the astronomical clock of Su Sung. One must therefore respectfully disagree with Joseph Needham's belief that 'the idea of the escapement came over to Europe at the end of the 13th century as a kind of stimulus diffusion from China' (Needham 2004:20).

46. The culmination was the famous astronomical clock of de Dondi. The development of springs and the balance wheel in the fifteenth century made portable time-pieces possible, and the replacement of the verge-and-foliot by the pendulum in the seventeenth century was the basis of a completely new level of accuracy. In Chinese towns, public time was also kept by use of water clocks and the beating of drums.

47. Kepler said: 'My aim in this is to show that the celestial machine is to be likened not to a kind of divine living being but rather to a clockwork. . . in so

far as nearly all the manifold movements are carried out by means of a single simple magnetic corporeal force, just as in a clockwork all motions come from a simple weight. Moreover I show how this physical conception is to be presented through calculation and geometry'. (Crombie 1990:226-7).

48. Knight 1992:28.

49. From the *Opus Maius of* 1267. In the same work he also goes into one of his rhapsodies about the amazing possibilities of lenses, such as 'from an incredible distance we may read the smallest letters, and may number the smallest particles of dust and sand, by reason of the greatness of the angle under which we see them. . .', and that lenses could 'cause the sun, moon, and stars in appearance to descend here below'. But whether he actually experimented with combinations of lenses, and so had some practical idea of the telescope, is disputed. See Crombie1996:55-6 and Ronan 1991.

50. See Rosen 1956 for a very detailed account. The first reference to spectacles is in a sermon by the Dominican Friar Giordano, on 23rd February 1306: 'It is not yet twenty years since there was found the art of making eye-glasses which make for good vision, one of the best arts and most necessary that the world has', which therefore makes 1286 or a few years later the most likely date for their invention. (The theme of his sermon was Man's unfailing ingenuity, so that we can look forward to a constant flow of new inventions in the future.) Giordano also said that he personally knew the inventor. The Chronicle of St Catherine's Monastery in Pisa for 1313 tells us more. It says that 'Friar Alessandro della Spina was a modest and good man. Whatever had been made, when he saw it with his own eyes he too knew how to make it. Eyeglasses were first made by someone else, who was unwilling to share them [i.e. the secret of making them, because he wanted to keep the profits for himself]. Spina made them and shared them with a cheerful and willing heart.' Giordano was also a member of St Catherine's Monastery, and we can be fairly sure that eye-glasses were invented in Pisa, not by a member of the Dominican order, but probably by a layman who was nevertheless well known to them.

By 1300 spectacles were being produced by the glass-makers of Venice, and soon became familiar conveniences, particularly among elderly members of the literate classes such as the clergy and merchants. These first lenses were convex, for correcting the presbyopia or long sight of advancing years, and it took about another two hundred years or so before concave lenses were made for correcting short sight, myopia.

51. He was the son of Leonard Digges, mathematician, surveyor, and astrologer/astronomer, and close friend of Dr John Dee, also famous as astrologer, alchemist and mathematician, who had a large library of manuscripts, including some by Roger Bacon. Thomas says that his father

had been stimulated to work on the telescope by reading Bacon, and had used combinations of concave and convex lenses and mirrors to magnify distant images. Thomas had been adopted by Dr Dee on his father's death and had continued his work. In 1571 he published his father's *Pantometria*, with a description of a reflecting telescope, or 'perspective glass', as telescopes were first called, and the reality of the Digges telescope was confirmed in a report made for Lord Burghley in about 1580, which describes an actual reflecting telescope. (A successful replica of this has recently been made.) For a full account of all this see Ronan 1991. In a revised edition of his father's *Prognostication Everlastinge* of 1576, Thomas also refers to the use of the telescope for observing the heavens, and the support it provides for the Copernican theory.

52. Crombie 1990:326.

53. 'He described how he had melted a piece of very clear Venice glass so that it ran down a thread, then caught the globule that formed at the bottom of the thread and ground and polished it with a whetstone and fine earth to make a lens of less than one tenth of an inch across.' Inwood 2003:66.

54. 'Kepler solved the problem of the formation of the retinal image by first isolating the geometrical optics of the eye from the questions of causation and perception, inherited within the package of ancient and medieval theories of vision, which inhibited a purely geometrical physical analysis. He treated the eye as a *camera obscura* containing a lens.' (Crombie 1996:105).

55. See in particular Taylor 1942.

56. It is often claimed that Galileo invented the thermometer, but his books do not refer to it and there is apparently no mention of it in his unpublished papers either. While he did claim to have invented it, in a letter to a friend, this was after he had been told of Santorio's demonstration of the use of the thermometer.

57. Each cannon was cast from a different mould, the bore might not be parallel to the gun itself, and the ball did not fit the bore closely because a gap of quarter of an inch or so was left as 'windage'. Powder varied widely in quality, the flight of the shot depended on air pressure, temperature, and humidity, and the resistance of the air depended on the shot's velocity.

58. Hall 1952:16.

59. A number of scientists, in the best tradition of Mr Gradgrind, see Tartaglia as a no-nonsense practical experimentalist, basically concerned with facts, not theoretical speculation. For example, 'When I think about which one event I would pick to mark the birth of modern science, I often come to a date in 1537. The Duke of Milan had just acquired some cannon. . .and had some questions about how to use his new toys. He summoned his chief engineer, a mathematician named Niccolo Tartaglia. . . and asked him a simple question:

At what angle should we set the barrel to insure maximum range for the cannonball? What happened next typified the new spirit of inquiry that was sweeping though Europe. Tartaglia didn't consult books written by ancient philosophers, nor did he retire to his study to ponder the question. Instead, he took some cannon to a field outside of Milan and started firing them. His result – that a barrel elevated at an angle of 45° gave the maximum range – would later be derived from Newton's Laws of Motion, but for Tartaglia it was simply a rule-of-thumb, seat-of-the-pants piece of engineering'. (Trefil 2002:89).

Tartaglia's own account is rather different: 'When I dwelt at Verona in 1531 . . . I had a very close and cordial friend, an expert bombardier at Castel Vecchio. . .He asked me about the manner of aiming a given artillery piece for its farthest shot. Now I had no actual practice in that art (for truly, . . . I have never fired artillery, arquebus, mortar, or musket); nevertheless, desiring to serve my friend, I promised to give him shortly a definite answer. And after I had chewed over and ruminated on this matter, I concluded and proved to him by physical and geometrical reasoning how the mouth of the piece must be elevated in such a way as to point at an angle of 45 degrees above the horizon.' (Drake and Drabkin 1969:63-4). Tartaglia goes on to say that his theory was tested successfully in the following year by some gunners at Padua, but he himself always remained a pure theorist, greatly influenced by Aristotle, and does not ever seem to have carried out any experiments with artillery. (Professor Trefil has thanked me for drawing these facts to his attention, and tells me that this myth of Tartaglia as practical experimenter occurs in a number of modern histories of science.)

60. See Drake and Drabkin 1969:63-143 for translations of the *Nova Scientia* and the *Questions*, where Tartaglia's ideas are set out very clearly. Aristotle's trajectory had been challenged by medieval philosophers, and the idea that part of it curved was not new. See Crombie 1979 (II):86-7.

61. Digges was very critical of Tartaglia's claim that an elevation of 45° attains maximum range, and of his theory of the path of a projectile. While Digges still adhered to Aristotle's idea that this is made up of natural and violent motion, he proposed that the path will always be a conic section: at 45° it will be a parabola, at a higher elevation a hyperbola, and below 45°, an elliptical arc, suggestions that are much closer to the truth.

62. Though cannon were first used to knock down castle walls, as soon as they began to be used to defend castles against the besiegers, an entirely new kind of fortification could be developed, based on geometrical principles. Whereas the medieval castle was essentially defensive, relying on high walls to deter scaling ladders, thick ones to deter sapping, and arrows and boiling oil to repel infantry, the new artillery castles could use cannon to go on

the offensive against the besiegers. Since cannon shoot in straight lines for hundreds of yards, especially when they have plunging fire, a whole set of ingenious geometrical constructions could be developed, whereby the new artillery bastions could protect one another, and create killing zones to trap the besiegers. From the sixteenth century, military engineering became, like gunnery, a theoretical as well as a practical topic, and many books were written on it. 'In the sixteenth century one can see emerging the idea that war called for an acquaintance with mathematics; many engineers such as Stevin, Pagan and Francois Blondel wrote treatises on mathematics, and others, who were primarily mathematicians, such as Galileo, Marolois and Bélidor, concerned themselves with the art of fortification.' (Hoppen 1999:21-2) These developments in fortification did not stand alone, however, but had a very important impact on the whole idea of rational town planning: 'The military treatises of the Renaissance, which inextricably linked the new defences with theories of town planning, which themselves reflected the current ideas of harmonious and rational proportion, undoubtedly promoted the concept of the ideal city.' (Hoppen 1999:25-6).

63. Crombie 1996:473.

64. Crombie 1996:89-90.

65. While Leonardo da Vinci published no books, he was the supreme example of the artist-engineer, whose insights into physical nature covered everything from the nature of flight to military machines and anatomy. The machine, in particular, was central to Leonardo's imagination: 'Since the world is a living organism, then even machines must be considered "living" things. . .because man, like God who in the beginning set the great machine of the Universe in motion, gives inert mechanisms and cold gears a "Spiritual force" which animates them "as if they were alive"'. . . 'How else can the Last Supper be defined if not as a "machine" based on mathematics (its perspective) and set in motion from the centre by Christ's gesture which activates the gears of the feelings and of the reactions of the Apostles.' (Cianchi n.d.:16).

66. Crombie 1994:204-5.

67. We know a good deal about his pioneering work on perspective from his biography by Antonio Manetti, although the first book on perspective was by Leon Alberti, *De Pictura*, in 1436, and dedicated to Brunelleschi.

68. Crombie 1994:432-3.

69. 'We must imagine [Brunelleschi in a room] standing before a flat mirror considering how, as his own visual axis struck the centre of the mirror surface perpendicularly , the mirror itself served as the basis of his "visual pyramid". He must then have noticed how the edges of all things parallel to each other in the room where he was standing, were reflected in the mirror as converging in perspective to points precisely level with his own optical plane (that is,

with his own eyes as reflected in the mirror). Furthermore, if the mirror were centred on one wall of a perfectly quadrangular room, and he looked squarely into it, he would notice that all the receding edges of floor and ceiling (the lines perpendicular to the wall on which the mirror was hung) appeared to be converging directly on a point identical to his reflected eyes themselves. Accordingly, if he raised or lowered himself, the locus of the converging lines always rose or declined, remaining identical with his changing eye level in the reflection.

Noting this, Brunelleschi could have concluded that the visual pyramid was actually reversed in the mirror; the apex – his eyes – of the first pyramid on the viewer's side of the mirror was connected by the visual axis and *cathetus* to the apex of the second pyramid as reflected in reverse. . .The apex on the mirror side also marked the centric vanishing point. Brunelleschi would thus have come upon the fundamental logic of what came to be known as "one point" or "frontal perspective". (Edgerton 1975:136-7). Here it is worth pointing out that the Romans, too, had large silver mirrors, and Seneca, for example, could have observed the same thing.

70. Brunelleschi painted a perspective view of the Baptistery from the door of the Cathedral, on a wooden panel, showing landmarks on either side and some of the piazza in front. He then bored a peep-hole in the picture, through the *back* of which an observer could look at the *reflection* of the picture in a mirror held in front in the other hand. When picture and mirror were properly lined up with the actual scene, 'it seemed as if the real thing was seen', said Manetti, who had himself tried it out many times. (The illusion was particularly effective because Brunelleschi had represented the sky with burnished silver-leaf, so that real clouds were reflected in it.) The whole point of the mirror (which otherwise seems a curiously elaborate contrivance) was to demonstrate experimentally the optical theory that Brunelleschi had worked out:

'The viewer, as he observed his hand-held mirror through this hole from the back of the picture . . . would have noted that all the lines of the piazza and architectural setting, which the artist had painted extending into the distance around the Baptistery, converged on a point identical with his own viewpoint. This hole thus provided scientific proof of the basic vanishing point principle in a most ingenious way. It dramatized for the viewer the optical fact that the visual axis, when directed straight ahead, defines the horizon – and therefore the single, unifying vanishing point.' (Edgerton 1975:137).

71. See David Hockney's very interesting study (2001) of the impact of optical aids on Renaissance painting.

72. Ancient church music, Gregorian chant, had consisted of a single melody closely tied to the words of the text that was being sung. But the development

of multiple melodies, polyphony, from the twelfth century onwards (a unique invention of Western culture) had meant that each note now had to be given a definite time-value to organize the different parts of the singers into a coherent whole. This in turn involved the idea of the bar, a fixed time-interval into which notes of different lengths had to be fitted, and in relation to different rhythms. Polyphony also, of course, involved complex harmonies, and the result was that, especially with increasingly sophisticated instruments, music by the fifteenth and sixteenth centuries had become extremely elaborate and mathematical.

73. Vincenzo Galilei, father of Galileo, was one of the most important musical experimenters and theorists of the sixteenth century. He was a skilled lutanist, mathematician, and musical preceptor to the Florentine musical academy, the Camerata. He translated the work of the ancient Greek musical theorist Aristoxenus, and explicitly followed Aristoxenus in trying to build musical science up from auditory sensation, instead of imposing on it a rigid mathematical scheme in the style of the Platonists and Pythagoreans. He used experimental methods, and measurements, and one of his discoveries was that the traditional Pythagorean ratio of 2:1 for the octave only applied to the lengths of strings, and that for the tensions of strings the ratio was 4:1.

74. 'Theology made Mersenne an apostle of science. When he came to offer, in a series of works published from 1633 to 1637, a programme for natural philosophy and its implementation, he did this in the first place in response on the one hand to magic and the occult and on the other to scepticism, and he did it with a combination of scientific with theological goals.' (Crombie 1990:403).

75. For example, to investigate the harmonic properties of vibrating strings he stretched two strings on a frame. One acted as the control, while in the other he kept all the relevant quantities (length, tension, density) constant except one, which he adjusted until it sounded in unison with, and hence vibrated with the same frequency as, the control. To discover the actual frequencies of notes of different pitch, he counted the slow vibrations of very long strings against time measured by pulse beats or a seconds pendulum. Then he used some general laws he had discovered, relating frequency to the length, tension and density of strings, to calculate the frequencies that were too rapid to count. On this he based the first experimental proof that musical intervals of octaves, fourths, and so on were determined by the frequencies of their vibrations, and that a harmonious sound is produced in the ear by two matching sound waves, while dissonance increases as coincidence of the waves decreases. This finally explained why the Pythagorean harmonies had to be whole tones, because these corresponded to whole sound waves. Again, he measured the upper and lower limits of audible frequencies, and

investigated the speed of sound, first by timing the difference between the flash or smoke and the report of a gun fired at a given distance. Later he timed the return of echoes, and the variations in these led him to realise the effect of atmospheric conditions on the speed of sound. He showed that the speed of sound was, however, independent of volume and pitch, and that the volume of a sound (like the intensity of light) is inversely proportional to the square of the distance from its source. (Crombie 1990: 408-9).

76. Crombie 1996:103.
77. Crombie 1990:323.
78. Knight 1992:29.
79. Crombie 1990:342.

## Chapter XVI Modern Science and Industrialism

1. e.g. Hessen 1931.
2. The Flemish Simon Stevin is an excellent example of this social type. He probably began as a merchant's clerk, and rose to become commissioner of public works and quartermaster-general to Prince Maurice of Nassau. He did fundamental research in statics, taking up the work of Archimedes and Hero in studies of the parallelogram of forces, buoyancy, and the nature of pressure in fluids; introduced the first form of decimals, especially for merchants; published comprehensive tide-tables, devised the first system of equal temperament in music, as well as designing windmills, sluices, and other public-works projects. He wrote all his books in Dutch, not Latin.
3. See Cooper 1935.
4. In *Two New Sciences* he also refutes the Aristotelians by a simple thought-experiment. Imagine, he says, a light and a heavy body tied together by a rope. In theory, the lighter body A should retard the heavier body B, because A falls more slowly than B but, on the other hand, the combination of the two bodies, being heavier than B, should fall even faster, an obvious contradiction.
5. To do this he first proved geometrically that a ball rolling down a slope with a vertical drop of, say, ten feet, will have reached the same speed at the bottom of the slope as if it had been dropped vertically for ten feet. He then took a twenty-foot plank of wood, with a groove along the edge lined with parchment, very smooth, regular, and highly polished. The plank was set up at a variety of different gradients, and a perfectly regular bronze ball was allowed to roll down from the top. Different intervals were marked off, and the times that the ball took to reach the different markers were measured. To time the ball's descent, he used a water-clock, a tank of water with a tube at the outlet that could be opened and closed. As the ball was allowed to roll, the water began to flow, until the ball passed one of the markers, when the flow was

stopped. The water that had been collected in a glass was then weighed in a very accurate balance, to give a measurement of the time taken. His notebooks recording these experiments have survived, and recent work duplicating his experiments has confirmed that they were extremely accurate.

6. '. . .inertia is a dynamic concept, whereas Galileo's published physics avoided the notion of forces in nature and remained purely kinematic.' (Drake 1990:68) Although he was quite familiar with impetus theory, in his mature work on motion, especially in the *Two New Sciences*, Galileo ignored the whole problem of force and how it related to motion, because he had no means of measuring it. 'Having no measure of force, he was content to discuss it only in general terms without trying to reduce it to a science. Force remained for him an indefinable term, as did energy . . .' (Drake 1990:221).

7. He devised another set of experiments to test this, which confirmed his geometrical theory that its path should be a parabola, not the combinations of straight lines that Aristotle had supposed, nor the trajectory of Tartaglia, but the angle of maximum distance would indeed be 45°. Since this did not take air resistance into account, the actual flight of cannon balls in the air is vastly more complicated, and it had no practical relevance to the art of gunnery.

8. 'Earth has intercourse with the Sun, and is impregnated for its yearly parturition', meaning its annual growth of vegetation and life of all kinds.

9. 'Now it seems to me gravity [weight] is but a natural inclination, bestowed on the parts of bodies by the Creator so as to combine the parts in the form of a sphere'.

10. 'The ear equipped to discern geometric harmony could detect a new neatness and coherence in the sun-centred astronomy of Copernicus, and if that neatness and coherence had not been recognized, there might have been no revolution.' (Kuhn 1957:172.)

11. The stellar parallax was finally discovered by Bessel only in 1838 – and there was no hard experimental evidence that the earth rotates on its axis. While it was known that the earth was flattened at the poles, and bulged at the equator, which was consistent with its rotation, the first actual experiment that proved the earth's rotation was performed by Foucault in Paris in 1851. He used a long pendulum and heavy weight to show that the arc of the pendulum's swing rotated in the course of a day, and that this could only be explained in relation to the earth's rotation. In the next year he proved the same thing by use of the gyroscope which he invented.

12. The whole dispute could, in theory at least, have been avoided. St Augustine, especially in his book *On the Literal Interpretation of Genesis*, had taken the view that the Bible was fundamentally about faith and morals,

not science. 'It is also customary to ask what one should believe about the shape and arrangement of heaven according to our Scripture. In fact, many people argue a great deal about such things, which with greater prudence our authors [of the books of the Bible] omitted... For what does it matter to me whether heaven, like a sphere, completely surrounds the earth, which is balanced in the centre of the universe, or whether like a discus it covers the earth on one side and from above?... briefly, it should be said that our authors did know the truth about the shape of heaven, but the Spirit of God, which was speaking through them, did not wish to teach men these things which are of no use for salvation.' He specifically warned the Church that, since unbelievers might know more about astronomy, for example, than Christians did, it was folly to interpret the Bible in ways that conflicted with science, because that would only bring the whole Bible into ridicule among unbelievers. Augustine therefore avoided the entire dispute between science and Scripture, and it was to Augustine, in particular, that Galileo appealed in his *Letter to the Grand Duchess Christina of Tuscany*, which is a masterly and entirely modern argument that the Books of Scripture and Nature are quite separate, and have to be interpreted by the completely different methods of theology and natural science.

While the Church did not reject Augustine, it did believe, however, that when the Bible referred to the facts of nature it should be presumed to be describing them accurately unless it could be proved that the literal meaning was untrue, when the passages would have to be reinterpreted. Until a scientific theory that contradicted Scripture was definitely proved, it could only be defended as a theory, not as a fact. For example, Cardinal Bellarmine, an admirer of Galileo and a leading figure in his dispute with the Church, wrote that it is one thing to suppose that the sun may be at the centre, and quite another to prove that it really is, 'for I believe the first demonstration may be available, but I have very great doubts about the second, and *in case of doubt one must not abandon the Holy Scripture as interpreted by the Holy Fathers*' [my emphasis]. He agreed that if it could be proved, the Church would have to accept it. 'If there were a true demonstration [of this] then one would have to proceed with great care in explaining the scriptures that appear contrary, and say rather that we do not understand them than that what is demonstrated is false. But I will not believe that there is such a demonstration, until it is shown me.' (cited by Finocchiaro 1989:68).

The problem for Galileo was that many astronomers did not accept the Copernican theory, and Galileo certainly could not provide the rock-solid proof demanded by the Church. Indeed, some of his attempts to do so, such as his theory of tides, and his use of sun-spot evidence, were justifiably rejected by many astronomers. Galileo, however, took the different view that it was not for him to have to provide a complete proof of the Copernican theory, but that the Church had to disprove it: 'whoever wants to condemn it judicially must first demonstrate it to be physically false by collecting the reasons against it'. (Ibid., 81) Bellarmine had had to admit that Scripture might not always mean what it appeared to mean, and Galileo used this to argue that, to discover the truth about some aspect of nature, the correct procedure was therefore to establish the facts scientifically, and then see what the Bible had to say. 'When the factual and physical truth has been found in this manner, then, and not before, can one be assured of the true meaning of Scripture and safely use it. . . If the earth *de facto* moves, we cannot change nature and arrange for it not to move. But we can rather easily remove the opposition of Scripture with the mere admission that we do not grasp its true meaning. Therefore, the way to be sure not to err is to begin with astronomical and physical investigations, and not with scriptural ones.' (Ibid., 82) (The Church, needless to say, was not impressed by this attempt to use science to interpret Scripture.) He admitted that the Copernican view was not fully proved, but it was nevertheless a far better explanation of the facts than the traditional view. 'One is not asking that in case of doubt the interpretations of the Fathers should be abandoned, but only that an attempt be made to gain certainty regarding what is in doubt, and that therefore no one disparage what attracts and has attracted very great philosophers and astronomers. Then, after all necessary care has been taken, the decision may be made.' (Ibid., 85.) This was an entirely reasonable position, but it was at odds with his claim that Copernican theory was true.

Copernicus had put forward his theory with the pretence that it was only a mathematical device 'to save the appearances'. Although neither he nor most of his readers believed this, and he was fiercely denounced by Luther and Calvin, and by the Aristotelian professors in the universities, the Church itself adopted no formal position on his theory until 1615-16, when a complaint was made to the Congregation of the Index of Prohibited Books that Copernican theory was heretical because it contradicted the Bible, and Galileo was linked with this complaint as a notorious defender of Copernicus. The Congregation appointed a committee to examine the matter, and it found that Copernicus had, in some passages, claimed that his theory was literally true, and that Galileo had also taught this. The report concluded that the theory was both scientifically false, and contrary to Scripture, and the Congregation placed Copernicus' book on the Index until the offending passages had been rewritten (in 1620). Galileo was ordered 'to abandon completely the above-mentioned opinion that the sun stands still at the centre of the world and the earth moves, and henceforth not to hold, teach, or defend it in any way whatever, either orally or in writing.' (cited in Finocchiaro 1989:147). This, rather than his later trial by the Inquisition, was the really fateful decision that finally broke with the Augustinian tradition, because it made further research on a condemned theory virtually impossible. When Galileo published his *Dialogue on the Two World Systems* in 1632, defending the Copernican view against that of Ptolemy, he pretended that it was really a defence of the Church's position, but no one was deceived, and he was summoned before the Inquisition, where his condemnation was inevitable in the circumstances. Pope Urban VIII was personally enraged with Galileo, whose friend he had once been, because he thought Galileo had deceived him about the nature of the book, and had held him up to ridicule in the person of Simplicio, an idiot defender of Ptolemy in the *Dialogue*. In any case, at this point in the Thirty Years War against the Protestants, and when the Papacy was particularly vulnerable, he considered the interests of the Church more important than astronomy: 'In fact, the Pope believes that the Faith is facing many dangers and that we are not dealing with mathematical subjects here but with Holy Scripture, religion, and Faith.' (Letter of the Tuscan Ambassador at the Vatican, cited in Finocchiaro 1987:232.)

The Cardinals of the Congregations of the Index and the Inquisition were well aware of Galileo's Augustinian arguments for the separation of science and Scripture, but ignored them. Not only did they fail to prevent the ultimate vindication of Copernicus, but they did so in a way that brought the maximum discredit on the Church. 'It was not until 1757 that Pope Benedict XIV annulled the anti-Copernican decree. At length in 1893 Pope Leo XIII made the *amende honorable* to Galileo's memory by basing his encyclical *Providentissimus Deus* on the principles of exegesis that Galileo had expounded...' (Crombie 1979 (II):225).

13. Kuhn 1957:233.

14. Mars has the greatest irregularities because, in fact, its orbit is the most elliptical, and although he did not of course realise this, it is the most suitable of the planets to demonstrate elliptical orbits. Gascoigne's micrometer, a very accurate measuring device attached to a telescope, showed in 1639 that the moon's image changes slightly in size during the month, proving experimentally that it moves in an elliptical orbit.

15. In the *Mysterium Cosmographicum* he had attempted to prove that there had to be six planets because there were five intervals between them, and these five intervals could be exactly filled by the five Platonic solids: Saturn – cube – Jupiter – tetrahedron – Mars – dodecahedron – Earth – icosahedron – Venus – octahedron – Mercury. This therefore seemed to explain beautifully why God had created only six planets, and why the distances between their orbits were as they were.

16. '[T]here exists only one moving soul [which he later changed to 'force']

in the centre of all the orbits, that is the sun, which drives the planet the more vigorously the closer the planet is, but whose force is quasi-exhausted when acting on the outer planets because of the long distance and the weakening of the force which it entails.' (*Mysterium Cosmographicum*, Chapter 20.) The discovery of sun-spots made it possible for astronomers to detect through the telescope that the sun revolves on its axis. Galileo, like Kepler, thought that it might be some attractive force from the revolving sun that swept the planets round in their orbits.

17. The third law says that the more distant a planet's orbit from the sun, the slower it will move, so that its year will not only be longer, obviously, because it has farther to go, but also because it is actually travelling more slowly around its orbit. More precisely, the square of a planet's period of revolution is inversely proportional to the cube of its average distance from the sun.

18. Stephenson 1987:204-5. Kepler published his *Epitome* in 1621, which is an essentially modern model of the solar system, with all the cycles and epicycles of Copernicus removed, and the *Rudolphine Tables* in 1627, based on Tycho's observations, which provided the essential data for Newton's final synthesis.

19. Inwood 2003:83.

20. He also guessed, on the analogy of light, that attraction decreased by the 'inverse square' law: at twice the distance it would only be a quarter as strong, and so on.

21. '[Newton] did not stumble into alchemy, discover its absurdity, and make his way to sober, "rational" chemistry. Rather he started with sober chemistry and gave it up rather quickly for what he took to be the greater profundity of alchemy.' (Westfall 1993:112).

22. McGuire & Rattansi 1966:136-37.

23. Momentum, however, is mass times velocity, the 'quantity of motion', (which Newton had had considerable difficulty in distinguishing from inertia).

24. The force of gravity is proportional to the combined masses of, say, the earth and the moon, and inversely proportional to the distance between them. So if the moon were twice as far away, it would only have one quarter of its pull on the tides. 'According to Newton himself, the "notion of gravitation" came to his mind "as he sat in a contemplative mood," and "was occasioned by the fall of an apple". He postulated that, since the moon is sixty times as far away from the centre of the earth as the apple, by an inverse square relation it would accordingly have an acceleration of free fall $1/(60)^2 = 1/3600$ that of the apple.' (Cohen 1974:61).

25. Alcohol vaporizes at 160° F, so had hitherto boiled away without being noticed, and it was only the innovation of cooling the still-head that allowed

its distillation. (Initially known as *aqua vitae* (90%) or *aqua ardens* (60%), it was only called 'alcohol' by Paracelsus much later.)

26. A good example of the effect that the search for the philosopher's stone had on alchemical research is the discovery of phosphorus by Hennig Brand of Hamburg in 1669. He accumulated large quantities of urine – on the basis that it, like gold, is yellow, and it was also an old alchemical maxim that gold was to be found in the dregs of the body. He allowed this to become very stale over several months until it was a black, decomposing mass and then added sand to it (his sand, like urine, was yellow). He then distilled the mixture with the mouth of the retort under water. The result was phosphorus, the first element to be discovered that cannot occur in nature because it is highly reactive with air and so only exists as a compound.

27. Paracelsus had added the principle of 'salt' to the old alchemical principles of mercury and sulphur, and while the traditional four elements were still accepted, some believed that salt, representing the body, mercury, representing water or spirit, and sulphur, representing fire or the soul, were even more important.

28. Pierre Gassendi (1592-1655) French priest, scientist, and professor of mathematics at the Collège Royal in Paris, who was also a general supporter of Galileo's ideas, in fact showed that atoms could easily be assimilated into Christian doctrine simply by assuming that God had created them, and that only God could destroy them.

29. Some scientists such as Boyle and Newton still regarded alchemical transmutation as possible, although others denied this. Boyle dismissed the traditional elements of air, water, and earth as merely matter at different densities, and fire as particles in motion. But he also thought that the atoms, or corpuscles, could combine into larger groups or 'mixts', and that these groups could act in a unitary way, as the new elements, during chemical reactions. Why, then, should transmutation not be possible by changing the composition of these mixts during alchemical processes? This problem of whether the individual atoms or their mixts were the basic units in chemical reactions was to remain a central issue for the next two hundred years, until it was finally established in the nineteenth century that atoms are indeed elements.

30. He heated chalk in a retort and collected the resulting carbon dioxide in the pneumatic trough: in this process chalk, calcium carbonate, $CaCO_3$, becomes slaked lime $CaO$, calcium oxide, and carbon dioxide, $CO_2$ but when, using an extremely precise balance, he weighed the chalk and then the slaked lime and the carbon dioxide, 'fixed air', he founds that their weights were identical.

31. Leicester 1965:32.

32. Becher had similar mystical ideas to van Helmont, and believed that in the plan of the Creator organic compounds were of central importance, while metals were merely a bye-product. Combustion, then, had to be explained by the burning of organic substances. In his view of the elements, there were three types of earth: the vitreous, the fatty, and the fluid (corresponding to salt, sulphur, and mercury), and the principle of combustion was therefore 'fatty earth', *terra pinguis*, an extension of the old view that sulphur was the principle of combustion, and no advance on Aristotelianism. The physician Georg Stahl (1660–1734) took up Becher's principle, which he called *phlogiston*, the Greek for 'something inflammable'. It is hard for us to get our heads around this idea, as it works in exactly the opposite way to oxygen: so when metals were heated, instead of gaining oxygen they *los*t phlogiston, which was carried away by the air; this purely mechanical function of the air therefore explained why it was necessary for combustion. Plants could obtain phlogiston from the air, and animals from plants, and when charcoal, which was very rich in phlogiston, was burned with metals it restored phlogiston to them. The most obvious objection to the theory is that since metals gain weight when they are oxidised how, then, can they *lose* phlogiston?

33. Leicester 1965:123.

34. He heated tin in a sealed vessel, but there was no gain in weight until he opened the vessel, when air could be heard rushing in. At this point oxidation became complete, and the oxide increased in weight, proving that the function of the air was not just to carry away the phlogiston, but contributed part of its volume to combustion.

35. Priestley used the burning glass because it produced pure heat without any form of combustion, which might have contaminated the experiment. Since air that would not support combustion was dephlogisticated air, this type of gas had to be phlogisticated air.

36. 'Lavoisier did not disbelieve in atoms, but he believed that chemists must adopt limited objectives; we could not know what the ultimate particles of things were like, and we should therefore found a science upon what we could know rather than upon what we could not.' (Knight 1992:52.)

37. Here we have to remember that, since Boyle, it had been supposed that the air particles are more or less static, (not in rapid motion as we know them to be), and that they repel each other, which is why it takes the pressure of a pump to compress them. Dalton reasoned that if the particles of nitrogen and oxygen all repelled one another, in time the heavier oxygen particles must eventually move to the bottom. But if only *similar* particles repelled one another, and had no effect on different particles, then there could be no tendency for layering to occur. He later came to realise that heat played a crucial part here, because under constant pressure, gases expand with heat.

He altered his theory so that the sizes/weights of the oxygen and nitrogen particles were different, and the mutual repulsion of similar particles was an effect produced by heat. While this work had no further direct effect on his atomic theory, (though he made fundamental contributions to the laws of gases), we can see that the relative weights of different types of atom had therefore become a crucial factor in his ideas about how they interacted.

38. 'Those [gases] whose particles are light and simple being least absorbable, and the others more as they increase in weight and complexity.' (Thackray 1972:75). This idea set him off on a new enquiry into how ultimate particles combine by relative size and weight: 'Other bodies besides elastic fluids [gases], namely liquids and solids, were subject to investigation . . . Thus a train of investigation was laid for determining the *number* and *weight* of all chemical elementary principles [elements, as we would say] which enter into any sort of combination one with another.' (Thackray 1972:83.)

39. His fundamental, philosophical, assumption was that these combinations must take the simplest possible forms. So if the atoms of two different elements, A and B, can combine in one or more ways, they will always do so in the order of simplicity:

| 1A + 1B | produces | 1 | compound | C (binary) |
|---|---|---|---|---|
| 1A + 2B | " | 1 | " | D (ternary) |
| 2A + 1B | " | 1 | " | E (ternary) |
| 1A + 3B | " | 1 | " | F (quaternary) |
| | **and so on** | | | |

In other words, if we find compounds of quaternary form, the same elements must also combine in ternary and binary forms as well.

40. The Law of Multiple Proportions says that if 2 elements, A and B, combine to form different compounds, then the various weights of A which combine with a fixed weight of B, e.g. 2A + 1B, 3A + 1B, bear simple ratios to one another – here, 2 : 3. The Law of Reciprocal Proportions says that if 2 elements, A and B, combine separately with a third element, C, then the weights of A and B which combine with a fixed weight of C have a simple ratio to one another. (Greenaway 1966:133.)

41. Lavoisier was chiefly responsible for supplying the French government with explosives of good quality, and solved many other technical problems for his countrymen. Berthollet was active in the French dyeing and bleaching industries, Gay-Lussac contributed his tower to the sulphuric acid manufacturers, Davy invented the miner's safety lamp, and many important industries expanded from the laboratory to the apothecary shop under the

influence of the Napoleonic Wars and the Continental Blockade. Production of beet sugar increased, and the Leblanc process for producing soda was also discovered. In fact, large-scale chemical industry may be said to have begun at this time. (See Hall 1952.)

42. Roger Bacon described him as 'knowing natural things by experience, medicine, alchemy, and the secrets of nature', in other words, a fellow natural magician. (Hackett 1997:312.)

43. 'It is the very obvious combination of animism and empiricism in Gilbert's *De Magnete* which has caused different historians to make conflicting claims as to his significance in the history of science. Because of the subsequent extreme divergence of experimentalism and animism in western culture, historians seem to have had difficulty in recognizing the simple fact that Gilbert worked at a time before that divergence had begun, when animistic thinking and the experimental method were both to be found in the tradition of natural magic.' (Henry 2001:117.)

44. Otto von Guericke belonged to a patrician family of Magdeburg, and was a university-educated engineer, who also represented Magdeburg at peace conferences and princely courts during the Thirty Years' War, being Mayor of Magdeburg 1646-76. He spent his leisure in scientific research.

45. See Williams 2000.

46. See Dibner 1964:40-50.

47. In around 1610 we have the Huguenot Salomon de Caus at the court of King James I, demonstrating his fountain by which a boiler could project water forty feet in the air by steam power. (He was a general wonder-worker, in the tradition of Hero, and moved on to the Court of Heidelberg.) David Ramsay, clockmaker to James I and Charles I, and a prolific inventor, was familiar with his work, and was awarded a patent in 1631 for raising water by fire, although it is not known how he intended to do this, or if it led to any practical results. Raising water was a very fashionable pre-occupation: 'of the fifty-five patents granted for inventions granted during the reign of Elizabeth, 1561-99, one in seven is for the raising of water, and of the 127 similar patents granted between 1617 and 1642, the same proportion is observable' (Dickinson 1963:16). The Marquis of Worcester (1601–67) was a young courtier and keen inventor who took up this work, receiving a patent for an engine to raise water by fire in 1663. This device actually operated at Vauxhall Gardens, probably to run an ornamental waterfall, but no adequate description of it survives, and his description of it published in his *Century of Inventions* (1663) is also vague. But there was a strong tradition that it was the direct inspiration of Savery's engine.

48. In this, steam from a boiler was fed into a large copper vessel, where it condensed. At the bottom of the vessel, a pipe led down to water at most

16 feet beneath it, because this was the practical limit of its suction, and as the steam condensed it sucked up this water into the vessel. More steam was then fed into the vessel, which now forced the water up another pipe for as much as forty feet where it discharged, and the water was prevented from falling back down the pipes by non-return valves. Savery's engine had two such vessels, working alternately, and they were cooled by cold water being poured over them.

49. Steam to force the water up the pipe had to be at several atmospheres of pressure, which, together with the excessive heat involved frequently burst the soldered joints, while the repeated cooling and heating of the vessels, and the amount of steam needed to pressurize the cold water involved an extravagant consumption of fuel. Savery's engine was only used, in fact, for supplying water for some water-wheels, and filling the water-tanks of one or two country mansions, so that this line of development was therefore a dead end in the use of steam power.

50. In France Pascal, who had heard of Torriccelli's work, took a tube and tied a bag of, then sank the tube into a deep vessel of water with the mouth of the tube above the surface. He showed that the deeper the bag of mercury was sunk in the water, the more mercury flowed up the tube, so confirming that pressure increases with depth. He then gave his brother-in-law the task of carrying one of Torriccelli's barometers up a neighbouring mountain, and he found, as Pascal had predicted, that the higher up the mountain he went the lower the mercury sank in the tube. One is struck by two things about these experiments: how readily men of the seventeenth century used them to settle theoretical questions, and how technically simple they were to perform.

51. Papin had been assistant to the great scientist Christiaan Huygens (1629–1695). They knew all about von Guericke and Torricelli, and spent some years in the early 1670s in Paris, trying to create vacuums in pistons by exploding gunpowder. To us it seems obvious that they should have used the condensing power of steam, but in the minds of scientists of the day this was narrowly associated with the suction of water, and not immediately thought of as something that could produce a vacuum anywhere.

52. Plate taken from Rhys Jenkins 1936:92.

53. Rolt & Allen 1977:34, who give a very good account of Newcomen's social background.

54. Indeed, Newcomen was specifically said by a contemporary who knew him (the coal-owner Stonier Parrott) to have derived his basic idea for his steam engine from Papin. (Musson 1963:vi.)

55. The major engineering problems that Newcomen had to overcome were: (a) The vertical motions of piston and pump rods had to be connected to a rotating beam – see picture.

(b) It was very difficult in those days to fabricate large cylinders with a true bore, so that air would easily leak past the piston and destroy the vacuum. Newcomen solved this with a leather washer on the piston, reinforced by a water seal on top of the piston that was essential and highly effective.

(c) Efficient working needed the vacuum to be created as quickly as possible. Newcomen began with a cold-water jacket around the cylinder, but this does not seem to have been very effective. It is likely that an accidental leakage of this water into the cylinder proved the importance of actually injecting cold water inside the cylinder to produce an instant vacuum. (See Rolt & Allen 1977:41-2.) Water injection was one of the most important features of the Newcomen engine.

(d) The steam entering the cylinder would also have drawn air in with it at each stroke. Since, unlike steam, this air could not condense, after a few strokes the engine would have stopped ('wind-logging'), so Newcomen had to install a 'snifting valve' through which this air could be blown off by the incoming steam.

(e) The injected water, and the water from the condensed steam, had to be drawn off by an 'eduction valve' at the bottom of the cylinder.

(f) The various valves to control the operation of the engine had to be controlled automatically by the engine itself. (Hills 1989:23-8 gives a good account of all these problems.)

56. This table was compiled in 1721 by Henry Beighton FRS, a collaborator of Newcomen. From Dickinson 1963:44.

*A Physico-Mechanical Calculation of the Power of an Engine.*

[table of engine power calculations showing Diameter of the Pump in Inches (4 to 10), Draws at a 6 Foot stroke in Ale Gallons, Draws at 16 Strokes in a Minute in Hogshead Gallons, and Diameter of the Cylinder values against The Depth to be Drawn in Yards (15 to 100)]

57. This historic engine is still preserved in the Hunterian Museum of Glasgow University. In his experiments Watt was puzzled by the amount of cold water needed for this, and found that when steam is passed through ice-cold water it can raise about six times its own weight of that water to boiling point. He discussed this result with his colleague, Professor Joseph Black, who was

making fundamental contributions to the study of heat. Thermometers were now extremely accurate, and just as he had used the balance in chemical experiments, so Black used the thermometer to investigate 'the distinction between *heat* and *temperature*, or *quantity of heat* and *intensity of heat*. In experiments with ice cubes, he realised that raising the temperature of different materials to a given degree could require different amounts of heat, regardless of the actual amount of matter, so that each substance has its own 'capacity for heat', or specific heat. In addition, 'definite quantities of heat disappear during certain changes of physical states, such as melting and evaporation. He also demonstrated that the same quantities of heat reappear during the reverse changes, freezing and condensation. Black called this disappearing and reappearing factor the "latent heat"...' (Singer 1943:299), and see Cardwell 1963. Specific and latent heat were the sources of Watt's problems, and the steam engine was therefore closely involved with the development of the new science of thermodynamics.

58. John Graunt, in particular, while working on mortality tables, made the fundamental discovery (1662) that statistical patterns or regularities, which are hidden in small numbers, will emerge when numbers are sufficiently large. Instead of merely looking at small changes in death rates week by week, he took the period 1603–23, with details of the causes of death of 229,250 individuals, and these revealed many factors in mortality that had not been suspected.

59. Crombie 1996:392.

60. In an Appendix to his book *Naval Timber and Arboriculture*, published in 1831, the same year that Darwin sailed on HMS *Beagle*. (An excellent account of Matthew and his work can be found in W.J.Dempster's *Patrick Matthew and Natural Selection*, 1983.) I think it is fair to say that if Darwin had been able to read Matthew's book before his voyage on the *Beagle* he would have saved himself a great deal of unnecessary work. When the *Origin* was published in 1859 (instigated by a communication from A.R.Wallace who had also independently worked out the theory of natural selection), Matthew drew Darwin's attention to these passages in *Naval Timber and Arboriculture*. Reasonably enough, Darwin replied that he had never heard of the book, but was obliged to concede that 'I freely admit that Mr Matthew has anticipated by many years the explanation which I have offered of the origin of species under the name of natural selection', (Dempster 1983:15) and in a subsequent edition of the *Origin* stated, 'he gives precisely the same view on the origin of species as that ...propounded by Mr Wallace and myself'. Matthew was ignored mainly, of course, because he published his theory in the obscure and casual way that he did, because he was an outsider in the scientific world, and was a political radical and dilettante who went

off to write on other topics such as emigration and the Schleswig-Holstein question. His book was also before its time: before Lyell's discoveries had been fully assimilated, and before writers like Herbert Spencer had popularised the whole idea of evolution.

61. Dempster 1983:46.
62. Matthew 1831:385.
63. Matthew 1831:381.
64. Medawar 1982:53.

## Conclusions

1. Ayer 1964:3.
2. 'That there is indeed a limit upon science is made very likely by the existence of questions that science cannot answer and that no conceivable advance would empower it to answer. These are the questions that children ask – the "ultimate questions" of Karl Popper. I have in mind such questions as:
>How did everything begin?
>What are we all here for?
>What is the point of living?

Doctrinaire positivism – now something of a period piece – dismissed all such questions as nonquestions or pseudoquestions such as only simpletons ask and only charlatans of one kind or another profess to be able to answer. This peremptory dismissal leaves one empty and dissatisfied because the questions make sense to those who ask them, and the answers, to those who try to give them; but whatever else may be in dispute, it would be universally agreed that it is not to science that we should look for answers.' (Medawar 1985:66.)
3. Leach 1982:52.

# REFERENCES

*Material quoted from these sources is used by kind permission of the copyright holders.*

Aberle, D.F. 1961. 'Matrilineal descent in cross-cultural perspective', in *Matrilineal Kinship*. eds. Schneider, D.M. & Gough, K. 655–727. University of California Press.

Ames, R. 1993. Translation of *Sun Tzu: The Art of Warfare*. With Intro. and comment. New York: Ballantine Books.

Arens, W. 1979. *The Man-Eating Myth*. New York: Oxford University Press.

Aristotle. *Complete Works* (2 vols.) 1984. ed. J.Barnes. Bollingen Series Princeton University Press.

Armstrong, A.H. 1968. *An Introduction to Ancient Philosophy*. London: Methuen.

Ayer, A.J. 1964. *Man as a Subject for Science*. Auguste Comte Memorial Lecture no. 6. London: Athlone Press.

Baldry, H.C. 1965. *The Unity of Mankind in Greek Thought*. Cambridge University Press.

Balikci, A. 1970. *The Netsilik Eskimo*. New York: Natural History Press.

Barnes, R. 2004. 'Cloistered bookworms in the chicken-coop of the Muses: the ancient library of Alexandria', 61–78, in Macleod, R. 2004. (ed.) *The Library of Alexandria. Centre of learning in the ancient world*. London: Tauris.

Barry, H., Child, I., & Bacon, M.K. 1959. 'The relation of child training to subsistence economy', *American Anthropologist*, 61, 51 – 63.

Basalla, G. 1988. *The Evolution of Technology*. Cambridge University Press.

Baxter, P. 1978. 'Boran age-sets and generation-sets: *gada*, a puzzle or a maze', in *Age, Generation and Time*, eds. P.Baxter & U.Almagor, 151-82. Cambridge University Press.

Beck, H.C. 1928. 'Early magnifying glasses', *The Antiquaries Journal*, 327-330.

Befu, H., and Ekholm, G.F. 1963. 'The true arch in pre-Columbian America ?', *Current Anthropology*, 5, 328.

Bellah, R.N. 1970. 'Religious evolution' in *Beyond Belief. Essays on religion in a post-traditional world*. 20-45. New York: Harper & Row.

Benveniste, E. 1973. *Indo-European Language and Society*. Tr. E.Palmer. University of Miami Press.

Bernal, J.D. 1969. *Science in History* (vol.1). 3$^{rd}$ ed. London: Pelican.

Bernardi, B. 1985. *Age Class Systems. Social institutions and polities based on age*. Tr. D.I.Kurtzer. Cambridge University Press.

Best, E. 1922. *The Astronomical Knowledge of the Maori*. Wellington: Dominion Museum.

Bodde, D. 1991. *Chinese Thought, Society, and Science*. University of Hawaii Press.

Boesche, R. 2002. *The First Great Political Realist. Kautilya and his Arthashastra*. Lanham, Md. :Lexington Books.

Bohannan, P. 1959. 'The impact of money on an African subsistence economy', *Journal of Economic History*, 19, 502.

Bohannan, P. 1967. 'Concepts of time among the Tiv of Nigeria', in *Myth and Cosmos. Readings in Mythology and Symbolism*. ed. J.Middleton, 315-30. New York: Natural History Press.

Bohannan, P., and Dalton, G. (eds.) 1962. *Markets in Africa*. Northwestern University Press.

Bose, D.M. et al. 1971. eds. *A Concise History of Science in India*. New Delhi.

Boyce, M. 1975. *A History of Zoroastrianism* (vol.1). Leiden: Brill.

Bray, F. [and Needham, J.] 1984. *Science and Civilisation in China*, vol. VI, Part 2: Agriculture. Cambridge University Press.

Brett, M. 1983. 'Middle Ages', in *Encyclopaedia Britannica*, 12, 138–64. Chicago.

Brown, D.E. 1991. *Human Universals*. New York: McGraw-Hill.

Brown, P., and Tuzin, D. (eds.) 1983. *The Ethnography of Cannibalism* Washington: Society for Psychological Anthropology.

Cardwell, D.S.L., 1963. *Steam Power in the Eighteenth Century. A case study in the application of science*. London: Sheed & Ward.

Chattopadhyaya, D.P. 1986. *History of Science and Technology in Ancient India*. Calcutta

Childe, G. 1954. *What Happened in History*. 2nd ed. London: Pelican.
Chomsky, N. 1980. *Rules and Representations*. Columbia University Press.
Chrisomalis, S. 2003a. *The Comparative History of Numerical Notation* Ph.D Thesis, McGill University.
Chrisomalis, S. 2003b. 'The Egyptian origin of the Greek alphabetic numerals', *Antiquity*, 77(Sept), 485-96.
Chrisomalis, S. 2004. 'A cognitive typology for numerical notation', *Cambridge Archaeological Journal*, 14(1), 37-52.
Cianchi, M. n.d. *Leonardo da Vinci's Machines*. Tr. L.G.Stoppato. Florence: Becocci Editore.
Claessen, H.J.M., and Skalnik, P. (eds.) 1978. *The Early State*. The Hague: Mouton.
Clark, W.E. 1937. 'Science' in *The Legacy of India*, ed. G.T.Garratt, 334 – 368. Oxford: Clarendon Press.
Coe, M.D. 1999. *The Maya*. 6th ed. London: Thames & Hudson.
Cohen, I.B. 1974. 'Newton', in *Dictionary of Scientific Biography*, X, 42–86. New York: Charles Scribner's Sons.
Cohen, M.N. 1977. *The Food Crisis in Prehistory. Overpopulation and the origins of agriculture*. Yale University Press.
Cohen, M.R., and Drabkin, I.E. 1958. *A Source Book in Greek Science*. 2nd ed. Harvard University Press.
Cohen, R. 1978 'State origins: a reappraisal', in *The Early State*, eds. H.J.M.Claessen & P.Skalnik, 31-75. The Hague: Mouton.
Cole, M., *et al*. 1971. *The Cultural Context of Learning and Thinking. An exploration in experimental anthropology*. London: Methuen.
Cole, M. & Scribner, S. 1974. *Culture and Thought: a psychological introduction*. New York: Wiley.
Constable, G. 1995. 'The Three Orders', in *Three Studies in Religious and Social Thought*. Cambridge University Press.
Cooper, L. 1935. *Aristotle, Galileo and the Tower of Pisa*. Ithaca.
Corlett, E. 1990. *The Iron Ship. The story of Brunel's SS Great Britain*. London: Conway Maritime Press.
Cornford, F.M. 1931. *The Laws of Motion in Ancient Thought*. Inaugural Lecture. Cambridge University Press.

Cornford, F.M. 1950. *The Unwritten Philosophy and Other Essays.* Cambridge University Press.

Crombie, A.C. 1953. *Robert Grosseteste and the Origin of Experimental Science 1100 – 1700.* Oxford: Clarendon Press.

Crombie, A.C. 1961. 'Quantification in medieval physics', *Osiris*, LII, 143-60.

Crombie, A.C. 1979. *From Augustine to Galileo* (2vols). London: Heinemann.

Crombie, A.C. 1990. *Science, Optics and Music in Medieval and Early Modern Thought.* London: Hambledon Press.

Crombie, A.C. 1994. *Styles of Scientific Thinking in the European Tradition.* 3 vols. London: Duckworth.

Crombie, A.C. 1996. *Science, Art and Nature in Medieval and Modern Thought.* London: Hambledon Press.

Crosby, A.W. 1997. *The Measure of Reality. Quantification and Western Science 1250 –1600.* Cambridge University Press.

Curtin, P.D. 1984. *Cross-Cultural Trade in World History.* Cambridge University Press.

Dalton, G. 1968. Introduction to *Primitive, Archaic, and Modern Economies. Essays of Karl Polanyi.* ed G.Dalton. NewYork Anchor Books.

David, F.N. 1955. 'Dicing and gaming (a note on the history of probability)', *Biometrika*, 42, 1 – 15.

Dasgupta, S. 1932. *History of Indian Philosophy.* I. Cambridge University Press.

Debus, A.G. 1967.'Alchemy and the historians of science', *History of Science*, 6, 128-38.

Dempster, W.J. 1983. *Patrick Matthew and Natural Selection.* Edinburgh: Harris.

Denny, J.P. 1986. 'Cultural ecology of mathematics: Ojibway and Inuit hunters', in *Native American Mathematics.* ed. M.P.Closs. 129-80. University of Texas Press.

Diamond, J. 1997. *Guns, Germs, and Steel.* London: Vintage.

Dibner, B. 1964. *Alessandro Volta and the Electric Battery.* New York Watts.

Dickinson, H.W. 1963. 2$^{nd}$ ed. *A Short History of the Steam Engine.* Introduction by A.E. Musson. London: Frank Cass.

Diener, P., Moore, K., & Mutaw, R. 1980. 'Meat, markets, and mechanical materialism: the great protein fiasco', *Dialectical Anthropology*, 5, 171-92.

Diocles *On Burning Mirrors*. Tr. and ed. G.J.Toomer. 1976. Berlin, Heidelberg, New York: Springer-Verlag.

Dodds, E.R. 1963. *The Greeks and the Irrational.* University of California Press.

Douglas, M. 1966. *Purity and Danger. An analysis of concepts of pollution and taboo.* London: Routledge & Kegan Paul.

Drachmann, A.G. 1948. *Ktesibios, Philon, and Heron. A Study in Ancient Pneumatics.* Copenhagen: Ejnar Munksgaard.

Drachmann, A.G. 1958a. 'The screw of Archimedes', *Acts of 8th International Congress of the History of Science, Florence-Milan*, 1956, vol.3, 940-43.

Drachmann, A.G. 1958b. 'How Archimedes expected to move the earth', *Centaurus*, 5, 278-82.

Drachmann, A.G. 1961. 'Heron's windmill', *Centaurus*, 7, 145 – 51.

Drachmann, A.G. 1963. *The Mechanical Technology of Greek and Roman Antiquity*. Copenhagen: Munksgaard.

Drake, Stillman. 1990. *Galileo: Pioneer Scientist.* University of Toronto Press.

Drake, Stillman, and Drabkin, I.E. 1969. *Mechanics in Sixteenth Century Italy. Selection from Tartaglia, Benedetti, Guido Ubaldo, & Galileo.* Translated and annotated. University of Wisconsin Press.

Drover, C.B. 1954. 'A mediaeval monastic water clock', *Antiquarian Horology*, 1, 54.

Duby, G. 1980. *The Three Orders. Feudal Society Imagined.* University of Chicago Press.

Eamon, W. 1994. *Science and the Secrets of Nature*. Princeton University Press.

Edgerton, R.B. 1992. *Sick Societies. Challenging the myth of primitive harmony*. New York: Free Press.

Edgerton, S. Y. 1975. *The Renaissance Rediscovery of Linear Perspective.* New York: Harper & Row.

Edwards, C. 1904. *The Hammurabi Code*. Reprinted 1971. Texas Kennikat Press.

Einzig, P. 1966. *Primitive Money*. 2nd ed. Oxford: Pergamon Press.
Eisenstadt, S.N. (ed.) 1986. *The Origins and Diversity of Axial Age Civilizations*. State University of New York Press.
Eliade, M. 1978. *The Forge and the Crucible: the origins and structures of alchemy*. 2nd ed. Chicago.
Elvin, M. 2004. 'Vale atque ave', in *Science and Civilisation in China*, vol VII (2), J.Needham, ed. K.Robinson, xxiv–xliii. Cambridge University Press.
van Ess, J. 1987. "Mu'tazilah", *Encyclopaedia of Religion*, ed. M.Eliade. 10, 220-29 New York: Macmillan.
Euclid, *Optics*. 1999. The Arabic version of Euclid's Optics, ed. and tr. with historical introd. and commentary by Elaheh Kheirandis Sources in the history of mathematics and early science no.16. New York: Springer.
Evans-Pritchard, E.E. 1940. *The Nuer*. Oxford: Clarendon Press.
Fakhry, M. 1970. *A History of Islamic Philosophy*. Columbia University Press.
Feil, D.K. 1987. *The Evolution of Highland Papua New Guinea Societies*. Cambridge University Press.
Findley, C.V. 2005. *The Turks in World History*. Oxford University Press.
Finley, M.I. 1973. *The Ancient Economy*. London: Chatto & Windus.
Finocchiaro, M. 1989. *The Galileo Affair. A documentary history*. University of California Press.
Firth, R. 1957. *We, the Tikopia*. 2nd ed. London: Allen & Unwin.
Fischer, S.R. 1997. *Rongorongo, the Easter Island Script: History, Traditions, Texts*. Oxford University Press.
Fisher, H.A.L. 1936. *A History of Europe*. London: Arnold.
Flynn, J.R. 2007. *What is Intelligence? Beyond the Flynn effect*. Cambridge University Press.
Forbes, R.J. 1964. *Studies in Ancient Technology*. 2nd ed. Vol. I. Alchemy. 125-49. Leiden: E.J.Brill.
Forbes, R.J. 1966. *Studies in Ancient Technology*. 2nd ed. Vol. V. Glass. 112-241. Leiden: E.J.Brill.
Forbes, R.J. 1971. *Studies in Ancient Technology*. 2nd ed. Vol. VIII. Metallurgy (gold, silver, lead). Leiden: E.J.Brill.

Forbes, R.J. 1972. *Studies in Ancient Technology.* 2nd ed. Vol. IX. Metallurgy (copper, bronze, iron). Leiden: E.J.Brill.

Forde, D. 1954. (ed.) *African Worlds.* International African Institute. London.

von Fürer-Haimendorf. 1967. *Morals and Merit. study of values and social controls in South Asian societies.* London: Weidenfeld & Nicolson.

Freeman, L.C., and Winch, R.F. 1957. 'Societal complexity: an empirical test of a typology of societies'. *American Journal of Sociology* 62, 461-6.

Gadd, C.J. 1948. *Ideas of Divine Rule in the Middle East.* Schweich Lectures. London: British Academy.

Galilei, G. [1991]. *Dialogues Concerning Two New Sciences.* Tr. H.Crew and A. de Salvio. New York: Prometheus Books.

Gardner, P.M. 1966. 'Symmetric respect and memorate knowledge', *Southwestern J. Anthrop.* 22, 389-415.

Garelli, P. 1975. 'The changing facets of conservative Mesopotamian thought'. *Daedalus,* 104, 47-55.

Gay, J., and Cole, M. 1967. *The New Mathematics and an Old Culture.* New York: Holt, Rinehart, Winston.

Gelb, I.J. 1963. *A Study of Writing.* 2nd ed. University of Chicago Press.

Gell, A. 1975. *The Metamorphosis of the Cassowaries. Umeda society, language and ritual.* London: Athlone Press.

Ginsberg, M. 1944. *Moral Progress.* Frazer Lecture. Glasgow University Press.

Gluckman, M. 1954. 'Political institutions', in *The Institutions of Primitive Society,* 66-80. Oxford: Blackwell.

Gluckman, M. 1967. *Politics, Law, and Ritual in Tribal Society.* Oxford: Blackwell.

Goldman, I. 1970. *Ancient Polynesian Society.* Chicago University Press.

Golson, J. 1977. *The Ladder of Social Evolution: archaeology and the bottom rungs.* Annual Lecture of the Australian Academy of the Humanities. Sydney University Press.

Gomme, G.L. 1880. *Primitive Folk-Moots, or open-air assemblies in Britain.* London: Samson Low.

Gorst, J.E. 2001. *The Maori King*. (First pub. 1864.) Auckland: Reed.

Graham, A.C. 1978. *Later Mohist Logic, Ethics, and Science*. Hong Kong: Chinese University Press.

Graunt, J. [1662]. *Natural and Political Observations made upon the Bills of Mortality*. ed. W.F.Wilcox, 1939. Baltimore: Johns Hopkins Press.

Greenaway, F. 1966. *John Dalton and the Atom*. London: Heinemann.

Grimm, T. 1977. 'Academies and urban systems in Kwantung', in G.W.Skinner ed. *The City in Late Imperial China*. 475-98. Stanford University Press.

Grube, G.M.A. 1958. *Plato's Thought*. London: Methuen.

Hackett, J. 1997. 'Roger Bacon on *scientia experimentalis*', in *Roger Bacon and the Sciences. Commemorative essays*. ed. J.Hackett, 277–315. Leiden: Brill.

Hall, A.R. 1952. *Ballistics in the Seventeenth Century. A study in the relations of science and war with reference principally to England*. Cambridge University Press.

Hallpike, C.R. 1973 'Functionalist interpretations of primitive warfare'. *Man* (n.s.), 4, 51–70.

Hallpike, C.R. 1977. *Bloodshed and Vengeance in the Papuan Mountains. The generation of conflict in Tauade society*. Oxford: Clarendon Press.

Hallpike, C.R. 1978. 'Social hair' (revised), in *The Social Aspects of the Human Body*. ed. T. Polhemus, 134 – 146. London: Penguin.

Hallpike, C.R. 1979. *The Foundations of Primitive Thought*. Oxford: Clarendon Press.

Hallpike, C.R. 1984. 'The relevance of the theory of inclusive fitness to human society', *Journal of Social and Biological Structures*, 7(2), 131-44.

Hallpike, C.R. 1986. *The Principles of Social Evolution*. Oxford: Clarendon Press.

Hallpike, C.R. 1999. Comment on Greek hoplites. *Journal of Royal Anthropological Institute*, 5, 627-9.

Hallpike, C. R. 2004 *The Evolution of Moral Understanding*. Prometheus Research Group.

Hallpike, C.R. 2007. 'Time-reckoning and calendars', in *New Encyclopaedia of Africa.* ed. J.Middleton. 5, 52-9. New York: Charles Scribner's Sons.

Hallpike, C.R. 2008. *The Konso of Ethiopia. A study of the values of an East-Cushitic people.* (Revised ed.) AuthorHouse UK. (First published 1972, Oxford: Clarendon Press.)

Harris, M. 1980. *Cultural Materialism. The struggle for a science of culture.* New York: Vintage.

Harris, W.V. 1989. *Ancient Literacy.* Harvard University Press.

Heath, T. 1981. *A History of Greek Mathematics* (2 vols.). (First pub. 1921). New York: Dover.

Heather, P. 1998. *The Goths.* Oxford: Blackwell.

Heather, P. 2005. *The Fall of the Roman Empire.* London: Macmillan.

Henry, J. 2001. 'Animism and empiricism: Copernican physics and the origins of William Gilbert's experimental method', *Journal of the History of Ideas, 61,* 99-122.

Hero, *Pneumatics.* The Pneumatics of Hero of Alexandria, Tr. and ed. Bennet Woodcroft, 1851, London.

Hertz, R. 1960. *Death and the Right Hand.* Tr. R.Needham. London Cohen & West.

Hessen, Boris. 1931. 'The social and economic roots of Newton's *Principia*', in N.Bukharin ed., *Science at the Crossroads.* London.

Hills, R.L. 1989. *Power from Steam: A history of the stationary steam engine.* Cambridge University Press.

Hockney, D. 2001. *Secret Knowledge. Rediscovering the lost techniques of the Old Masters.* London: Thames & Hudson.

Hodge, A.T. 1960. *The Woodwork of Greek Roofs.* Cambridge University Press.

Holmberg, A. 1969. *Nomads of the Long Bow. The Siriono of Eastern Bolivia.* New York: Natural History Press.

Holmes, G. 1992. Foreword to *The Oxford History of Medieval Europe,* ed. G.Holmes. Oxford University Press.

Hoppen, A. 1999. *The Fortification of Malta by the Order of St John.* Malta: Mireva.

Hourani, G.F. 1985. *Reason and Tradition in Islamic Ethics.* Cambridge University Press.

Howard, A., and Kirkpatrick, J. 1989. 'Social organization', in *Developments in Polynesian Ethnology.* eds. A. Howard & R. Borofsky. 47-94. University of Hawaii Press.

Howell, S. 1989. *Society and Cosmos. The Chewong of Peninsular Malaysia.* 2nd ed. Chicago University Press.

Huizinga, J. 1949. *Homo Ludens. A study of the play element in culture.* London: Routledge & Kegan Paul.

Humphreys, S.C. 1975. ' "Transcendence" and intellectual roles: the Ancient Greek case', *Daedalus,* 104, 91-118.

Hussey, E. 1972. *The Pre-Socratics.* London: Duckworth.

Inwood, S. 2003. *The Man Who Knew Too Much. The strange and inventive life of Robert Hooke 1635–1703.* London: Pan Books.

Jaspers, K. 1953. *The Origin and Goal of History.* London: Routledge & Kegan Paul.

Jenkins, Rhys. 1936 . *The Collected Papers.* The Newcomen Society: Cambridge University Press.

Johnson, A.W., and Earle, T. 2000. *The Evolution of Human Societies. From foraging group to agrarian state.* 2nd ed. Stanford University Press.

Kaplan, E. 1960. *Robert Recorde (c. 1510–1558): Studies in the Life and Works of a Tudor Scientist.* Ph. D Thesis, Dept. of History, New York University.

Kaye, J. 1988. 'The impact of money on the development of fourteenth century scientific thought', *Journal of Medieval History,* 14, 251-70.

Kendall, M.G. 1956. 'The beginnings of a probability calculus', *Biometrika,* 43, 1–14.

Kirch, P.V. & Green, R.C. 2001. *Hawaiki, Ancestral Polynesia. An essay in historical anthropology.* Cambridge University Press.

Kirk, G.S., Raven, J.E., and Schofield, M. 1983. 2nd ed. *The Presocratic Philosophers.* Cambridge University Press.

Kitcher, P. 1985. *Vaulting Ambition. Sociobiology and the quest for human nature.* Cambridge, Mass.: MIT Press.

Knight, D. 1992. *Ideas in Chemistry. A history of the subject.* London: Athlone Press.

Koch, K–F. 1974. *War and Peace in Jalémó. The management of conflict in Highland New Guinea.* Harvard University Press.

Kohlberg, L. 1984. *The Psychology of Moral Development: the nature and validity of moral stages.* San Francisco: Harper & Row.

Kroll, J.L. 1985-7. 'Disputation in early Chinese culture', *Early China*, 11-12, 118-45.

Kuhn, T.S. 1957. *The Copernican Revolution. Planetary astronomy in the development of Western thought.* Harvard University Press.

Landels, J.G. 2000. *Engineering in the Ancient World.* 2nd ed. London: Constable.

Leach, E.R. 1982. *Social Anthropology.* Oxford University Press.

Leaman, O. 1985. *An Introduction to Medieval Islamic Philosophy.* Cambridge University Press.

Lee, H.D.P. 1936. *Zeno of Elea.* Cambridge University Press.

Lee, R.B. 1979. *The !Kung San. Men, women and work in a foraging society.* Cambridge University Press.

Lee, R.B. 1984. *The Dobe !Kung.* New York: Holt, Rinehart, Winston.

Leicester, H.M. 1965. *The Historical Background of Chemistry.* New York: Wiley.

Lewis, B. 1982. *The Muslim Discovery of Europe.* London: Weidenfeld & Nicolson.

Lewis, B. 1993. 6th ed. *The Arabs in History.* Oxford University Press.

León-Portilla, M. 1964. 'Philosophy in the culture of Ancient Mexico', in *Cross-Cultural Understanding: Epistemology in Anthropology*, eds. F.S.C.Northrop & H.H.Livingston, 35-54. New York: Wenner-Gren Foundation.

Lienhardt, G. 1961. *Divinity and Experience. The Religion of the Dinka.* Oxford: Clarendon Press.

Littlejohn, J. 1963. 'Temne space', *Anthropological Quarterly*, 36, 1 – 17.

Lloyd, G.E.R. 1966. *Polarity and Analogy.* Cambridge University Press.

Lloyd, G.E.R. 1970. *Early Greek Science. Thales to Aristotle.* London: Chatto & Windus.

Lloyd, G.E.R. 1973. *Greek Science after Aristotle.* London: Chatto & Windus.

Lloyd, G.E.R. 1979. *Magic, Reason, and Experience. Studies in the origins and development of Greek science.* Cambridge University Press.

Lloyd, G.E.R. 2002. *The Ambitions of Curiosity. Understanding the World in Ancient Greece and China.* Cambridge University Press.

Long, A.A. 1986. *Hellenistic Philosophy. Stoics, Epicureans, Sceptics.* 2nd ed. London: Duckworth.

Lucretius, *De Rerum Natura.* 1975. Tr. W.H.D Rouse, ed. M.F.Smith Loeb Classical Library. London: Heinemann.

Luria, A.R. 1976. *Cognitive Development: its cultural and social foundations.* Trs. M.Lopez-Morillas & S.Solotaroff. Harvard University Press.

Machinist, P. 1986. 'On self-consciousness in Mesopotamia', in Eisenstadt, S.N. (ed.) *The Origins and Diversity of Axial Age Civilizations.* 183-202. State University of New York Press.

Mahony, W.K. 1987. 'Soul: Indian concepts', in *The Encyclopaedia of Religion,* ed. Mircea Eliade. 13, 438—43. New York: Macmillan.

Makdisi, G. 1980. 'On the origin and development of the college', in *Islam and the West,* ed. Khalil I. Semaan. 26-49. State University of New York Press.

Malinowski, B. 1922. *Argonauts of the WesternPacific.* London: Routledge & Kegan Paul

Marsden, E. 1969. *Greek and Roman Artillery. Their historical development.* Oxford: Clarendon Press.

Marshall, L. 1976. 'Sharing, talking, and giving: relief of social tension among the !Kung', in *Kalahari Hunter-Gatherers,* eds. R.B.Lee & I. De Vore, 349-71, Harvard University Press.

McGuire, J.E., & Rattansi, P.M. 1966. 'Newton and the "Pipes of Pan"'. *Notes and Records of the Royal Society of London,* 21(2), 108 43.

Medawar, P.B. 1982. *Pluto's Republic.* Oxford University Press.

Medawar, P.B. 1985. *The Limits of Science.* Oxford University Press.

Mesoudi, A., Whiten, A., & Laland, K.N. 2004. 'Perspective: Is human cultural evolution Darwinian? Evidence reviewed from the perspective of The Origin of Species.' *Evolution* 58, 1-11.

Misra, G.S.P. 1984. *Development of Buddhist Ethics*. Delhi: Munshiram? Mandharlal Publishers.

Mitchiner, M. 1973. *The Origins of Indian Coinage*. London: Hawkins Publications.

Momigliano, A. 1975. *Alien Wisdom. The limits of Hellenization*. Cambridge University Press.

Morris, B. 1976. 'Whither the savage mind? Notes on the natural taxonomies of a hunting and gathering people', *Man* (n.s.), 11, 542–57.

Munro, D.S. 1969. *The Concept of Man in Early China*. Stanford University Press.

Murdock, G.P., and Provost, C. 1973. 'Measurement of cultural complexity.' *Ethnology*, 12, 379–92.

Murray, A. 1978. *Reason and Society in the Middle Ages*. Oxford: Clarendon Press.

Musson, A.E. 1963. 'Introduction' to H.W.Dickinson *A Short History of the Steam Engine*. London: Frank Cass.

Naroll, R. 1956. 'A preliminary index of social development'. *American Anthropologist*, 59, 664–87.

Nasr, S.N. 1987. (2$^{nd}$ ed.) *Science and Civilization in Islam*. Cambridge: Islamic texts Society.

Needham, J. 1956. *Science and Civilisation in China*. Vol. 2. *History of scientific thought*. Cambridge University Press.

Needham, J. 1959. *Science and Civilisation in China*. Vol. 3. *Mathematics and the sciences of the heavens and the earth*. Cambridge University Press.

Needham, J. 1962. *Science and Civilisation in China*. Vol. 4, Part 1: Physics. Cambridge University Press.

Needham, J. 1965. *Science and Civilisation in China*. Vol. 4, Part 2: Mechanical Engineering. Cambridge University Press.

Needham, J. 1971. *Science and Civilisation in China*. Vol.4, Part 3. Physics and Physical Technology. (Nautical technology). Cambridge University Press.

Needham, J. 1980 'The guns of Kaifeng Fu', *Times Literary Supplement*, 11 January, 39–42.

Needham, J. 1986. *Science and Civilisation in China*. Vol. 5, Part 7. *Military Technology: the Gunpowder Epic*. With Ho Ping-Yü, Lu Gwei-Djen, and Wang Ling. Cambridge University Press.

Needham, J. *et al.* 2004. *Science and Civilisation in China*. Vol.7 (2). General Conclusions and Reflections. Cambridge University Press.

Needham, R. 1967. 'Right and left in Nyoro symbolic classification', *Africa*, 37(4), 423 – 51.

Needham, R. 1973. (ed.) *Right and Left. Essays on dual symbolic classification*. University of Chicago Press.

Needham, R. 1978. *Primordial Characters*. Virginia University Press.

Neugebauer, O. 1969. *The Exact Sciences in Antiquity*. 2$^{nd}$ ed. New York: Dover.

Newman, W.R. 1997. 'An overview of Roger Bacon's alchemy' in *Roger Bacon and the Sciences*. Commemorative essays. ed. J.Hackett, 317–335. Leiden: Brill.

Nigosian, S.A. 1993. *Zoroastrian Faith: tradition and modern research*. Montreal: McGill-Queen's University Press.

North, J. 2005. *God's Clockmaker. Richard of Wallingford and the invention of time*. London: Hambledon & London.

Obeyesekere, R. 1998. 'Cannibal feasts in nineteenth-century Fiji: seamen's yarns and the ethnographic imagination', in *Cannibalism and the Colonial World*, ed. F.Barker, 63-86. Cambridge University Press.

Onians, R.B. 1954. *The Origins of Modern European Thought*. 2$^{nd}$ed. Cambridge University Press.

Oppenheim, A.L. 1975. 'The position of the intellectual in Mesopotamian society', *Daedalus*, 104, 37–46.

Peerenboom, R.P. 1993. *Law and Morality in Ancient China. The silk manuscripts of Huang-Lao*. State University of New York Press.

Peregrine, P.N., Ember, C.R., and Ember, M. 2004. 'Universal patterns in cultural evolution. An empirical analysis using Guttman scaling.' *American Anthropologist* 101(1), 145-49.

Perrett, R.W. 1998. 'History, time, and knowledge in ancient India', *History and Theory*, 38(3), 307-21.

Piaget, J. 1929. *The Child's Conception of the World*. London: Routledge & Kegan Paul.

Piaget, J. 1930. *The Child's Conception of Physical Causality*. London: Routledge & Kegan Paul.

Piaget, J. 1932. *The Moral Judgment of the Child*. London: Routledge & Kegan Paul.

Piaget, J. 1952. 'The child and moral realism', in *Moral Principles of Action*. ed. R.N.Anshen, 417 – 35. New York: Harper.

Piaget, J. & Inhelder, B. 1975. *The Origin of the Idea of Chance in Children*. London: Routledge & Kegan Paul.

Plantzos, D. 1997. 'Crystals and lenses in the Graeco-Roman world'. *American Journal of Archaeology*, 101, 451-64.

Plato, 1985. *Collected Dialogues*. eds. E.Hamilton & H.Cairns. Bollingen Series. Princeton University Press.

Polanyi, K. 1957. *The Great Transformation*. Boston: Beacon Press.

Popper, K.R. 1957. *The Poverty of Historicism*. London: Routledge & Kegan Paul.

Price, D.J. de Solla. 1964. 'Automata and the origin of mechanics', *Technology and Culture*, 5, 9-23.

Price, D.J. de Solla. 1990. 'Philosophical mechanism and mechanical philosophy', *Annali dell'Istituto e Museo*, 5, 75–85.

Prince, J.R. 1969. *Science Concepts in a Pacific Culture*. Sydney: Angus & Robertson.

Ptolemy, *Almagest*. 1998. 2$^{nd}$ed. Tr. and annotated G.J.Toomer. Princeton University Press.

Ptolemy, *Optics*, see Smith 1996.

Rabinovitch, N.L. 1969. 'Probability in the Talmud', *Biometrika*, 56, 15 – 19.

Rabinovitch, N.L. 1970.'Combinations and probability in rabbinic literature', *Biometrika,* 57, 21 – 23.

Randall, J.H. 1962. *Aristotle*. Columbia University Press.

Rappaport, R.A. 1968. *Pigs for the Ancestors. Ritual in the ecology of a New Guinea People.* Yale University Press.

Read, K.E. 1955. 'Morality and the concept of the person among the Gahuku-Gama', *Oceania*, 25(4), 233-82.

Reed, C.A. (ed.) 1977a. *The Origins of Agriculture*. The Hague: Mouton.

Reed, C.A. 1977b. 'Origins of agriculture; discussion and some conclusions' in *The Origins of Agriculture*, 879-956.

Rhys Davids, T.W. 1955. *Buddhist India*. Calcutta: Susil Gupta.

Rienstra, M.H. 1974. 'Porta', in *Dictionary of Scientific Biography*. XI, 95-8. New York: Charles Scribner's Sons.

Ritter, E.A. 1955. *Shaka Zulu. The rise of the Zulu empire*. London: Allen Lane.

Robertson, J.D. 1931. *The Evolution of Clockwork*. London: Cassell.

Roller, D.H.D. 1959. *The* De Magnete *of William Gilbert*. Amsterdam: Menno Hertzberger.

Rolt, L.T.C., & Allen, J.S. 1977. *The Steam Engine of Thomas Newcomen*. New York: Science History Publications.

Ronan, C. A. 1991. 'The origins of the reflecting telescope', *Journal of the British Astronomical Association*. 101, 335-42.

Rosen, E. 1956. 'The invention of eyeglasses', *Journal of the History of Medicine and Allied Sciences*. 11, 13-46, and 183-218.

Rossi, P. 1978. *Francis Bacon: From Magic to Science*. Chicago University Press.

Rostovtzeff, M. 1957. *The Social and Economic History of the Roman Empire*. 2 vols. 2nd ed. Revised P.M.Fraser. Oxford: Clarendon Press.

Rudgley, R. 1999. *Lost Civilisations of the Stone Age*. London: Arrow Books.

Sahlins, M. 1974. *Stone Age Economics*. London: Tavistock.

Sahlins, M. 2003. 'Artificially maintained controversies. Global warming and Fijian cannibalism'. *Anthropology Today*, 19(3), 3-5.

Sahlins, M., and Service, E.R. 1960. *Evolution and Culture*. University of Michigan Press.

Sambursky, S. 1956. 'On the possible and the probable in ancient Greece', *Osiris*, 12, 1–14.

Sambursky, S. 1963. *The Physical World of the Greeks*. London: Routledge & Kegan Paul.

Sampson, G. 1985. *Writing Systems*. Stanford University Press.

Sanderson, S.K. 2001. *The Evolution of Human Sociality. A Darwinian conflict perspective*. Lanham, Mld.: Rowman & Littlefield.

Schapera, I. 1956. *Government and Politics in Tribal Societies*. London: Watts.

Scharf, P.M. 1996. *The Denotation of Generic Terms in Ancient Indian Philosophy: Grammar, Nyaya, and Mimamsa.* Transactions of the American Philosophical Society 86 (3), Philadelphia.
Schmandt-Besserat, D. 1992. *Before Writing.* (2 vols) University of Texas Press.
Schwartz, B.I. 1985. *The World of Thought in Ancient China.* Harvard University Press.
Scribner, S., & Cole, M. 1981. *The Psychology of Literacy.* Harvard University Press.
Sen, M.K. 1991. *Hinduism.* With Foreword by Amartya Sen. London: Penguin.
Seneca, *Quaestiones Naturales.* Tr. J.Clarke 1910 as *Physical Science in the Time of Nero.* London: Macmillan.
Sharma, J.P. 1968. *Republics in Ancient India c. 1500 BC – 500 BC.* Leiden: E.J.Brill.
Simoons, F.J. 1967. *Eat Not This Flesh. Food avoidances in the Old World.* University of Wisconsin Press.
Sines, G. 1992. 'Precision in engraving of Etruscan and Archaic Greek gems', *Archaeomaterials*, 6, 53-67.
Sines, G., and Sakellarkis, Y. 1987. 'Lenses in antiquity', *American Journal of Archaeology*, 91, 191-96.
Singer, C. 1943. *A Short History of Scientific Ideas to 1900.* Oxford University Press.
Skinner, A. 1999. 'Analytical Introduction' to Adam Smith's *Wealth of Nations.* London: Penguin Classics.
Smith, G.Elliott. 1933. *The Diffusion of Culture.* London: Watts.
Smith, M. 1996. 'Ptolemy's Theory of Visual Perception', *Transactions of the American Philosophical Society*, 86(2), 1-300.
Smith, P.H. 2004. *The Body of the Artisan. Art and experience in the scientific revolution.* University of Chicago Press.
Stahl, W.H. 1962. *Roman Science.* University of Wisconsin Press.
Stephenson, B. 1987. *Kepler's Physical Astronomy.* Berlin, Heidelberg, New York: Springer-Verlag.
Sun-Tzu, *The Art of Warfare*, see Ames 1993.

Sylla, E. 1973. 'Medieval concepts of the latitude of forms: the Oxford calculators', *Archives d'histoire doctrinale et litéraire du moyen age,* 40, 223-83.

Tambiah, S.J. 1976. *World Conqueror and World Renouncer. A study of Buddhism and polity in Thailand against a historical background.* Cambridge University Press.

Taylor, F.S. 1942, 'The origin of the thermometer', *Annals of Science,* 5, 129-156.

Taylor, F.S. 1951. *The Alchemists: Founders of modern chemistry.* London: Heinemann.

Thackray, A. 1972. *John Dalton. Critical assessments of his life and science.* Harvard University Press.

Thapar, R. 1975. 'Ethics, religion, and social protest in the first millennium BC in Northern India', *Daedalus,* 104, 119–32.

Thapar, R. 2002. *Early India. From the origins to AD 1300.* University of California Press.

Thornton, R.J. 1980. *Space, Time, and Culture among the Iraqw of Tanzania.* New York: Academic Press.

Toomer, G.J. 1978. 'Hipparchus', *Dictionary of Scientific Biography,* XV, 207-24.

Trefil, J. 2002. *Cassell's Laws of Nature.* London: Cassell.

Trigger, B. 1972. 'Determinants of urban growth in pre-industrial societies', in *Man, Settlement and Urbanism,* eds. P.J.Ucko, Tringham, & G.W.Dimbleby, 576–99. London: Duckworth.

Trigger, B. 1990. 'Maintaining economic equality in opposition to complexity: an Iroquoian case study', in *The Evolution of Political Systems,* ed. S.Upham, 119-45. Cambridge University Press.

Trigger, B. 2003. *Understanding Ancient Civilizations.* Cambridge University Press.

Trigger, B. 2004 'Writing systems: a case study in cultural evolution'. In *The First Writing. Script invention as history and process,* ed. S.D.Houston, 39-68. Cambridge University Press.

Trowbridge, M.L. 1928. 'Philological studies in ancient glass', *University of Illinois Studies in Language and Literature,* 13 (3-4), 3-206. (Published 1930).

Tuden, A., and Marshall, C. 1972. 'Political organization: cross-cultural codes 4', *Ethnology*, 11, 436–64.

Turnbull, C. 1965. *The Mbuti Pygmies: an ethnographic survey.* New York: American Museum of Natural History.

Turney-High, H.H. 1971. *Primitive War. Its practice and concepts.* 2nd ed. University of South Carolina Press.

Ucko, P.J., and Dimbleby, G.W. (eds.) 1969. *The Domestication and Exploitation of Plants and Animals.* London: Duckworth.

Ucko, P.J., Tringham, R., & Dimbleby, G.W. (eds.) 1972. *Man, Settlement and Urbanism.* London: Duckworth.

Ulansey, D. 1991. *The Origins of the Mithraic Mysteries: Cosmology and salvation in the ancient world.* Oxford University Press.

Vinette, F. 1986. 'In search of Mesoamerican geometry', in *Native American Mathematics*, ed. P.Closs. 387–407. University of Texas Press.

Vitruvius. *Ten Books on Architecture.* 1999. Tr. I.Rowland, Comm. T.N.Howe. Cambridge University Press.

Wade, N. 2007. *Before the Dawn. Recovering the lost history of our ancestors.* London: Duckworth.

Wallace, W.A. 1971. 'Mechanics from Bradwardine to Galileo', *Journ. Hist. Ideas* 32, 15-28.

Weisheipl, A. 1984. 'Science in the thirteenth century', in *The History of the University of Oxford*, ed. J.I.Catto. Oxford: Clarendon Press.

Westfall, R. 1993. *The Life of Isaac Newton.* Cambridge University Press.

White, J. 1957. *The Birth and Rebirth of Pictorial Space.* London: Heinemann.

White, K.D. 1984. *Greek and Roman Technology.* Cornell University Press.

White, L. 1978. *Medieval Religion and Technology.* University of California Press.

Whitelock, D. 1952. *The Beginnings of English Society.* London: Pelican.

Whybray, R.N. 1973. *The Intellectual Tradition in the Old Testament.* Berlin: de Gruyter.

Williams, B.I. 2000. *The Matter of Motion and Galvani's Frogs*. Bletchingdon: Rana.

Williams, S.J. 1997. 'Roger Bacon and the Secret of Secrets', in *Roger Bacon and the Sciences. Commemorative essays*. ed. J.Hackett, 365–391. Leiden: Brill.

Williams, S.J. 2003. *The Secret of Secrets: the scholarly career of a pseudo Aristotelian text in the Latin Middle Ages*. University of Michigan Press.

Wilson, C. 1995. *The Invisible World. Early modern philosophy and the invention of the microscope*. Princeton University Press.

Wilson, P. 2003. *Sacred Signs. Hieroglyphs in ancient Egypt*. Oxford University Press.

Wittfogel, K.A. 1957. *Oriental Despotism. A comparative study of total power*. Yale University Press.

Woodburn, J. 1968. Discussion in *Man the Hunter*. eds. R.B.Lee & I. De Vore, p. 91, Chicago: Aldine.

Yates, F. 1972. *The Rosicrucian Enlightenment*. Routledge & Kegan Paul.

Zipf, G.K. 1949. *Human Behavior and the Principle of Least Effort*. Harvard University Press.

# INDEX

*A sequence of pages in which a subject is discussed in detail is shown as, e.g. '162–69'*

*A sequence of pages in which a subject is mentioned is shown as, e.g. '162...69'*

abacus (see also number systems) 222, 225, 226, 400, 539
Abbasid dynasty 360, 362, 364, 510
Aberle, D.F. 478, 565
Aborigines, Australian 29, 35, 50, 107, 140, 474, 495
accumulation of necessary conditions 18, 20, 24, 116, 188, 205, 287, 394, 465, 467, 498
acids, strong mineral 442
Adam 334, 399, 439, 513
Adaptation 4, 6, 7, 8, 12, 16, 17, 18, 28, 43, 49, 50, 51, 53, 71, 91, 92, 107, 113, 114, 122, 133, 179, 189, 196, 204, 264, 273, 275, 362, 418, 457, 459, 464, 466, 470, 472, 473, 475, 481, 487, 521
Adelard of Bath 400
aether 305, 306, 441, 448
Africa 20, 21, 30, 52, 56, 61, 62, 63, 73, 98, 134, 158, 162, 174, 180, 181, 185, 203, 207, 211, 373, 390, 401, 470, 476, 477, 482, 483, 490, 491, 497, 499, 500, 511, 539
age-grouping systems 10, 72, 73, 102, 119, 153, 180, 185, 478, 484
Agricola 400, 411, 539
agriculture 18, 19, 23, 26, 29, 31, 52-58, 61, 63, 64, 65, 75, 78, 100, 153, 154, 158, 159, 188, 191, 208, 227, 230, 251, 256, 368, 372, 381, 399, 471, 475, 476, 477, 479, 497, 498, 501, 515, 543
Ainu, the 31, 45,
Alberti, Leone 424, 548
alchemy 320, 330-5, 359, 361, 367, 372, 397, 398, 411, 433, 442, 461, 462, 539, 535 556, 560
alcohol 390, 418, 442, 484, 556-7
Alexander the Great 239, 258, 260, 262, 343, 538
Alexandria 281, 321, 323, 327, 330, 334, 343, 346, 352, 356, 357, 418, 461, 521, 525-6, 534, 565, 573
algebra 234, 235, 371, 403, 535, 539
al-Ghazzali 361-2, 399, 536

al-Hazen 406, 535
al-Kindi 525, 535
al-Khwarazmi 400, 535
Almagest, the 327-8, 527
al-Mamun, Caliph 361
alphabet 214, 215, 216, 219, 220, 223, 500, 507, 518, 524
al-Rashid, Caliph 361
altruism 29, 40, 41, 42, 48, 481
amber 335, 448
Ames, R. 509, 565
Anaxagoras 283, 322, 531
Anaximenes 519
ancestor worship 70, 71, 163, 183, 184
animals, domestication of 1, 18, 26, 45, 52, 53, 54, 57, 58-61, 62, 64, 67, 100, 205, 477-8
Apollonius 325
Aquinas, St Thomas 361, 389, 404, 405, 510
arch, the 205, 206, 345, 369, 495, 497
Archimedes 331, 343...6, 348, 349, 351, 352, 355, 425, 503, 522, 529, 533, 551
area, concept of 33, 54, 127...32, 221, 231...4, 438, 487, 504
Arens, W. 483, 564
Aristarchus 323, 324, 430
Aristotle 244, 245, 248, 258, 260, 263, 274, 278, 290, 294, 298, 299, 302-11, 312, 321...5, 328, 332, 342, 357, 361, 371, 389, 405, 406, 409, 420, 425, 435, 438, 440, 448, 452, 461, 515, 517, 519, 521, 522, 524, 525, 528, 530, 535, 536, 538, 547, 552
Armstrong, A.H. 507, 511, 565
artillery (see also guns) 370, 410, 412, 418-21, 539, 547, 548
artist-engineers 522-5, 538, 548
asceticism 241, 243, 257, 270
Ashoka, Emperor 251
assemblies, popular 57, 172, 180-3, 238, 239, 243, 250, 251, 253, 354, 377, 378, 380, 381, 382, 384, 385, 425, 494, 517
astrology 12, 15, 227, 228, 230, 289, 296, 320, 321, 328-30, 331, 333, 334, 359, 361, 367, 397, 398, 403, 411, 412, 436, 447, 461, 462, 503, 528, 538, 540, 545

astronomy (see also calendars) 15, 79, 163, 205, 212, 218, 225, 226-31, 234, 235, 242, 283, 285, 286, 295, 298, 300, 308, 320, 321-8, 330, 337, 338, 352, 361, 363, 366, 368, 370-1, 373, 388, 390, 396, 400, 405, 412, 413, 414, 416, 429, 430-41, 447, 497, 503, 514, 520, 521, 527, 534, 535, 536, 544, 545, 552…6

Athens 229, 243, 260, 283, 285, 321, 357, 359, 387, 507, 512, 531

atmosphere 306, 349, 435, 445, 450, 534, 551

atmospheric pressure 348-51, 428, 450-53, 455, 457, 534

atomism 208, 292-7, 298, 302, 303, 305, 307, 311, 318, 321, 328, 333, 348, 372, 405, 423, 427, 429, 430, 436, 441-6, 461, 462, 464, 472, 517-18, 524, 525, 530, 557, 558, 559

Augustine, Saint 282, 358, 361, 404, 552-3

Augustus, Emperor 261, 355

Axial Age 236, 246, 248, 266, 267, 273, 277, 360, 505, 507

Ayer, A.J. 564, 565

Aztecs, the 14, 26, 159, 161, 164, 166, 168, 174, 177, 204, 210, 231, 322, 496, 499, 507

Babylonians 164, 215, 225, 229, 230, 232, 234, 281, 321, 322, 324, 326, 328, 370, 371, 504, 505, 506, 508

Bacon, Francis 399-400, 539

Bacon, Roger 396-7, 398, 406, 410, 415, 538, 545, 546, 560

Baghdad 334, 360, 361, 364, 365, 366, 400, 442, 510, 535

balance, the 444, 535, 542, 552, 557, 563

Baldry, H.C. 510, 565

Balikci, A. 475, 565

ballistics (see also artillery, guns) 344, 419, 429

bands 33-5, 39, 40, 41, 46, 50, 67, 76, 152, 478

banking 2, 239, 349

barley 31, 52, 55, 475, 490, 494, 501

Barnes, R. 525, 565

Barry, H. 475, 489, 565

barter 100, 101, 103, 480

Basalla, G. 471, 495, 496, 565

Baxter, P. 480, 565

Beck, H.C. 530, 565

Befu, H. 497, 566

Bellah, R.N. 506, 510, 511, 566

Benveniste, E. 512, 566
Bernal, J.D. 513, 534, 566
Bernardi, B. 478, 566
Best, E. 487, 566
Big Men (Papua New Guinea) 60, 71, 81, 82, 96, 98, 102, 107, 108, 117, 151, 169, 480
Black, Joseph 443, 445, 456, 562, 563
Bodde, D. 509, 512, 536, 537, 566
Boesche, R. 508, 509, 566
Bohannan, P. 481, 482, 487, 566
Bolton, Matthew 393, 456
Bose, D.M. 515, 566
boundaries 87, 88, 89, 133, 186
Boyce, M. 505, 566
Boyle, Robert 443, 444, 539, 557, 558
Brahe, Tycho 435, 436
Bráhman 278-9, 280, 281, 286, 512
Brahmins 181, 232, 234, 238, 243, 251, 252, 253, 254, 258, 509
Bray, F. 476, 566
Brett, M. 537, 566
bronze 196, 197, 198, 332, 336, 381, 423, 505, 506, 551
Brown, D.E. 470, 566
Brunel, Isambard 190, 494, 495
Brunelleschi, F. 392, 423, 424, 425, 538-9
Buddha 238, 243, 505
bureaucracy 19, 23, 29, 163, 186, 227, 242, 252, 256, 258, 354, 355, 364, 365, 367, 373, 375, 376, 379, 393, 394, 467, 499, 537
Buridan, Jean 408, 544
Bushmen, the 30, 32, 37, 39, 41, 42, 131
Byzantine Empire 354, 358, 376, 386, 388, 393, 423
calendars 47, 86, 130, 135, 163, 212, 225, 226, 227-30, 285, 286, 368, 370, 432, 487, 488
cannibalism 88, 109-10, 483
capitalism 2, 6, 20, 25, 42, 96, 354, 364, 375, 391, 393, 394, 429, 455, 463, 464, 468, 474, 480, 481, 541
Cardwell, D.S.L. 563, 566
cartography 341-2, 361, 390, 422, 423

castration 90, 110, 111, 112, 167, 274
catapults (see also artillery) 202, 343-5, 346, 352, 418, 421, 532
de Caus, S. 560
causality, concepts of, 145-9, 303-5, 308, 371, 415, 426, 536
Cavendish, Henry 445
Celts, the 180, 217, 377, 378, 378, 537
ceremonial exchange 87, 95, 101, 114, 117, 479, 481, 482
Chandragupta Maurya, Emperor 238, 252, 254
Chattopadhyaya, D.P. 503, 504, 515
Cheng Ho, Admiral 373, 537,
Chewong, the 41, 574
chiefdoms 92, 114, 116, 118, 119, 157, 158, 160, 165, 166, 174, 188, 237, 377, 483, 493, 502
Childe, Gordon 471, 473, 503, 504, 567
China 16, 19, 26, 55, 141, 156, 159, 162, 163-4, 166, 172, 174, 175, 181, 183-4, 193, 197, 198, 201-3, 204, 205, 206, 210, 211, 212, 215, 218, 236-7, 239, 241, 242, 243-4, 245, 248, 250, 254-8, 279-82, 285, 289, 290, 296, 314, 332, 336, 354, 359, 367-75, 376, 390, 393, 474, 476, 490, 494, 501, 506, 510, 512, 513, 518, 523, 524, 536, 537, 544
Chomsky, N. 295, 518, 567
Chrisomalis, S. 223, 501, 502, 503, 567
Christ 285, 358, 406, 548
Christianity (see also Church) 48, 178, 227, 267, 270, 271, 272, 281, 282, 285, 302, 358, 360, 378, 386-9, 404-6, 485, 500, 510, 511, 552-5
Church, the 285, 359, 373, 386-9, 394, 398, 404-5, 436, 552-5
Cianchi M. 558, 567
Cicero 248, 261, 339, 355, 533
cities 2, 18, 19, 22, 24, 26, 52, 65, 155, 156, 161, 170-6, 180, 185, 207, 238, 240, 259, 365, 381, 383, 386, 387, 388, 394, 450, 475, 494
city-state 159, 166, 172-6, 180, 185, 238, 239, 249, 258, 259, 263, 283, 355, 365, 394, 467, 505, 512, 521
Claessen, H.J.M. 491, 493, 567
Clark, W.E. 515, 567
classification 46, 47, 83, 127, 130, 138-41, 155, 246, 262, 311

clock, mechanical 357, 370, 371, 390, 398, 402, 410, 413-15, 423, 426, 534
clock, water 346, 352, 371, 413, 544, 551
Coe, M.D. 498, 503
cognatic descent 34, 35, 69, 70, 116, 181, 378, 389
Cohen, I.B. 556, 567
Cohen, M.N. 476, 567
Cohen, M.R. 541, 567
Cohen, R. 490, 491, 567
coinage, origins (see also money) 186, 236, 239, 331, 379, 381, 402, 505-6
Cole, M. 487, 488, 499, 500, 567
colours, of spectrum (see also light) 342, 439, 532
colours, symbolic 87, 88, 330, 331, 333, 334, 397, 493, 518
competition 4, 5, 6, 12, 24, 32, 43, 60, 66, 87, 103, 106, 109, 157, 166, 169, 182, 189, 272, 288, 375, 392, 428, 458, 459, 461
concrete operational stage of thought 134, 138, 142, 152, 154, 155, 476
Confucius 237, 254, 258, 280, 283, 505, 508, 514
conscience 29, 48, 152, 154, 249, 264, 508, 509
Constable, G. 538, 567
Constantine, Emperor 272, 284, 358, 359
Constantinople 386, 511
construction versus selection 12, 16, 18, 24, 465
Cooper, L. 551, 567
Copernicus 321, 322, 323, 429, 430, 432-6, 437, 439, 447, 527, 542, 552, 554, 555, 556
copper 14, 195, 196-7, 198, 199, 330, 331-2, 449, 518
core principles 115-20, 180-5, 187, 375, 484, 494
Corlett, E. 494, 567
Cornford, F.M. 508, 518, 522, 567
councils 2, 39, 72, 73, 77-8, 97, 106, 119, 165, 174, 175, 180...6, 355, 374, 376, 382, 383
craft knowledge 179, 220, 392, 394, 410, 422, 429, 458
craftsmen, status of 178-9
Cree, the 46,
Crombie, A.C. 513, 524, 525, 535, 538, 540...3, 545...8, 550, 551,

555, 563, 568
Crosby, A.W. 539, 568
Ctesibius 343, 347, 352, 357
Curtin, P.D. 494, 568
Cusa, Nicholas of 542
Cynics 241, 262
Dahomeans, the 139, 211
Dalton, G. 480, 481, 482, 566, 568
Dalton, J. 445-6, 558, 572, 582
Darwin, Charles 458, 459, 563
Darwinism (see also adaptation, competition) 4-20, 23-5, 31, 66, 109, 169, 189, 192, 264, 273, 288, 318, 396, 460-5, 471, 495
Dasgupta, S. 507, 512, 568
David, F.N. 523, 568
Davy, H. 449, 559
debate 143, 182, 219, 238, 239, 240, 242-7, 269, 277, 355, 360, 375, 428, 507-10, 511
Debus, A.G. 539, 568
decoration of body 3, 14, 195, 196, 231, 470
Democritus 292, 296, 517, 531
Dempster, W.J. 563, 564, 568
Denny, J.P. 475, 568
Descartes, R. 403, 417
Diamond, J. 475, 568
Dibner, B. 560, 568
Dickinson, H.W. 560, 562, 568
Diener, P. 583, 569
diffusion 205-7, 498
Digges, T. 416, 421, 436, 546, 547
Dinka, the 143, 144, 148, 161, 575
Diocles 530, 569
Diocletian, Emperor 359
distillation 194, 333, 390, 399, 442, 557
divination 49, 83, 84, 122, 150, 155, 211, 212, 314, 315, 336, 359, 491, 499, 502
division of labour 11, 19, 64, 75, 95, 482, 484
Dodds, E.R. 511, 513, 569

Douglas, M. 479, 569
Drachmann, A.G. 532, 533, 534, 569
Drake, S. 547, 552, 569
Drover, C.B. 544, 569
Duby, G. 569
Dürer, A. 411
Durkheim, E. 521
dynamics 295, 343, 344, 408, 521
Eamon, W. 513, 536, 538, 543, 569
East Cushitic societies 118-20, 484
Easter Island 217
eclipses 283, 308, 322, 526
economic surplus 18, 19, 75, 100, 156, 160, 163, 169
Edgerton, R.B. 472, 569
Edgerton, S.Y. 549, 569
Edwards, C. 49, 569
Egypt 16, 156, 159, 162, 164, 166, 174, 175, 178, 188, 199, 210, 211, 214, 221, 231, 298, 331, 337, 344, 359, 364, 477, 490, 491, 498, 501, 511, 529, 532, 536
Einzig, P. 482, 570
eldest son, status of 37, 91, 93, 117
electricity (see also magnetism) 14, 18, 194, 200, 208, 288, 335-6, 393, 427, 446-50, 457, 461, 534
elements 290-1, 293, 294, 297, 300, 301, 305, 306, 307-11, 332, 345, 397, 430, 436, 441-6, 461, 515-17, 518, 528, 557, 558
elliptical orbits 434, 437, 438, 440, 461, 462, 547, 555
Elvin, M. 524, 536, 570
energy, harnessing of 8, 9, 64, 162, 455
Epicureans (see also atomism) 241, 269, 485
epicycle 325, 326, 527
equant 327, 433, 527, 528
Eratosthenes 323, 526, 532
Eskimo, the 34, 39, 42, 43, 565
van Ess, J. 511, 570
Ethiopia 53, 71, 76, 81, 103, 477, 478, 483, 490, 511
Euclid 298, 339, 340, 346, 352, 372, 488, 519, 520, 531, 570
Eudoxus 324, 325

Evans-Pritchard, E.E. 488, 570
evolution, biological, see Darwinism, adaptation, competition, selection
evolutionary potential 12, 13, 14, 17, 24, 61, 113, 194, 199, 203-5, 220, 244, 295, 299, 449, 461, 462, 464, 473, 479
evolutionary psychology 470-1
experimental science, nature of 26, 27, 208, 287, 335, 354, 367, 371, 372, 373, 394, 395, 396, 400, 410, 422, 424, 425, 429, 430, 439, 455, 462, 463, 465, 467
eye-beams 338-9
Fakhry, M. 509, 570
Faraday, Michael 449
Feil, D.K. 483, 484, 570
feudalism 237, 239, 240, 379, 380...5, 394, 474
Fibonacci, Leonardo 401, 520, 539
Fijians, the 110, 116, 483, 578, 580
Findley, C.V. 536, 570
Finley, M.I. 496, 509, 535, 570
Finocchiaro, M. 553, 554, 555, 560
Firth, R. 478, 482, 570
Fischer, S.R. 501, 570
Fisher, H.A.L. 25, 474, 570
fishing 31, 54, 57, 62
Flynn, J.R. 389, 570
focal activities 171, 172, 176, 365
Forbes, R.J. 195, 496, 528, 570
Forde, D. 488, 571
formal operational stage of thought 127, 128, 140, 142, 150, 152, 154, 155, 236, 263, 264, 274, 277, 287, 288, 296, 318, 335, 489
Franklin, Benjamin 393
Franks, the 376, 385, 386
Freeman, L.C. 374, 571
Frontinus 343, 346, 532
Gadd, C.J. 494, 499, 509, 571
Gahuku-Gama, the 144, 579
Galen 417
Galileo 288, 344, 346, 396, 398, 408, 414, 416, 421, 424, 425, 427,

429, 431, 432, 435, 438, 440, 442, 451, 461, 518, 539, 541, 546, 548, 550, 552-5, 556, 567
Galvani, L. 448-9
gambling (see also probability) 313-17, 502, 524, 540, 541
Gardner, P.M. 475, 571
Garelli, P. 506, 571
Gases (in atomic theory) 290, 333, 442-6, 461, 558, 559
Gassendi, P. 557
Gat, A. 483
Gay, J. 487, 488, 571
Gelb, I.J. 499, 571
geometry 219, 221, 231, 232-4, 235, 296, 298, 300-2, 313, 315, 318, 320, 322-3, 337, 338-42, 343, 346, 352, 371, 372, 388, 390, 391, 403, 406, 432, 437, 441, 519, 526, 532, 533, 539, 545
Gerbert of Aurillac 400
Gilbert, William 437, 447, 448, 543, 560, 573
Gilgamesh, Epic of 235, 504
Ginsberg, M. 571
Glasgow University 392, 443, 445, 562
glass 191, 194, 199-200, 205, 337-8, 341, 342, 397, 398, 415-18, 423, 443, 445, 448, 451, 496, 529-30, 532, 541, 545, 546
Gluckman, M. 482, 483, 571
God 82, 83, 142, 161, 183, 267, 271, 273, 275-8, 280-6, 288, 329, 334, 338, 361-2, 364, 387, 389, 399, 404-6, 408-9, 411, 414, 432, 433, 439, 441, 447, 448, 473, 486, 511, 513, 524, 548, 553, 555, 557
gods 38, 46, 83, 116, 117, 122, 150, 161-3, 164, 166, 176, 211, 243, 247, 249, 251, 259, 260, 266, 269, 270, 274-5, 277, 279, 283, 286, 295, 314, 347, 377, 470, 485, 499, 507, 512, 517
gold 14, 15, 17, 101, 102, 162, 167, 170, 195-7, 199, 201, 304, 330-4, 345, 379, 383, 501, 505, 506, 518, 528, 529, 533, 538, 557
Golden Rule 48, 152, 153, 249, 264
Golden Section 520
Goldman, I. 482, 484, 571
Golson, J. 476, 571
Gomme, G.L. 571
Gorst, J.E. 493, 572

Goths, the 376, 386, 573
Graham, A.C. 507, 572
grammar 218, 245, 246, 260, 261, 357, 367, 376, 507, 581
Graunt, J. 563, 572
gravity 306, 309, 313, 322, 324, 329, 345, 409, 419, 431, 436, 440, 441, 460, 533, 535, 542, 552, 556
Greece, ancient 16, 141, 162, 166, 174, 177, 180, 183, 193, 218, 236, 237, 238-9, 241, 242-3, 244, 245, 246, 250, 260, 269, 270, 277, 290, 296, 313, 328, 387, 512, 513, 529
Greenaway, F. 559, 572
Gresham College 392, 455
Grimm, T. 536, 572
Grosseteste, Robert 405, 406, 568
Grube, G.M.A. 521, 572
von Guericke, Otto 448, 452, 560, 561
guilds, craft and merchant 252, 365, 366, 382, 383, 388, 392,
gunpowder 194, 200-3, 205, 206, 370, 390, 398, 410, 421, 443, 496, 561, 578
guns (see also artillery) 202, 418-21, 455, 543, 546, 547, 548, 551
Hackett, J. 560, 572, 578, 584
Hadza, the 42
Hall, A.R. 546, 560, 572
Hammurabi, King 164, 569
Han Fei-tzu 257
Harris, M. 473, 483, 573
Harris, W.V. 501, 573
Heath, T. 520, 573
Heather, P. 537, 573
heliocentric theory (see also Copernicus) 323, 327, 328, 535
van Helmont, J.B. 443, 558
Henry, J. 543, 573
Heraclides 323
Heraclitus 278, 279, 511
Hermes Trismegistus 178, 334, 433
Hero of Alexandria 343, 345, 347-52, 353, 418, 534, 551, 560, 573
Hertz, R. 479, 573
Hessen, B. 551, 573

hierarchical structures  19, 75, 78, 152
Hills, R.L. 562, 573
Hinduism (see also India) 110, 228, 238, 267, 271-2, 278, 279, 509, 515, 581
Hipparchus 324, 326, 327, 328, 528, 582
historical inevitability 25, 27, 463
Hittites, the 180, 494
Hobbes, Thomas 255, 258
Hockney, D. 549, 573
Hodge, A.T. 534, 573
Hokhma 281-2
Holmberg, A. 489, 573
Holmes, G.538, 573
Hooke, Robert 400, 416, 417, 438, 443, 452, 539, 574
Hoppen, A. 548, 573
horse 58, 59, 60, 63, 109, 180, 204, 379, 397, 455, 476-8
Hourani, G.F. 507, 573
Howard, A. 484, 574
Howell, S. 475, 489, 574
Huizinga, J. 479, 574
humours, the four  297, 330, 334, 397
Humphreys, S.C. 507, 574
Hundred Flowers period 244, 467
Huns, the 63, 376
hunting and gathering 18, 31, 32, 33, 52, 54, 56, 57, 62, 118, 476, 484, 577
Huron, the 209, 210
Hussey, E. 511, 574
Huygens, C. 414, 561
ideograms 209
impetus theory 409, 420, 433, 542, 552
Inca, the164, 174, 175, 178, 204, 211
India 19, 26, 141, 162-4, 166, 174, 180-3, 193, 198, 206, 218, 228, 231-4, 236, 237-8, 239, 241-6, 251-4, 257, 278-9, 289, 290, 296, 328, 354, 364, 373, 376, 390, 394, 474, 490, 497, 501, 505, 509-15, 517, 523, 524, 535, 538
Indo-Europeans, the 63, 164, 180-3, 184, 377, 483, 494, 566

Indus Valley civilisation 159, 162, 237, 490
Industrial Revolution 155, 206, 288, 393, 396, 456
inequality 2, 30, 38, 65, 71, 72, 97, 115, 157, 189, 251
inertia 431, 439, 440, 522, 542, 552, 556
Inquisition, the 389, 398, 436, 555
instrument makers 391, 411, 455, 534
intention 146, 151, 152, 154, 249, 510
intuitive stage of thought 26, 125-6, 127, 128, 130, 131, 133, 135-8, 141, 146-50, 151, 154, 274, 296, 308, 309, 339, 389
invention, and necessity 190-4
Inwood, S. 549, 546, 556, 574
iron 195, 197-8, 205, 238, 296, 330, 331, 334, 345-6, 352, 356, 368, 369, 409, 445, 518, 528
Iroquois, the 117-18
irrigation 65, 92, 161, 170, 252, 367, 372, 490, 497
Islam 226, 235, 269, 270, 281, 321, 351, 352, 354, 359-67, 373, 374, 376, 384, 388, 389, 393, 394, 396, 410, 410, 491, 496, 500, 510, 511, 535
Israel, ancient 164, 236, 269, 271, 277, 281-2, 404
Jamnitzer, A. 411, 520
Japanese, the 31, 177, 215, 216, 219, 438, 477, 490,
Jaspers, K. 505, 574
Jenkins, Rhys 561, 574
Johnson, A.W. 489, 574
Judaism see Israel
Justinian, Emperor 285, 359, 387
Kaplan, E. 539, 540, 574
Kautilya 252-4, 255, 256, 258, 259, 508, 509, 566
Kaye, J. 402, 539, 542, 574
Kendall, M.G. 541, 574
Kepler, Johannes 417, 426, 427, 435, 436-8, 439, 440, 520, 544, 546, 556, 581
kin groups 35, 58, 67-72, 73, 74, 76, 94, 186, 375, 378, 482
kings 120, 156, 157, 158, 160, 161-7, 168, 172-7, 178, 179, 182-4, 187, 230, 250-7, 259, 262, 263, 269, 273, 321, 345, 377-85, 491, 493, 454, 498, 499, 509, 515, 525, 537
Kirch, P.V. 484, 574

Kirk, G.S. 519, 574
Kitcher, P. 470, 574
Knight, D. 529, 545, 551, 558, 574
Koch, K-F. 483, 575
Kohlberg, L. 476, 575
Konso, the 61, 71, 76-7, 80 81, 85, 89, 91, 96…9, 102, 103, 105, 110-12, 113, 116, 118-20, 149, 164, 171, 173, 299, 478, 481, 483, 484, 487, 495, 497, 573
Kosala 174, 238
Kpelle, the 132
Kroll, J.L. 507, 575
Kshatriyas 181, 238, 243, 509
Kuhn, T.S. 526, 527, 528, 541, 542, 552, 555, 565
Lactantius 284, 513
Landels, J.G. 496, 534, 575
latitude of forms 407, 582
Lavoisier, A. 444, 445, 558, 559
Leach, E.R. 564, 575
leadership 20, 37-8, 39, 43, 46, 58, 65, 70, 72, 73, 74, 78, 82, 96, 99, 105, 106, 113…17, 119, 120, 153, 158, 160, 161, 166, 185, 186, 377, 378, 483, 484, 493
Leaman, O. 510, 575
Lee, H.D.P. 517, 575,
Lee, R.B. 475, 575, 576, 584
Legalists 255-8, 367, 514, 516
legitimacy, political 158, 160, 169, 272, 285, 387
Leicester, H.M. 557, 558, 575
lenses 200, 337, 338, 339, 352, 391, 410, 415-17, 496, 529-30, 545, 546, 579, 581
Léon-Portilla, M. 507, 575
Leucippus 517
Lewis, B. 536, 575
Lienhardt, G. 488, 575
light, nature of 147-8, 289, 294, 338-9, 342, 406, 438, 441, 530, 532, 541, 551
literacy (see also writing) 18, 19, 142, 179, 214, 217, 218, 237, 376, 387, 403, 491, 501, 573

Littlejohn, J. 487, 575
Lloyd, G.E.R. 507, 516, 517, 521, 522, 525, 534, 535, 575, 576
logic 139, 140, 141, 155, 218, 244-6, 260, 287, 292, 296, 298, 366, 464, 510, 511, 514, 535
Logos 278-9, 280, 281, 286, 298, 329, 525
Lombards, the 376, 386
London 381, 382, 384, 392, 411, 452, 454
Long, A.A. 511, 576
Lucretius 327, 336, 341, 517, 521, 576
Lunar Society 393
Luria, A.R. 488, 576
Lyceum 302, 321, 521
Machiavelli, N. 253, 259
machine 14, 22, 155, 200, 204-5, 294, 342-3, 346-7, 350-1, 352, 356-7, 369, 371, 397, 398, 413, 414-15, 419, 422, 424, 425, 426, 456, 544, 548
Machinist, P. 499, 506, 576
Maghada 174, 238
magic 6, 7, 9, 12, 13, 15, 22, 83, 122, 146, 155, 162, 193, 199, 269, 289, 299, 330, 334, 397-400, 416, 418, 425, 430, 446-7, 450, 461, 470, 473, 485, 538, 550
magnetic compass 206, 336, 370, 390, 398, 410, 411, 447, 449, 536
magnetism 335-6, 352, 369, 372, 398, 410, 412, 429, 437, 446-8, 449, 461, 529, 543, 545, 560
Mahony, W.K. 512, 576
Makdisi, G. 536, 576
Malinowski, B. 488, 576
Maori, the 99, 210, 482, 493, 566, 572
de Maricourt, P. 437, 447
market economy (see also subsistence economy) 94, 480
markets 95, 100…103, 113, 114, 119, 132, 160, 171…4, 221, 240, 256, 365, 374, 381, 401, 478, 480, 481, 505
marriage 29-30, 34-7, 50, 69, 74-5, 79, 80, 84, 87…90, 299, 475
Mars 230, 322, 518, 555, 436
Marsden, E. 532, 533, 576
Marshall, L. 41, 475, 489, 494, 576, 583
Marx, Karl 25, 103, 193, 473

Marxism 429, 458, 466, 468
Masai, the 98
materialism 4, 7-16, 59, 86, 107, 109, 198, 207-8, 233-4, 241, 288, 289, 318, 354, 374, 420, 432, 450, 460, 462, 473, 480, 499, 504, 513, 573
matrilineal descent 34, 68-9, 70, 118, 478, 560
Matthew, P. 458-9, 563-4, 568
de Maupertuis, Baron 457, 458
Mauryan Empire 174, 238, 252
Maya, the 159, 164, 166, 174, 204, 210, 211, 212, 225, 229, 231, 322, 494, 497, 499, 567
McGuire, J.E. 556, 576
measurement 23, 129, 130, 131, 135, 137, 221, 222, 231, 299, 302, 311, 313, 318, 323, 342, 352, 396, 400, 402, 407, 409, 412, 417, 424, 425, 426, 430, 431, 488, 506, 534, 550
Medawar, P.B. 460, 564, 576
medicine 201, 260, 297, 330, 333, 334, 352, 361, 366, 368, 372, 388, 397, 398, 411, 412, 447, 450, 521, 535, 560,
memes 471-2
Mencius 248, 255, 508
merchants, status of 103, 157, 169, 170, 173, 176-8, 179, 184, 237, 238, 239, 250, 253, 258, 260, 263, 355, 356, 357, 365, 369, 373, 374, 375, 381, 382-4, 385, 394, 401, 402, 410, 414, 454, 478, 505, 506, 509, 536, 539, 540, 545, 551
mercury 331, 332, 442, 444, 451, 518, 527, 528-9, 555, 557, 558, 561
Mersenne, M. 424, 425, 429, 540
Mesopotamia 16, 156, 159, 162, 163, 164, 166, 172, 174, 176, 191, 199, 203, 206, 211, 214, 215, 217, 222, 231, 235, 242, 337, 494, 498, 499, 501, 505, 506, 508, 533, 571
Mesoudi, A. 471, 576
metal (see also bronze, copper, gold, iron, alchemy) 65, 159, 179, 195-9, 220, 330-5, 356, 496, 505, 515, 516, 518, 528, 529
microscope 200, 336, 390, 416-7, 584
Minoans, the 217, 337, 491
Misra, G.S.P. 508, 577
Mitchiner, M. 501, 506, 577

Mohists 244, 245, 280, 381, 514, 572
moieties 87, 89, 90, 474
momentum 542, 556
Momigliano, A. 505, 577
money 2, 17, 23, 40, 94, 95, 97, 98, 101, 102, 103, 219, 317, 355, 374, 377, 378, 401, 402, 479, 482, 505, 506, 539
Montagnais-Naskapi, the 49
Morris, B. 475, 577
Muhammad 172, 359, 365, 510
Munro, D.S. 508, 509, 513, 577
Murdock, G.P. 474, 577
Murray, A. 549, 577
Museum of Alexandria 343, 347, 348, 350, 356, 357, 461, 521, 525, 534
music 310, 352, 388, 422…5, 427, 466, 521, 549, 550, 551
Musson, A.E. 561, 568, 577
mutation, (see also variation) 188, 203, 277, 396, 461, 462
myth 83, 89, 90, 178, 216, 217, 235, 270, 274, 286, 288, 491, 504, 514

names 22, 46, 141-2, 143, 167, 246
Naroll, R. 474, 577
Nasr, S.N. 535, 536, 577
Natufians, the 31
navigation 135, 230, 390, 402, 410, 411, 422, 425, 432, 526
Needham, Joseph 279, 280, 286, 476, 495, 496, 497, 498, 510, 512, 513, 514, 518, 522, 536, 537, 544, 566, 570, 577, 578
Needham, R. 479, 573, 578
Neugebauer, O. 330, 502, 503, 525, 528, 578
Newcomen, Thomas 351, 453-5, 456, 457, 561-2, 580
Nigosian, S.A. 505, 578
Norman, R. 411-12, 447
North, J. 527, 540, 544, 578
North-West Coast Indians, the 31
Nuer, the 79, 136, 570
number, conceptions of 137, 131-3, 299-301, 519-20
number notation, systems of 220-31
Obeyesekere, R. 483, 578

Ogham script 217
Onians, R.B. 488, 511, 578
Oppenheim, A.L. 506, 578
optics 321, 336-42, 352, 361, 399, 406, 415-17, 424, 426, 439, 539-40, 541, 546, 548-9
orders of society 152, 184, 250, 258, 260, 261, 538
Oresme, N. 402, 408, 409, 542
Orphism 269, 270, 511
Oxford University 321, 396, 406, 407
Paliyans, the 41, 47,
Panini 218
Papacy, the 279, 385-8, 394, 555
paper 219, 366, 370, 390, 391
Papin, D. 452, 453, 454, 561
Papua New Guinea 9, 21, 56, 76, 82, 86, 95, 107, 108, 131, 132, 483
Paracelsus 411, 442, 443, 557
parallax, stellar 323, 324, 434, 552
Paris 386, 396, 408, 444, 538, 552, 557, 561
Pascal, B. 541, 561
pastoral nomadism 43, 62-3, 73, 85, 91, 158, 181, 183, 484
patents 392, 416, 450, 560
patrilineal descent 34, 68, 69, 70, 76, 91, 116, 181, 183, 379, 474, 478
Peerenboom, R.P. 513, 578
pendulum 414, 438, 444, 550, 552
Peregrine, P.M. 474, 578
Perret, R.W. 515, 578
Persians, the 164, 174, 180, 239, 271, 277, 281, 359, 360, 361, 365, 478, 505, 511, 535, 538
perspective 336, 340-2, 352, 402, 423-4, 531, 548-9, 569
petroglyphs 209, 210, 217
Pharaoh, the 15, 164, 166, 174, 211, 491
Philip of Macedon, King 260, 343
Philo of Alexandria 281, 283, 361
phlogiston 444, 558
Phoenicians, the 216, 217

Piaget, J. 21, 150, 338, 341, 415, 488, 489, 530, 578, 579
pigs 10, 57...60, 68, 82, 95, 96, 98, 108, 109, 132, 138, 169, 196, 476, 481, 483
Plantzos, D. 530, 579
Plato, Platonism 246, 248, 258, 260, 278, 289, 294, 300, 301, 302, 314, 358, 389, 398, 430, 432, 436, 437, 461, 509, 512, 518, 519, 521, 550
Platonic Solids 300, 411, 555
play, and ritual 14, 84, 85
Pleiades 86, 227, 487
Plutarch 524, 533
pneumatics 321, 347-51, 352, 418
Polanyi, K. 480, 579
Polynesians, the 66, 69, 82, 100, 116-17, 135, 181, 217, 484, 490, 571, 574
Popper, K.R. 25, 460, 474, 564, 579
population pressure 53, 56, 57, 58, 66, 158, 188, 189, 492
della Porta, G. 398, 399, 416, 418, 425, 580
pottery 57, 64, 65, 101, 179, 191, 194, 199, 203, 206, 496, 497
Pre-Socratic philosophers 290, 305, 574
prestige goods 98, 101, 118, 160
Price, D.J.de Solla 534, 543, 579
Priestley, Joseph 393, 444, 455, 558
primary and secondary properties 293, 294, 297, 407, 426, 427, 430, 463, 464
primitive thought 20-3, 84, 122-55
Prince, J.R. 487, 579
printing 206, 219, 364, 370, 390, 391, 393, 410, 411, 514
private property 30, 67, 97, 251
probability 128, 149, 150, 313-17, 352, 404, 457-8, 489, 524, 540-1
projectile motion 310-11, 342, 408, 418-21
Ptolemies of Egypt 321, 344, 356, 357, 525
Ptolemy, Claudius 227, 230, 323, 324, 327-30, 335, 341, 342, 352, 423, 432, 433, 434, 527, 531, 532, 534, 555, 589
pumps 17, 346, 347, 352, 372, 415, 450-5, 534, 558, 561
purity and impurity 87-8, 236, 249, 267, 334, 479, 569
Pygmies, the 42, 583

Pythagoras, Pythagoreans 233, 298-301, 398, 425, 432, 546, 439, 461, 504, 512, 516, 519, 520, 531, 550
Rabinovitch, N.L. 524, 579
Randall, J.H. 510, 511, 524, 579
Read, K.E. 144, 488, 579
Recorde, Robert 403, 539, 574
Reed, C.A. 476, 579
religion 9, 12, 38, 43-6, 82-4, 85-90, 110-13, 117, 141, 157, 158, 161-5, 166, 193, 236, 244, 245, 249, 266-87, 305, 358-9, 360-7, 386-9, 393, 404-6, 430, 433, 439, 466, 468, 473, 485-6, 537, 552-5
Rhys Davids, T.W. 507, 509, 580
Ricci, M. 285
Rienstra, M.H. 538, 580
right and left 87, 134, 157, 578
rites of passage 84, 86, 88
Ritter E.A. 493, 496, 580
Robertson, J.D. 544, 580
Roller, D.H.D. 529, 580
Rolt, L.T.C. 561, 562, 580
Romans and Roman Empire 63, 167, 171, 173, 180, 193, 204, 206, 219, 222, 247, 260-1, 263, 343, 345, 346, 354-9, 373, 374, 375-6, 384
romantic love 29, 36, 42, 48
Ronan, C.A. 545, 546, 580
Rosen, E. 545, 580
Rossi, P. 539, 580
Rostovtzeff, M. 580
Royal Society of London 283, 357, 392, 400, 449, 453, 339
Rudgley, R. 498, 580
Rudolph II, Emperor 436
Runic script 217
sacrifice 16, 45, 60, 80, 81, 84, 85, 98, 106, 111, 112, 160, 162, 163, 164, 166-7, 219, 228, 266, 279, 334, 382, 486, 491, 506, 507, 509
Sahlins, M. 370, 373, 374, 378, 380, 382, 383, 580
Sambursky, S. 517, 521, 524, 580
Sampson, G. 499, 580

Sanderson, S.K. 471, 473, 474, 580
Santorii, S. 418
Savery, Thomas 450, 454, 560, 561
Saxons, the 166, 376, 380, 383, 386, 387, 537
Schapera, I. 482, 580
Scharf, P.M. 507, 581,
Schmandt-Besserat, D. 501, 581
Schwartz, B.I. 491, 494, 508, 510, 581
Scribner, S. 488, 499, 500, 567, 581
Secret of Secrets, the 397-8, 584
selection 4, 6, 7, 9, 11-13, 16, 24, 71, 104, 113, 122, 192, 208, 214, 264, 288, 295, 457, 458, 459-61, 464, 465, 470, 473, 496
self-awareness 488
Sen, M.K. 503, 581
Seneca 261, 289, 337, 357, 530, 532, 534, 535, 549, 581
sensori-motor stage of thought 124
settlements, permanent 30, 31, 52, 57, 58, 66
Shang kingdom 165, 166, 174, 175, 212, 491, 506
Sharma, J.P. 494, 581
shifting cultivation 47, 57, 65, 66, 86, 91
silk 102, 170, 205, 368, 369, 512, 537, 578
Simoons, F.J. 476, 581
simplification 186, 187, 214, 215, 216, 223, 277, 434
Sines, G. 529, 530, 581
Singer, C. 526, 533, 563, 581
Skinner, A. 498, 581
slavery, slaves 17, 31, 72, 102, 120, 157, 160, 167, 178, 181, 182, 189, 237, 239, 260, 262, 355, 356, 364, 377, 378, 465, 474, 491, 506, 509, 537
Smith, Adam 100, 193, 207, 458, 581
Smith, G. Elliott 498, 581
Smith, M. 581
Smith, P.H. 543, 581
social classes, see orders of society
socio-biology 470
Socrates 246, 261, 283, 505, 512, 514
Sophists 243, 258

soul 83, 148, 162, 247, 248, 261, 262, 266, 267, 268, 270, 278, 279, 284, 290, 301, 407, 447, 485, 510…13, 518, 530, 555, 557
space, concepts of (see also Void) 21, 47, 85, 88, 127, 130, 131, 133-5, 293, 294, 297, 307, 406, 441, 488, 515, 517, 519, 522, 529, 531, 542, 575
spectacles 200, 390, 391, 398, 415, 545
spokesmen 39, 96, 114, 118, 158
Stahl, W.H. 535, 581
state, the (see also city-state, territorial state) 156-190
statics 344, 352, 361, 535, 551
statistics (see also probability) 149, 150, 313, 315, 317, 457, 458, 518, 563
steam (see also atmospheric pressure, vacuum) 14, 17, 190, 194, 200, 207, 208, 288, 347-8, 350-1, 352, 393, 399, 443, 450-7, 461, 560-1, 562-3
Stephenson, B. 556, 581
Stephenson, George 457
Stevin, S. 403, 429, 548, 551
Stoics, the 241, 248, 262, 269, 276, 277, 278, 280, 314, 321, 328, 329, 330, 332, 333, 358, 511, 513, 514, 517, 525
stone technology 10, 11, 57, 63, 64, 98, 159, 190, 195
Strato 348, 521, 522
subsistence economy 10, 62, 95, 97, 103, 187, 491, 500, 565
sulphur 200, 201, 202, 331, 332, 446, 448, 528, 529, 557, 558
Sumerians, the 203, 211, 212, 213, 215, 225, 491, 494
sun 15, 22, 79, 129, 133, 228-31, 283, 306, 308, 318, 322…7, 329, 333, 338, 342, 368, 413, 418, 432-41, 479, 514, 525…8, 545, 552…6
Sun Tzu 256
Sung dynasty 368-9, 370, 373, 374
surveying 221, 226, 231, 232, 298, 402, 410, 411, 501, 526, 543
survival of the mediocre 7, 11, 13, 22, 461, 495
Sylla, E. 541, 582
syllabaries 215, 216, 219, 220, 500
Tambiah, S.J. 509, 582
Taoists, the 201, 202, 241, 267, 278-80, 281, 286, 368, 512, 514
Tartaglia, Niccolo 419-21, 541, 546-7, 552, 569

Tauade, the 9-10, 21, 68, 71, 77, 81, 82, 91, 96, 102, 105, 107-8, 109, 122, 129, 136, 146, 150-1, 169, 479-81, 572
taxation 102, 156, 157, 160, 167, 171, 178, 188, 189, 211, 221, 239, 259, 355, 365, 374, 377, 379, 381, 385, 491, 505
taxonomic classes 47, 138, 139, 246
Taylor, F.Sherwood 529, 546, 582
technology, development of 91, 151, 155, 159, 190-208, 233-4, 238, 288, 289, 318, 321, 343-53, 354, 356-7, 360, 368, 369-70, 373, 381, 387, 393, 412, 414, 419, 430, 450, 454, 465, 466, 467, 472, 487, 495, 497, 513-14, 534, 535
telescope 200, 324, 336, 337, 390, 415, 416, 417, 425, 429, 435, 438, 545-6, 556, 580
temples 19, 85, 110, 156, 160, 162...5, 167, 169, 170, 172, 188, 193, 211, 314, 343, 346, 347, 368, 369, 491, 594, 598, 599
territorial state 173, 174-6, 178, 185
Thackray, A. 559, 582
Thales 290, 298
Thapar, R. 509, 582
thermometer 351, 407, 417-18, 426, 546, 563, 582
Thornton, R.J. 488, 582
Three Estates 181, 537-8, 567
Tierra del Fuego Indians, the 42
time, concepts of 129, 135-8
time-reckoning, see calendars
Toomer, G.J. 527, 530, 569, 579, 582
Torricelli, E. 418, 451, 452, 561
town organization 355, 365, 374-5, 381-4
transport 11, 61, 159, 203-4, 208, 356, 368, 476-8, 491, 496
Trefil, J. 547, 582
tribute 18, 81, 93, 99-10, 102, 115, 119, 158, 159, 163, 167, 172, 174, 186, 211, 251, 491, 498, 499
Trigger, B. 484, 491, 493, 494, 496, 498, 499, 582
Trowbridge, M.L. 529, 582
Tsou-Yen 515-16
Tuden, A. 494, 583
Turks, the 364, 570
Turnbull, C. 475, 583

Turney-High, H.H. 482, 583
Ucko, P.J. 476, 482, 583
Ulansey, D. 528, 583
Umeda, the 89-90, 91, 133, 138, 479, 571
universities 357, 366, 368, 387, 388-9, 390, 392, 396, 398, 399, 403, 404, 406, 411, 422, 429, 443, 455, 456, 539, 554, 560, 562
urbanization 19, 208, 236, 243, 490
vacuum (see also Void) 294, 307, 348...51, 429, 441, 450-3, 456-7, 461, 561, 562
Vai script 216, 499-500
variation (see also mutation) 4-5, 24, 71, 192, 264, 288, 459, 464, 471-2
Venus 230, 322, 435, 518, 523, 527, 555
Viète, F. 403
da Vinci, Leonardo 422, 548, 567
Vinette, F. 503, 583
virtue(s) 153, 154, 240, 247-9, 261, 280, 508
Vitruvius 341, 343, 345, 346, 351, 531, 532, 583
Void, the 292, 307, 348, 436, 450, 451, 452, 522, 527
Volta, A. 449, 568
volume, concept of 127, 128, 130, 132, 221, 232, 298, 345, 446, 522, 533
voyages of discovery 373, 390, 536-7
Wade, N. 482, 583
Wallace, A.R. 563
Wallace, W.A. 541, 583
warfare, primitive 16, 104-15, 318, 482-3, 572
warfare, of states 165-8
Warring States, period of 237, 244, 257
water-wheel 193, 204, 346, 351, 356, 450, 456, 497, 561
Watt, James 393, 443, 455-7, 562-3
weaving 64, 65, 101, 103, 143, 179, 191, 206, 456, 478, 497
Weisheipl, A. 538, 583
West Cushitic societies 119-20, 483
Westfall, R. 556, 583
wheat 31, 52, 55, 477, 490, 494, 501
wheel 159, 203-5

wheel-barrow 191, 369
White, J. 531, 583
White, K.D. 496, 583
White, L. 513, 543, 583
Whitelock, D. 537, 583
Whybray, R.N. 513, 583
wild and the tame 28, 43, 86…90, 157, 479, 504
Williams, B.I. 560, 584
Williams, S.J. 538, 584
Wilson, C. 584
Wilson, P. 499, 584
windmill 200, 350, 351, 369, 551, 569
witchcraft 8, 22, 39, 118, 122, 123, 146, 148, 150, 155, 398, 485-6
Wittfogel, K.A. 490, 584
Woodburn, J. 42, 475, 489, 584
Worcester, Marquis of 454, 560
writing, origins of (see also literacy) 209-20, 498-501
Xenophanes 274-5
Yakuts, the 31
Yang and Yin 278, 299
Yates, F. 539, 584
Zeno of Citium 514, 525
Zeno of Elea 517, 575
zero, concept of 224-7, 235
Zipf, G.K. 472, 584
Zoroaster, Zoroastrianism 271, 277, 360, 505, 511, 566, 578
Zulu, the 174, 198, 491, 492, 580